Electrochemical Immunosensors and Aptasensors

Special Issue Editors

Paolo Ugo

Ligia M. Moretto

MDPI

Special Issue Editors
Paolo Ugo
University Ca' Foscari of Venice
Italy

Ligia M. Moretto
University Ca' Foscari of Venice
Italy

Editorial Office
MDPI AG
St. Alban-Anlage 66
Basel, Switzerland

This edition is a reprint of the Special Issue published online in the open access journal *Chemosensors* (ISSN 2227-9040) from 2016–2017 (available at: http://www.mdpi.com/journal/chemosensors/special_issues/EIA).

For citation purposes, cite each article independently as indicated on the article page online and as indicated below:

Author 1; Author 2; Author 3 etc. Article title. *Journal Name*. **Year**. Article number/page range.

ISBN 978-3-03842-406-2 (Pbk)
ISBN 978-3-03842-407-9 (PDF)

Table of Contents

Chapter 1: Articles

Chapter 2: Reviews

Chapter 3: Technical Note

About the Guest Editors

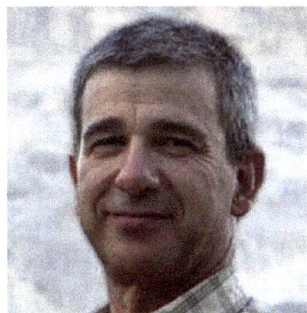

Paolo Ugo is full professor of Analytical Chemistry at the University Ca' Foscari of Venice (Italy) since 2006. He has been Visiting Associate at the California Institute of Technology (Pasadena, USA), Visiting Professor at the University Bordeaux (France) and at the University of Southampton (UK). His research interests focus on chemosensors and biosensors for environmental and food analysis, electrochemistry in nonaqueous solvents, ion-exchange voltammetry and polymer modified electrodes, nanoelectrochemistry, molecular diagnostics with nanoelectrode arrays and electrochemical biosensors. He is the author and co-author of over 138 scientific articles in international peer-reviewed journals, 16 book chapters, and approximately 150 scientific communications presented at national and international conferences. His bibliometric data are: citations: >2400; H-index: 28 (ISI-WOS-Feb 2017).

Ligia Maria Moretto graduated in Chemical Engineering at the Federal University of Rio Grande do Sul, Brazil, and received her PhD in analytical chemistry in 1994 from the University Ca' Foscari of Venice, Italy. Since 2015 she has held the position of associate professor of Analytical Chemistry at the University Ca' Foscari of Venice. She has been an invited professor and invited researcher at several institutions in Brazil, France, Argentina, Canada and the USA. Working at the Laboratory of Electrochemical Sensors, her research field has been the development of electrochemical sensor and biosensors based on modified electrodes, the development of gold arrays and ensembles of nanoelectrodes for environmental and food analysis application. She is a co-editor of two books, author and co-author of more than 65 full papers, seven book chapters, and has presented around 100 contributions at international conferences. Her bibliometric data are: citations >1500, h-index: 22 (ISI-WOS-March 2017).

Preface to "Electrochemical Immunosensors and Aptasensors"

Since the first electrochemical biosensor for glucose detection was pioneered in 1962 by Clark and Lyons, research and application in the field has grown at an impressive rate and, nowadays, we are still witnessing the continuing evolution of research on this topic. Initially, major research and applicative efforts were devoted to develop biocatalytic electrochemical sensors, aimed at exploiting the specificity of the reaction of an enzyme for its substrate. In the 1980s–1990s, the first examples of use of immunochemical reactions for electrochemical sensing were proposed by Heineman and Halsall, followed by the first examples of electrochemical immunosensors where antibodies were immobilized on the electrode surface. These biosensors exploit the specificity of the antigen (Ag)–antibody (Ab) interaction to detect one of the two partners as the analyte. In order to achieve electrochemical detection, a label is typically used which can be electroactive itself or able to generate or consume an electroactive molecule. Using this approach, many sensors have been developed to detect a number of disease markers.

Recent research trends in the field of affinity biosensors are indeed moving forward, trying to overcome some of the present limitations. On the one hand, many studies are aimed at developing novel capture agents, not necessarily belonging to the antibodies category. Aptamers represent a successful example of an efficient capture molecules' alternative to antibodies. On the other hand, the possibility of avoiding using enzyme labels appears very attractive in order to simplify detection schemes, thus avoiding complex functionalization procedures.

This Special Issue collates several contributions which offer an overview of recent developments in the field of electrochemical immunosensors and aptasensors, outlining future prospects and research trends.

The use of advanced capturing agents as an alternative to "classical" antibodies, such as affibodies and aptamers, constitutes the focus of three original research articles and three review papers. Another review presents and discusses the recent literature on immunosensors based on field effect transistors. Two research papers, one technical note and one review report on the development of novel immunosensors for food control, particularly the analytical capabilities of using non-conventional detection schemes. A final note highlights the results achieved and described in the special issue, outlining future prospects of research development in the field.

Paolo Ugo and Ligia M. Moretto

Guest Editors

Chapter 1:
Articles

chemosensors

MDPI

Article

Simultaneous Determination of the Main Peanut Allergens in Foods Using Disposable Amperometric Magnetic Beads-Based Immunosensing Platforms

Víctor Ruiz-Valdepeñas Montiel [1], Rebeca Magnolia Torrente-Rodríguez [1], Susana Campuzano [1,*], Alessandro Pellicanò [2], Ángel Julio Reviejo [1], Maria Stella Cosio [2] and José Manuel Pingarrón [1,*]

[1] Departamento de Química Analítica, Facultad de CC. Químicas, Universidad Complutense de Madrid, E-28040 Madrid, Spain; victor_lega90@hotmail.com (V.R.-V.M.); rebeca.magnolia@gmail.com (R.M.T.-R.); reviejo@quim.ucm.es (Á.J.R.)

[2] Department of Food, Environmental and Nutritional Sciences (DEFENS), University of Milan, Via Celoria 2, 20133 Milan, Italy; alessandro.pellicano@unimi.it (A.P.); stella.cosio@unimi.it (M.S.C.)

* Correspondence: susanacr@quim.ucm.es (S.C.); pingarro@quim.ucm.es (J.M.P.); Tel.: +34-91-394-4315 (J.M.P.)

Academic Editors: Paolo Ugo and Ligia Moretto
Received: 26 March 2016; Accepted: 24 June 2016; Published: 28 June 2016

Abstract: In this work, a novel magnetic beads (MBs)-based immunosensing approach for the rapid and simultaneous determination of the main peanut allergenic proteins (Ara h 1 and Ara h 2) is reported. It involves the use of sandwich-type immunoassays using selective capture and detector antibodies and carboxylic acid-modified magnetic beads (HOOC-MBs). Amperometric detection at -0.20 V was performed using dual screen-printed carbon electrodes (SPdCEs) and the H_2O_2/hydroquinone (HQ) system. This methodology exhibits high sensitivity and selectivity for the target proteins providing detection limits of 18.0 and 0.07 ng/mL for Ara h 1 and Ara h 2, respectively, with an assay time of only 2 h. The usefulness of the approach was evaluated by detecting the endogenous content of both allergenic proteins in different food extracts as well as trace amounts of peanut allergen (0.0001% or 1.0 mg/kg) in wheat flour spiked samples. The developed platform provides better Low detection limits (LODs) in shorter assay times than those claimed for the allergen specific commercial ELISA kits using the same immunoreagents and quantitative information on individual food allergen levels. Moreover, the flexibility of the methodology makes it readily translate to the detection of other food-allergens.

Keywords: Ara h 1; Ara h 2; dual determination; magnetic beads; SPdCEs; amperometric immunosensor; food extracts

1. Introduction

Food allergies, i.e., adverse immunologic (IgE and non-IgE mediated) reactions to food, have resulted in considerable morbidity and reached high proportions in the industrialized world, affecting up to 10% of young children and 2%–3% of adults [1]. Analysis for food allergens is required both for consumer protection and food fraud identification. The eight food major allergens are peanuts, wheat, eggs, milk, soy, tree nuts, fish, and shellfish [2]. Peanut allergy deserves particular attention because very small amounts of peanut proteins can induce severe allergic reactions. It persists throughout life and accounts for most of food-induced anaphylactic reactions with a prevalence that has doubled in a five-year time span [3,4]. Consequently, there is an increasing concern and need to protect food allergic consumers from acute and potentially life-threatening allergic reactions through detection of peanut trace contamination and accurate food labeling [5]. Although Regulation No. 1169/2011 established food allergen labelling and information requirements under the EU Food Information for Consumers [6], food allergic patients are still at high risk of consuming unintentional trace amounts of allergens that may have contaminated the food product at some point along the production line.

The detection of peanut allergens in food products is sometimes challenging since they are often present unintentionally and in trace amounts, or can be masked by compounds of the constituting food matrix. Moreover, since there are no established thresholds below which an allergen poses only a small risk of causing harm to an allergic consumer so far, there is general agreement in the analytical community and especially standardization bodies to look for validated methods that can detect food allergens in the low ppm range (1–10 mg allergenic ingredient kg^{-1} food product) [7,8].

Analytical techniques used to detect peanut allergens can be divided into protein-or DNA-based assays. The former detect specific peanut protein allergens, using enzyme-linked immunosorbent assays (ELISAs), or total soluble peanut proteins. Commercially available ELISA kits constitute the most widely used analytical tool by food industries and official food control agencies for monitoring adventitious contamination of food products by allergenic ingredients [4]. However, these methods are limited to providing only qualitative or semi-quantitative information and can suffer from unexpected cross-reactivity in complex food matrices [3]. On the other hand, DNA-based techniques allow the presence of allergens to be detected by PCR amplification of a specific DNA fragment of a peanut allergen gene. False positive results due to cross-reactivity with other nuts [9], significant differences regarding quantification with respect to ELISA kits [1,4] and the high number of replicates for samples required by the PCR methods are important limitations hindering their applicability to processed foods or complex food matrices [8,10]. Most importantly, these methods require different assays to detect each of the different food allergens [11].

Recently, liquid chromatography-mass spectrometry (LC-MS)/MS has emerged as an interesting alternative for food allergen analysis because it provides wide linear dynamic ranges and absolute identification and quantification of allergens. However, apart from a high level of expertise and costly equipment, multiple extraction and cleanup steps are necessary making this method laborious and time consuming [1]. Therefore, the development of accurate and simpler methods for performing highly sensitive and specific simultaneous detection of multiple food-product allergens is highly demanded.

In this context, electrochemical immunosensors constitute clear alternatives to the above-mentioned techniques due to their simplicity, low cost and easy use. However, their applications for the detection and quantification of allergens are still scarce [12]. Although some electrochemical immunosensors have been reported recently for the determination of peanut allergenic proteins [12–15], to the best of our knowledge, no electrochemical immunosensor has been so far reported for the multiplexed determination of food allergens. This paper describes the first electrochemical immunosensor for the simultaneous determination of the two major peanut allergenic proteins, Ara h 1 and Ara h 2, in one single experiment. More than 65% of peanut allergic individuals have specific IgE to Ara h 1 and more than 71% to Ara h 2 [16]. The implemented methodology involved the use of functionalized magnetic beads (MBs), a specific pair set of antibodies for sandwiching each target protein and amperometric detection at dual screen-printed carbon electrodes (SPdCEs) using the hydroquinone (HQ)/horseradish peroxidase (HRP)/H_2O_2 system. The dual immunosensor was successfully applied to the detection of both endogenous target proteins in food extracts and, in addition, to the detection of peanut traces (0.0001% or 1.0 mg· kg^{-1}) in wheat flour spiked samples.

2. Materials and Methods

2.1. Materials

Amperometric measurements were performed with a CHI812B potentiostat (CH Instruments, Austin, TX, USA) controlled by software CHI812B. Dual screen-printed carbon electrodes (SPdCEs) (DRP-C1110, Dropsens, Oviedo, Spain) consisting of two elliptic carbon working electrodes (6.3 mm^2 each), a carbon counter electrode and an Ag pseudo-reference electrode were employed as transducers.

A specific cable connector (ref. DRP-BICAC also from DropSens, S.L.) acted as interface between the SPdCEs and the potentiostat. Single screen-printed carbon electrodes (SPCEs) and their specific connector (DRP-C110 and DRP-CAST, respectively, Dropsens) were also used. All measurements were carried out at room temperature.

A Bunsen AGT-9 Vortex (Lab Merchant Limited, London, UK) was used for the homogenization of the solutions. A Thermomixer MT100 constant temperature incubator shaker (Universal Labortechnik GmbH & Co. KG, Leipzig, Alemania) and a magnetic separator Dynal MPC-S (Thermo Fisher Scientific Inc., Madrid, Spain) were also employed. Capture of the modified-MBs onto the SPCE surface was controlled by a neodymium magnet (AIMAN GZ S.L., Madrid, Spain) embedded in a homemade casing of Teflon. Centrifuges Cencom (J.P. Selecta S.A., Barcelona, Spain) and MPW-65R (Biogen Científica, Madrid, Spain) were used in the separation steps.

All reagents were of the highest available grade. Sodium di-hydrogen phosphate, di-sodium hydrogen phosphate, Tris-HCl, NaCl and KCl were purchased from Scharlab (Barcelona, Spain). Tween®20, N-(3-dimethylaminopropyl)-N′-ethylcarbodiimide (EDC), N-hydroxysulfosuccinimide (sulfo-NHS), ethanolamine, hydroquinone (HQ), hydrogen peroxide (30%, w/v), lysozyme (from chicken egg white) and albumin from chicken egg white (OVA) were purchased from Sigma-Aldrich (Madrid, Spain). 2-(N-morpholino)ethanesulfonic acid (MES) and bovine serum albumin (BSA Type VH) were purchased from Gerbu Biotechnik GmbH (Heidelberg, Alemania) and commercial blocker casein solution (a ready-to-use, phosphate buffered saline (PBS), solution of 1% w/v purified casein) was purchased from Thermo Fisher Scientific (Madrid, Spain). Carboxylic acid-modified MBs (HOOC-MBs, 2.7 μm Ø, 10 mg/mL, Dynabeads® M-270 Carboxylic Acid) were purchased from Invitrogen (San Diego, CA, USA). Peanut allergen Ara h 1 Enzyme-linked immunosorbent assay (ELISA) kit (EL-AH1, containing mouse monoclonal IgG1 (2C12) antiAra h 1 capture antibody, AbC-Ara h 1, purified Ara h 1 standard, and biotinylated mouse monoclonal IgG1 (2F7) antiAra h 1 detection antibody, b-AbD-Ara h 1) and Ara h 2 ELISA kit (EL-AH2, containing mouse monoclonal IgG1 (1C4) antiAra h 2 capture antibody, AbC-Ara h 2, purified peanut allergen Ara h 2 standard, and Polyclonal rabbit antiserum raised against natural purified Ara h 2 as detection antibody, AbD-Ara h 2) were purchased from Indoor Biotechnologies, Inc. (Charlottesville, VA, USA). Peroxidase-conjugated AffiniPure F(ab')₂ Fragment Goat anti-Rabbit IgG (F(ab')₂-HRP), Fc Fragment Specific was purchased from Jackson ImmunoResearch Laboratories, Inc. (West Grove, PA, USA). A high sensitivity Strep-HRP conjugate from Sigma Aldrich (Ref: 000000011089153001, 500 U/mL) (Madrid, Spain) was also used.

All buffer solutions were prepared with water from Milli-Q Merck Millipore purification system (18.2 MΩ cm) (Darmstadt, Germany). Phosphate-buffered saline (PBS) consisting of 0.01 M phosphate buffer solution containing 137 mM NaCl and 2.7 mM KCl; 0.01 M sodium phosphate buffer solution consisting of PBS with 0.05% Tween®20 (pH 7.5, PBST); 0.05 M phosphate buffer, pH 6.0; 0.1 M phosphate buffer, pH 8.0; 0.025 M MES buffer and 0.1 M Tris-HCl buffer, pH 7.2. Activation of the HOOC-MBs was carried out with an EDC/sulfo-NHS mixture solution (50 mg/mL each in MES buffer, pH 5.0). The blocking step was accomplished with a 1 M ethanolamine solution prepared in a 0.1 M phosphate buffer solution of pH 8.0.

2.2. Modification of MBs

Dual Ara h 1 and Ara h 2 determinations at SPdCEs were accomplished by simultaneously preparing two different batches of MBs each of them suitable for the determination of each protein receptor following slightly changed protocols (in order to rearrange the assay times) with respect to those described previously for the individual determination of each protein [14,15]. In brief, 3-μL aliquot of the HOOC-MBs commercial suspension was transferred into a 1.5 mL Eppendorf tube for each batch. MBs were washed twice with 50 μL MES buffer solution for 10 min under continuous stirring (950 rpm, 25 °C). Between washings, the particles were captured using a magnet and, after 4 min, the supernatant was discarded. The MBs-surface confined carboxylic groups were activated by incubation during 35 min at 25 °C under continuous stirring (950 rpm) in 25 μL of the EDC/sulfo-NHS mixture solution. The activated MBs were washed twice with 50 μL of MES buffer and re-suspended in 25 μL of the corresponding capture antibody solution (25 μg/mL AbC-Ara h 1 and 50 μg/mL AbC-Ara h 2, prepared in MES buffer) during 30 min at 25 °C under continuous stirring (950 rpm). Subsequently, the AbC-modified MBs were washed twice with 50 μL of MES buffer solution. Thereafter, the unreacted activated groups on the MBs were blocked by adding 25 μL of a 1 M ethanolamine solution in 0.1 M phosphate buffer, pH 8.0, and incubating the suspension under continuous stirring (950 rpm) for 60 min at 25 °C. After one washing step with 50 μL of 0.1 M Tris-HCl buffer solution (pH 7.2) and two more with 50 μL of the commercial blocker casein solution, the magnetic beads modified with the capture antibody (AbC-MBs) were re-suspended in 25 μL of the target analyte standard solution or the sample (prepared in blocker casein solution) and incubated during 45 min (950 rpm, 25 °C). Then, the modified MBs were washed twice with 50 μL of the blocker casein solution and immersed into the corresponding AbD solution (b-AbD-Ara h 1 and AbD-Ara h 2 1/10,000 and 1/1000 diluted, respectively, in blocker casein solution) during 45 min (950 rpm, 25 °C). After two washing steps with 50 μL of PBST buffer solution (pH 7.5), the resulting beads were incubated during 30 min (950 rpm, 25 °C) in the corresponding labeling reagent solution: Strep-HRP (1/1000) for Ara h 1 and F(ab')$_2$-HRP (1/10,000) for Ara h 2, both prepared in PBST, pH 7.5. Finally, the modified-MBs were washed twice with 50 μL of PBST buffer solution (pH 7.5) and re-suspended in 5 μL of 0.05 M sodium phosphate buffer solution (pH 6.0).

Total determination of Ara h 1 and Ara h 2 was performed at SPCEs. In this case, 3 μL of AbC-Ara h 1-MBs and 3 μL of AbC-Ara h 2-MBs (after the blocking step with ethanolamine) were commingled together into a 1.5 mL Eppendorf tube and incubated 45 min (950 rpm, 25 °C) in a 25 μL of the standard/sample solution (prepared in blocker casein solution). This MB mixture was washed twice with 50 μL of the blocker casein solution and immersed into a mixture solution containing both AbDs 1/10,000 (b-AbD-Ara h 1) and 1/1000 (AbD-Ara h 2) diluted in commercial blocker casein solution during 45 min (950 rpm, 25 °C). After two washing steps with 50 μL of PBST buffer solution (pH 7.5), the resulting beads were incubated during 30 min in a mixture solution containing the two labeling reagents: Strep-HRP (1/1000) and F(ab')$_2$-HRP (1/10,000), prepared in PBST, pH 7.5. Finally, the modified-MBs were re-suspended in 45 μL of 0.05 M sodium phosphate buffer solution (pH 6.0) to perform the amperometric detection.

2.3. Amperometric Measurements

The amperometric measurements at the SPdCEs were performed as follows: the 5 μL of the resuspended MBs modified for Ara h 1 determination were magnetically captured onto one of the working electrodes of the dual SPCE. Similarly, the 5 μL suspension of the modified MBs for the Ara h 2 determination were captured on the second working electrode by keeping the dual SPCE in a horizontal position after placing it in the corresponding homemade magnet holding block. Then, the magnet holding block was immersed into an electrochemical cell containing 10 mL of 0.05 M phosphate buffer of pH 6.0 and 1.0 mM HQ (prepared just before performing the electrochemical measurement). Amperometric measurements in stirred solutions were made by applying a detection potential of −0.20 V vs. Ag pseudo-reference electrode upon addition of 50 μL of a 0.1 M H$_2$O$_2$ solution until the steady-state current was reached at both working electrodes (approx. 100 s). The amperometric signals given through the manuscript corresponded to the difference between the steady-state and the background currents.

To perform the detection at SPCEs, the 45 μL of the MBs mixture solution were magnetically captured on the working electrode of the SPCE. The same protocol described before for the detection at SPdCEs was followed.

2.4. Analysis of Real Samples

The dual Ara h 1 and Ara h 2 amperometric magnetoimmunosensor was applied to the analysis of different food samples containing unknown endogenous amounts of both proteins and also samples free of peanuts (wheat flour) spiked at trace levels.

Different types of foodstuffs, purchased in local supermarkets, were analyzed: wheat flour, hazelnuts; peanuts (peanut flour, raw and fried); chocolate bars with roasted peanuts and peanut cream.

Regarding the analysis of spiked samples, peanut-free wheat flour (verified using the commercial Ara h 1 and Ara h 2 ELISA spectrophotometric kits) was spiked with different amounts of peanut flour that consisted of 100% raw peanut (unknown variety) from a commercial retailer (Frinuts). Accordingly, a series of mixtures containing 1.0%, 0.5%, 0.1%, 0.05%, 0.025%, 0.01%, 0.0075%, 0.005%, 0.001%, 0.0005% and 0.0001% (w/w) of peanut were prepared.

The following protocol was used for the extraction of proteins present in peanuts in all the food samples analyzed: 0.5 g of accurately weighted ground sample (previously blended) were introduced in plastic tubes and incubated in 5.0 mL of Tris-HCl (pH 8.2) overnight at 60 °C under continuous stirring (950 rpm). Regarding the chocolate sample, it was frozen at −20 °C before blending, and 0.5 g of skimmed milk powder (Central Lechera Asturiana®, Asturias, Spain) were added during the extraction in order to avoid masking of the target protein by tannins [7]. Subsequently, the aqueous phase was isolated by centrifugation involving a first step at 3600 rpm during 10 min and a second step at 10,000 rpm during 3 min (4 °C) for a 1 mL aliquot of the first supernatant [14,17,18]. The resulting supernatant appropriately diluted was used to perform the determinations with the MBs-based immunosensor. No significant differences between the Ara h 1 and Ara h 2 content determined was observed after one month storage of these food extracts at 4 °C.

In order to make comparison, the same food extracts were also analyzed by applying both ELISA methods involving the use of the same immunoreagents.

3. Results

Figure 1 shows schematically the principles on which the dual electrochemical magnetoimmunosensor is based. Similar to that previously reported for the individual determination of each allergen protein [14,15], sandwich immunoassays were performed onto HOOC-MBs. Target proteins were sandwiched between respective specific capture antibodies and a biotinylated detector antibody for Ara h 1 (b-AbD-Ara h 1) and a non-biotinylated detector antibody for Ara h 2 (AbD-Ara h 2). These detector antibodies were labeled in a latter step with a streptavidin-HRP (Strep-HRP) polymer in the case of Ara h 1 or an HRP-conjugated secondary antibody in the case of Ara h 2. The MBs bearing the sandwich immunocomplexes for each target protein were magnetically captured on the corresponding working electrode (WE 1 and WE 2) of the SPdCE and amperometric detection at −0.20 V of the catalytic currents generated upon H_2O_2 addition and using HQ as redox mediator in solution at each working electrode was employed to determine each target protein concentration. It is important to note that this methodology implied that the SPdCEs acted only as the electrochemical transducer while all the affinity reactions occurred on the surface of the MBs, thus minimizing unspecific adsorptions of the bioreagents on the electrode surfaces.

The working variables used in the assays are summarized in Table S1 (in the Supporting Information) and were the same as those optimized for the single determination of each target protein with the exception of the incubation time in the AbC-Ara h 2 solution, which has been extended from 15 to 30 min in order to finish the preparation of both MBs batches at the same time. The detection potential value was also previously optimized for the HQ/HRP/H_2O_2 system [19]. Moreover, the working conditions used in the HOOC-MBs activation procedure, the successive washings and the unreacted carboxylic groups blocking step were established according to the protocols provided by the MBs supplier.

Cross-talking between the adjacent working electrodes is considered a potential major drawback to be avoided in the design of electrochemical multisensory platforms [20]. In addition, cross-reactivity amongst antibody pairs selected should be evaluated to demonstrate the feasibility of the bioplatform to perform the simultaneous determination of Ara h 1 and Ara h 2. Figure 2 shows the amperometric measurements obtained with the dual MBs-based immunosensor in solutions containing different Ara h 1 and Ara h 2 mixtures. As it can be deduced, no significant cross-talking between electrodes was apparent and the selected antibody pairs gave rise to significant responses only for the target protein despite the similar structural motifs described in both proteins [21]. These results endorsed the viability of the dual MBs-based immunosensing platform for the simultaneous specific detection of both allergenic proteins. Furthermore, the currents measured in the absence of the target protein can be considered as the negative control to account for any nonspecific binding of the AbDs or the enzymatic labels on the functionalized MBs. As it is shown in Figure 2, the immunosensor responses were mostly due to the selective sandwich immunocomplexes attached to the MBs surface.

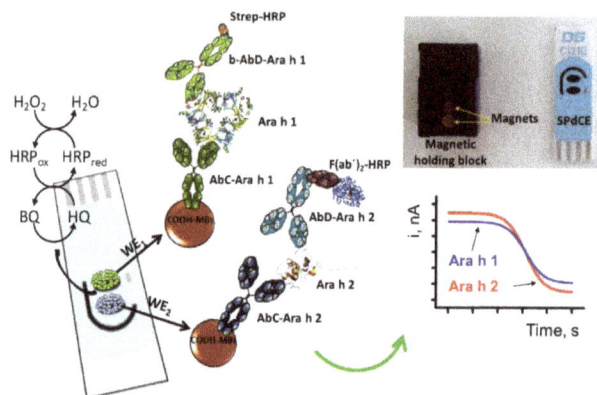

Figure 1. Schematic display of the fundamentals involved in the dual determination of Ara h 1 and Ara h 2 using screen-printed dual carbon electrodes (SPdCEs) as well as in the reactions implied in the amperometric responses. A real picture of the SPdCE and the homemade magnetic holding block is also shown.

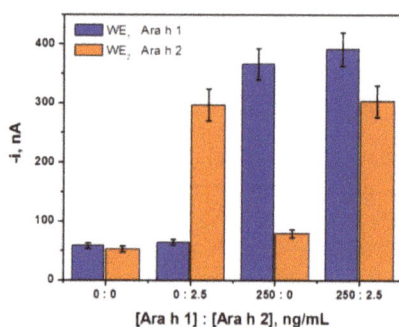

Figure 2. Simultaneous amperometric responses measured with the dual magnetic beads (MBs)-based immunosensor for mixtures containing: 0 ng/mL of both proteins; 0 ng/mL Ara h 1 and 2.5 ng/mL Ara h 2; 250 ng/mL Ara h 1 and 0 ng/mL Ara h 2; 250 ng/mL Ara h 1 and 2.5 ng/mL Ara h 2. $E_{app} = -0.20$ V vs. Ag pseudo-reference electrode. Error bars estimated as triple the standard deviation ($n = 3$).

3.1. Analytical Characteristics

The reproducibility of the simultaneous amperometric responses for 500 ng/mL Ara h 1 and 1.0 ng/mL Ara h 2 was checked using eight different dual MBs-based immunosensors. Relative standard deviation (RSD) values of 7.3% and 8.9% were calculated for Ara h 1 and Ara h 2, respectively, confirming that the whole dual immunosensor preparation process, including MBs modification, MBs magnetic capture on the surface of each working electrode and amperometric measurements, was reliable and that reproducible amperometric responses can be obtained with different immunosensors constructed in the same manner.

Figure 3 displays the calibration plots for both target protein standards with the dual immunosensor. The corresponding analytical characteristics are summarized in Table 1. It is worth to note the remarkably higher sensitivity obtained for the determination of Ara h 2, which is in agreement with that reported by other authors using the same immunoreagents [5], and attributed to a better affinity of the antibody pair used for this target protein. Low detection limits (LODs) of 18 and 0.07 ng/mL (450 and 1.75 pg in 25 μL) were calculated according to the $3 \times s_b / m$ criterion, where s_b was estimated as the standard deviation for 10 blank signal measurements and m is the slope value of the calibration plot. These low LODs are relevant from a clinical point of view since some patients exhibit strong allergic reactions against allergen levels as low as in the ng/mL range [5]. These LODs are slightly higher than those reported with the immunosensors developed for the individual determination of each proteins (6.3 and 0.026 ng/mL for Ara h 1 and Ara h 2, respectively), which is most likely due to the remarkably smaller active surface area of the dual SPCEs working electrodes when compared with the single SPCEs (6.3 vs. 12.6 mm^2). Nevertheless, the LOD values achieved with the dual immunosensor were shown to be sufficient to allow detecting both target proteins in food extracts as well as peanut traces, as it will be demonstrated below.

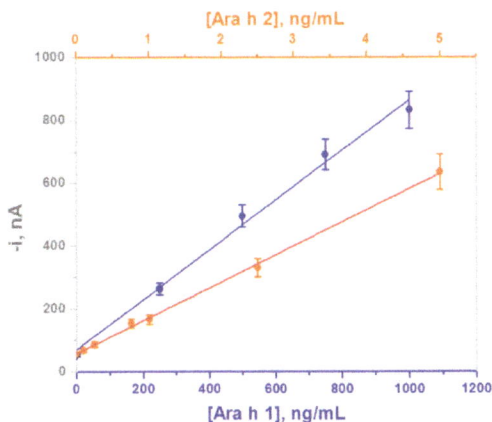

Figure 3. Calibration plots obtained with the dual immunosensing platform for Ara h 1 and Ara h 2 standards. Error bars estimated as triple of the standard deviation ($n = 3$).

Table 1. Analytical characteristics for the determination of Ara h 1 and Ara h 2 using the dual magnetic beads (MBs)-based immunosensing platform.

	Ara h 1	Ara h 2
Linear range (LR), ng/mL	60–1000	0.25–5
r	0.996	0.999
Sensitivity, nAmL/ng	0.79 ± 0.05	115 ± 2
LOD, ng/mL *	18	0.07
Limit of determination (LQ), ng/mL **	60	0.25

* Calculated as $3 \times s_b / m$ where s_b was the standard deviation for 10 blank signal measurements and m is the slope value of the calibration plot; ** Calculated as $10 \times s_b / m$

It is also important to note that the achieved LODs are better than those claimed with commercial ELISA kits for the individual detection of Ara h 1 and Ara h 2 (31.5 and 2 ng/mL, respectively) using the same immunoreagents employed in the dual immunosensor.

The storage stability of the AbC-MBs was tested by keeping them at 4 °C in Eppendorf tubes containing 50 μL of filtered PBS. Two replicates of the stored AbC-Ara h 1-MBs and AbC-Ara h 2-MBs conjugates were incubated each working day in solutions containing no target protein, 250 ng/mL Ara h 1 and 2.5 ng/mL Ara h 2. Control charts were constructed by setting the average current value calculated from 10 measurements made the first day of the study (when the AbC-Ara h 1-MBs and AbC-Ara h 2-MBs were prepared) as the central values, while the upper and lower limits of control were set at $\pm 3 \times$ SD of these central values. The obtained results (not shown) showed that the immunosensors prepared with the stored AbC-MBs provided measurements within the control limits during 25 and 50 days, for Ara h 1 and Ara h 2, respectively. This good storage stability suggests the possibility of preparing sets of AbC-Ara h 1-MBs and AbC-Ara h 2-MBs conjugates and storing them under the above-mentioned conditions until the dual bioplatform needs to be prepared.

3.2. Selectivity of the Dual Magnetoimmunosensor

The selectivity of the dual magnetoimmunosensor was evaluated towards non-target proteins such as BSA, lysozyme and OVA, which can coexist with the target proteins in food extracts. A comparison of the current values measured with the dual immunosensing platform for 0 and 500 ng/mL Ara h 1 and 0 and 1.0 ng/mL Ara h 2 in the absence and in the presence of these potential interfering compounds is shown in Figure 4. No significant effect in the measurements for Ara h 1 and Ara h 2 was apparent as a result of the presence of the three non-target proteins even at the large concentrations tested. Moreover, no noticeable cross-reactivity was observed between Ara h 1 and Ara h 2 despite these proteins showing similar structural motifs [21], and even although Ara h 1 was tested at a 500 times larger concentration than Ara h 2. The high selectivity of the developed platform against other Ara h, legumes and nuts proteins will be also evidenced in the analysis of different complex food extracts where other non-targeted proteins are present in a large extent.

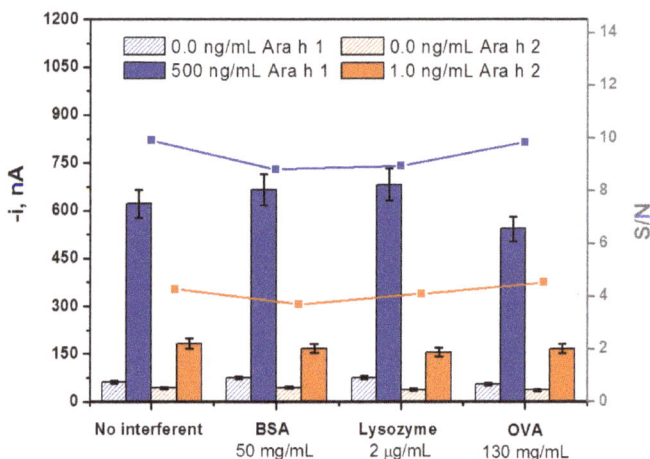

Figure 4. Current values measured for 0 and 500 ng/mL Ara h 1 and 0 and 1.0 ng/mL Ara h 2 in the absence or in the presence of 50 mg/mL bovine serum albumin (BSA), 2 μg/mL lysozyme and 130 mg/mL ovalbumin (OVA). Supporting electrolyte, 0.05 M sodium phosphate solution, pH 6.0; E_{app} = −0.20 V vs. Ag pseudo-reference electrode. Error bars estimated as triple of the standard deviation (n = 3).

3.3. Simultaneous Determination of Ara h 1 and Ara h 2 in Food Samples

The usefulness of the dual immunosensor for the analysis of real samples was verified by determining both target allergen proteins in different food extracts containing variable and unknown amount of endogenous Ara h 1 and Ara h 2 as well as in target-spiked protein-free samples. Most interestingly, no significant matrix effects were found once the sample extracts were appropriately diluted with blocker casein solution. Using the dilution factors summarized in Table 2, no statistically significant differences were observed between the slope value of the calibration plots constructed with Ara h 1 and Ara h 2 standards (Figure 3) and the slope values of the calibration graphs constructed from all the extracts by spiking with growing amounts of standards solution (up to 750 and 2.5 ng/mL for Ara h 1 and Ara h 2, respectively). It is worth remarking that the dilution factors mentioned in Table 2 correspond to those required to fit the target analyte concentration into the linear range of the calibration graphs. Therefore, Ara h 1 and Ara h 2 quantification could be accomplished by simple interpolation of the measured currents in the properly diluted samples at each working electrode of the SPdCE into the calibration plot constructed with standard solutions.

Table 2. Determination of the endogenous content of allergenic proteins Ara h 1 and Ara h 2 in different food extracts with the amperometric dual immunosensor and comparison with the results provided by ELISA spectrophotometric kits.

Extract	[Ara h 1], mg/g			[Ara h 2], mg/g		
	Dilution Factor	Dual Platform	ELISA	Dilution Factor	Dual Platform	ELISA
Fried peanuts	1/1000	(7 ± 2)	(7.3 ± 0.6)	1/250,000	(3.6 ± 0.8)	(3.4 ± 0.6)
Raw peanuts	1/1000	(2.3 ± 0.4)	(2.8 ± 0.3)	1/250,000	(3.8 ± 0.3)	(4.1 ± 0.5)
Chocolate bars with roasted peanuts	1/50	(0.18 ± 0.01)	(0.18 ± 0.03)	1/25,000	(0.23 ± 0.05)	(0.30 ± 0.08)
Peanut cream	1/50	(1.9 ± 0.2)	(1.8 ± 0.4)	1/500,000	(4.5 ± 0.8)	(4.3 ± 0.8)
Raw hazelnuts	—	ND	ND	—	ND	ND
Wheat flour	—	ND	ND	—	ND	ND

ND: non detectable.

The results obtained for all the analyzed food extracts are given in Table 2. In addition, these results were compared with those obtained by using the ELISA kits containing the same immunoreagents. As it can be deduced in Figure 5, excellent correlations were found for both proteins' concentrations determined both with the amperometric multiplexed platform and the single-plexed ELISA kits. RSD values obtained ($n = 3$) are between 2.3%–9.6% and 3.4%–10.5% using the developed dual platform and the conventional ELISA methodology, respectively. For both target proteins, the confidence intervals (at a significance level of $\alpha = 0.05$) for the slope and intercept included the unit and the zero values, respectively. These results demonstrated the great selectivity of the developed platforms against other legumes and nuts proteins since no detectable amperometric responses were obtained for undiluted raw halzenuts and wheat flour extracts.

Furthermore, wheat flour containing no detectable content of the target allergen was spiked with different increasing amounts of peanut flour and the corresponding extracts were obtained and analyzed as described in Section 2.4. Figure 6 shows as the dual immunosensor was able to discriminate samples contaminated with 0.0001%–0.01% peanut through Ara h 2 detection whereas samples contaminated with 0.01%–1.0% peanut could be identified by means of Ara h 1 detection. The ability to detect clearly 0.0001% (1.0 mg· kg^{-1}) peanut improves in a factor of 500–1000 the lowest detectable trace peanuts concentration reported previously by other authors based on Ara h 1 determination, 0.05 [14] and 0.1% [12,13]. This enhanced sensitivity means a major comparative advantage taking into account the serious public health problem that the contamination degree with peanuts of commercial food samples, whether fraudulent or accidental, may cause in sensitive individuals. Moreover, it is important to highlight that although a level of 10 mg· kg^{-1} is considered relevant for the detection of potentially hazardous residues of undeclared allergens in foods, the achievement of a 10 times lower detection limit, as it is the case of this work, may be highly helpful since minimal amounts of the target allergen can be critical [9]. In addition, it is also important to note that the detection level of about 1.0 mg/kg peanut is achieved without any amplification step and, even so, is lower than those reported so far with PCR-based approaches, which are in the 2 to 10 mg/kg range [9,22–24]. Furthermore, the developed immunosensor is suitable to allow allergen determination in a simple way without requirements of a high number of sample replicates and the use of a high precision thermocycler. These features make this methodology possess inherent advantages with respect to PCR-based assays for an easy implementation in analytical food quality and safety laboratories performing routine or decentralized analyses.

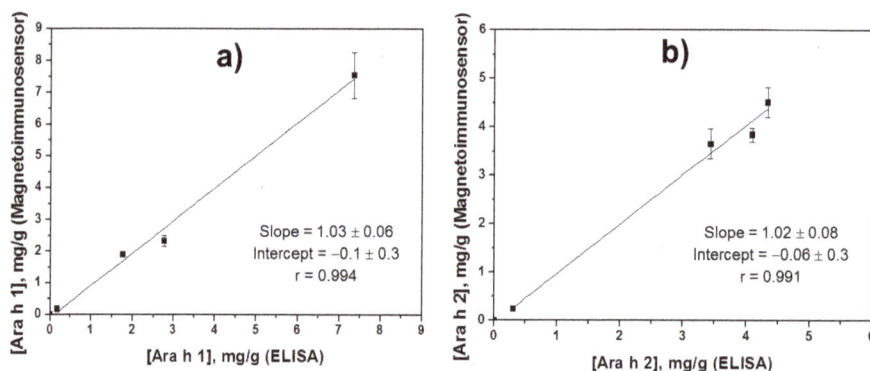

Figure 5. Correlation plots between the results obtained in the determination of Ara h 1 (**a**) and Ara h 2 (**b**) in food extracts using the developed dual immunoplatform and those provided by the individual ELISA spectrophotometric kits (data from Table 2).

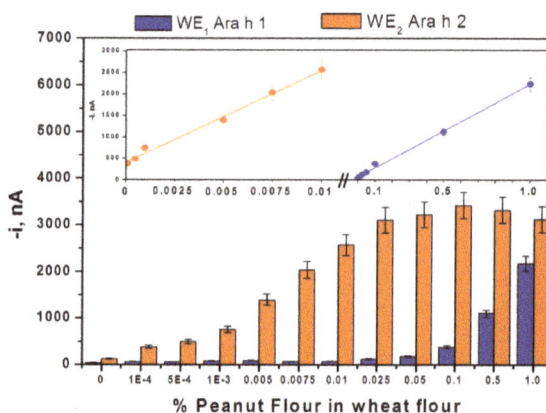

Figure 6. Amperometric responses obtained with the dual MBs-based immunosensing platform for extracts prepared from wheat flour spiked with increasing amounts of peanut flour (final concentrations: 0.0001%, 0.0005%, 0.001%, 0.005%, 0.0075%, 0.01%, 0.025%, 0.05%, 0.1%, 0.5% and 1.0% (*w/w*)). Error bars estimated as triple the standard deviation (*n* = 3).

4. Discussion

Apart from the better sensitivity achieved, it is important to note that the multiplexing capability of the developed platform provides higher levels of information from samples that are unavailable with current commercial ELISA detection kits. Some ELISA kits and detection methods are only capable of providing information on total allergen amounts, therefore lacking information and breakdown of the actual individual allergens in the sample, assuming that patients exhibit the same levels of allergic reaction against all peanut allergens, which is not the case [25,26]. Moreover, data about the individual content of food allergens could be important also in processed foods since some treatments have demonstrated having major roles in changing the allergenic characteristics of particular allergenic proteins (i.e., enhancing allergenic properties of Ara h 2 with roasting) [27,28].

The ability to perform multiplexed allergen detection using the developed platform would therefore provide a means to quantitatively detect specific peanut allergens like Ara h 1 and Ara h 2. This could potentially open up opportunities for patients who, provided with information from clinicians, understand their specific threshold levels for particular allergens to ingest foods that are labeled in detail.

Furthermore, the possibility of shortening the assay time and simplifying as much as possible the sample treatment was considered. With this purpose, measurements in spiked extracts performed according with the whole protocol described in Section 2.4, were compared with those carried out by omitting the centrifugation steps or replacing them by a 30 min natural decantation process. Figure S1 (in the Supporting Information) shows that although, as expected, the resulting S/N ratios were smaller when the extracts were not centrifuged, the amperometric responses were still sufficiently large to discriminate clearly between spiked and non-spiked samples. This relevant result outlined the potentiality of the developed methodology to be employed as a rapid method for alarm or screening purposes able to discriminate samples containing only 0.005% peanut through detection of Ara h 2 protein in a cloudy extract (see orange bars 3 in Figure S1a and tube 3) in Figure S1b).

Total Detection of Both Major Peanut Allergenic Proteins in Food Samples

In view of the results presented in Figure 6 and with the aim of developing a simple methodology for screening the presence of peanuts in foods in a wider range of concentrations, the total detection of both target proteins was evaluated by using a mixture of target specific modified magnetic beads (MBs) sets and amperometric detection at a single SPCE.

Figure 7a shows a schematic display of the approach. The modified MBs are commingled together and incubated sequentially in the sample and mixture of the two AbD solutions as well as in the labeling reagents' solutions and then captured on the surface of a SPCE. The results presented in Figure 7b demonstrate that this methodology allows a clear discrimination of samples contaminated with peanut over five orders of magnitude concentrations, between 0.0001% and 1.0%. Therefore, this approach, although not designed to identify the type of peanut allergenic protein detected, proves to be a user-friendly, attractive, effective and rapid tool (2 h) for detecting the presence in a wide concentration range or verifying the absence of peanuts in foods.

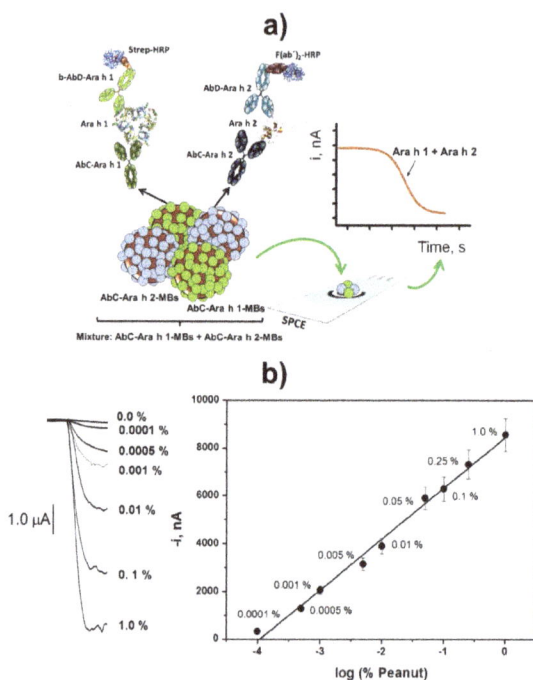

Figure 7. (**a**) schematic display of the methodology involved for total detection of Ara h 1 and Ara h 2 at a SPCE; (**b**) amperometric responses obtained by measuring the total content of Ara h 1 and Ara h 2 with the MBs-based immunosensing approach at SPCEs from extracts prepared for wheat flour unspiked and spiked with increasing amounts of peanut flour (from 0.0001% to 1.0% (w/w)). Error bars estimated as triple the standard deviation ($n = 3$).

5. Conclusions

In this work, a dual electrochemical immunoplatform for the simultaneous detection of peanut allergenic proteins Ara h 1 and Ara h 2 in one single experiment has been described for the first time. This platform provides quantitative information on individual food allergen levels, and the achieved LODs for both target proteins are lower than those claimed for standard well-based ELISAs using the same immunoreagents. Apart from its successful applicability for trace allergen contaminant detection and quantification, the flexibility of this MBs-based electrochemical design allows for further expansion to any allergen that patients are potentially allergic to and to produce a comprehensive array for determining the most important food allergens with a single assay on a single chip. Furthermore, the sensors can be mass produced, making them low cost and disposable. Given its demonstrated important advantages, this novel bioplatform provides food manufacturers and allergic patients or caretakers with an attractive solution to the need for highly sensitive and specific detection of any multiple trace allergen contaminants in food samples.

Supplementary Materials: The following are available online at http://www.mdpi.com/2227-9040/4/3/11/s1, Table S1: Optimized experimental variables affecting the performance of the electrochemical dual MBs-based immunosensing platform for the simultaneous determination of Ara h 1 and Ara h 2, Figure S1: (**a**) Current values measured with the dual immunosensing platform from extracts prepared with unspiked and spiked with 0.5 (Ara h 1) and 0.005 (Ara h 2) % (w/w) wheat flour using different sample treatments: protocol described in Section 2.4 (1); centrifugation steps were substituted by a 30-min natural decantation process (2); centrifugation steps were omitted (3). (**b**) Real picture of the extracts obtained after applying each sample treatment. Supporting electrolyte, 0.05 M sodium phosphate solution, pH 6.0; $E_{app} = -0.20$ V vs. Ag pseudo-reference electrode. Error bars estimated as triple of the standard deviation ($n = 3$).

Acknowledgments: The financial support of the Spanish Ministerio de Economía y Competitividad Research Projects, CTQ2015-64402-C2-1-R, and the NANOAVANSENS Program from the Comunidad de Madrid (S2013/MT-3029) are gratefully acknowledged. Financial support by COFIN 2010–2011 (Programme di Ricerca Scientifica di Relevante Interesse Nazionale, MIUR, 2010AXENJ8_002) is also acknowledged. R.M. Torrente-Rodríguez and V. Ruiz-Valdepeñas Montiel acknowledge predoctoral contracts from the Spanish Ministerio de Economía y Competitividad and Universidad Complutense de Madrid, respectively.

Author Contributions: V.R-V.M. and R.M.T-R. contributed equally to this work. V.R-V.M., R.M.T-R., S.C. and J.M.P. conceived and designed the experiments; V.R-V.M., R.M.T-R. and A.P. performed the experiments; V.R-V.M., R.M.T-R., S.C., M.S.C. and A.J.R. analyzed the data; S.C. and J.M.P. wrote the paper.

Conflicts of Interest: The authors declare no conflict of interest.

Abbreviations

The following abbreviations are used in this manuscript:

ELISAs	Enzyme-linked immunosorbent assays
HQ	Hydroquinone
LOD	Limit of detection
MBs	Magnetic beads
PCR	Polymerase chain reaction
SPdCEs	Screen-printed dual carbon electrodes

References

1. Walker, M.J.; Burns, D.T.; Elliott, C.T.; Gowland, M.H.; Clare Mills, E.N. Is food allergen analysis flawed? Health and supply chain risks and a proposed framework to address urgent analytical needs. *Analyst* **2016**, *141*, 24–35. [CrossRef] [PubMed]
2. U.S. Food and Drug Administration. Consumers-Food Allergies: What You Need to Know. 2015. Available online: http://www.fda.gov/Food/ResourcesForYou/Consumers/ucm079311.htm (accessed on 25 August 2015).
3. Van Hengel, A.J. Food allergen detection methods and the challenge to protect food-allergic consumers. *Anal. Bioanal. Chem.* **2007**, *389*, 111–118. [CrossRef] [PubMed]
4. Montserrat, M.; Sanz, D.; Juan, T.; Herrero, A.; Sánchez, L.; Calvo, M.; Pérez, M.D. Detection of peanut (Arachis hypogaea) allergens in processed foods by immunoassay: Influence of selected target protein and ELISA format applied. *Food Control* **2015**, *54*, 300–307. [CrossRef]
5. Ng, E.; Nadeau, K.; Wang, S.X. Giant magneto resistive sensor array for sensitive and specific multiplexed food allergen detection. *Biosens. Bioelectron.* **2016**, *80*, 359–365. [CrossRef] [PubMed]
6. Food Allergen Labelling and Information Requirements under the EU Food Information for Consumers, Regulation No. 1169/2011: Technical Guidance. Available online: http://www.food.gov.uk/sites/default/files/food-allergen-labelling-technical-guidance.pdf (accessed on 27 June 2016).
7. Poms, R.E.; Agazzi, M.E.; Bau, A.; Brohee, M.; Capelletti, C.; Nørgaard, J.V.; Anklam, E. Inter-laboratory validation study of five commercial ELISA test kits for the determination of peanut proteins in biscuits and dark chocolate. *Food Addit. Contam.* **2005**, *22*, 104–112. [CrossRef] [PubMed]
8. Pedreschi, R.; Nørgaard, J.; Maquet, A. Current Challenges in Detecting Food Allergens by Shotgun and Targeted Proteomic Approaches: A Case Study on Traces of Peanut Allergens in Baked Cookies. *Nutrients* **2012**, *4*, 132–150. [CrossRef] [PubMed]
9. López-Calleja, I.M.; de la Cruz, S.; Pegels, N.; González, I.; García, T.; Martín, R. Development of a real time PCR assay for detection of allergenic trace amounts of peanut (Arachis hypogaea) in processed foods. *Food Control* **2013**, *30*, 480–490. [CrossRef]
10. Holzhauser, T.; Kleiner, K.; Janise, A.; Röder, M. Matrix-normalised quantification of species by threshold-calibrated competitive real-time PCR: Allergenic peanut in food as one example. *Food Chem.* **2014**, *163*, 68–76. [CrossRef] [PubMed]
11. Cho, C.Y.; Nowatzke, W.; Oliver, K.; Garber, E.A.E. Multiplex detection of food allergens and gluten. *Anal. Bioanal. Chem.* **2015**, *407*, 4195–4206. [CrossRef] [PubMed]

12. Alves, R.C.; Pimentel, F.B.; Nouws, H.P.A.; Marques, R.C.B.; González-García, M.B.; Oliveira, M.B.P.P.; Delerue-Matos, C. Detection of Ara h 1 (a major peanut allergen) in food using an electrochemical gold nanoparticle-coated screen-printed immunosensor. *Biosens. Bioelectron.* **2015**, *64*, 19–24. [CrossRef] [PubMed]

13. Alves, R.C.; Pimentel, F.B.; Nouws, H.P.A.; Correr, W.; González-García, M.B.; Oliveira, M.B.P.P.; Delerue-Matos, C. Detection of the peanut allergen Ara h 6 in foodstuffs using a voltammetric biosensing approach. *Anal. Bioanal. Chem.* **2015**, *407*, 7157–7163. [CrossRef] [PubMed]

14. Ruiz-Valdepeñas Montiel, V.; Campuzano, S.; Pellicanò, A.; Torrente-Rodríguez, R.M.; Reviejo, A.J.; Cosio, M.S.; Pingarrón, J.M. Sensitive and selective magnetoimmunosensing platform for determination of the food allergen Ara h 1. *Anal. Chim. Acta* **2015**, *880*, 52–59. [CrossRef] [PubMed]

15. Ruiz-Valdepeñas Montiel, V.; Campuzano, S.; Pellicanò, A.; Torrente-Rodríguez, R.M.; Reviejo, A.J.; Cosio, M.S.; Pingarrón, J.M. Electrochemical detection of peanuts at trace levels in foods using a magnetoimmunosensor for the allergenic protein Ara h 2. *Sens. Actuator B Chem.* **2016**. [CrossRef]

16. Koppelman, S.J.; Vlooswijk, R.A.; Knippels, L.M.; Hessing, M.; Knol, E.F.; Van Reijsen, F.C.; Bruijnzeel-Koomen, C.A. Quantification of major peanut allergens Ara h 1 and Ara h 2 in the peanut varieties Runner, Spanish, Virginia, and Valencia, bred in different parts of the world. *Allergy* **2001**, *56*, 132–137. [CrossRef] [PubMed]

17. Pele, M.; Brohée, M.; Anklam, E.; Van Hengel, A. Presence of peanut and hazelnut in cookies and chocolates: The relationship between analytical results and the declaration of food allergens on product labels. *Food Addit. Contam.* **2007**, *24*, 1334–1344. [CrossRef] [PubMed]

18. Jollet, P.; Delport, F.; Janssen, K.P.F.; Tran, D.T.; Wouters, J.; Verbiest, T.; Lammertyn, J. Fast and accurate peanut allergen detection with nanobead enhanced optical fiber SPR biosensor. *Talanta* **2011**, *83*, 1436–1441.

19. Gamella, M.; Campuzano, S.; Conzuelo, F.; Reviejo, A.J.; Pingarrón, J.M. Amperometric magnetoimmunosensor for direct determination of D-dimer in human serum. *Electroanal.* **2012**, *24*, 2235–2243. [CrossRef]

20. Escamilla-Gómez, V.; Hernández-Santos, D.; González-García, M.B.; Pingarrón-Carrazón, J.M.; Costa-García, A. Simultaneous detection of free and total prostate specific antigen on a screen-printed electrochemical dual sensor. *Biosens. Bioelectron.* **2009**, *24*, 2678–2683. [CrossRef] [PubMed]

21. Bublin, M.; Breiteneder, H. Cross-reactivity of peanut allergens. *Curr. Allergy Asthma Rep.* **2014**, *14*. [CrossRef] [PubMed]

22. Watanabe, T.; Akiyama, H.; Maleki, S.; Yamakawa, H.; Iijima, K.; Yamazaki, F.; Matsumoto, T.; Futo, S.; Arakawa, F.; Watai, M.; et al. A specific qualitative detection method for peanut (*Arachis hypogaea*) in foods using polymerase chain reaction. *J. Food Biochem.* **2006**, *30*, 215–233. [CrossRef]

23. Scaravelli, E.; Brohee, M.; Marchelli, R.; Van Hengel, A. Development of three real-time PCR assays to detect peanut allergen residue in processed food products. *Eur. Food Res. Technol.* **2008**, *227*, 857–869. [CrossRef]

24. Köppel, R.; Dvorak, V.; Zimmerli, F.; Breitenmoser, A.; Eugster, A.; Waiblinger, H.U. Two tetraplex real-time PCR for the detection and quantification of DNA from eight allergens in food. *Eur. Food Res. Technol.* **2010**, *230*, 367–374. [CrossRef]

25. Koppelman, S.J.; Wensing, M.; Ertmann, M.; Knulst, A.C.; Knol, E.F. Relevance of Ara h 1, Ara h 2 and Ara h 3 in peanut-allergic patients, as determined by immunoglobulin E Western blotting, basophil-histamine release and intracutaneous testing: Ara h 2 is the most important peanut allergen. *Clin. Exp. Allergy* **2004**, *34*, 583–590. [CrossRef] [PubMed]

26. Mitchell, D.C. Peanut allergy diagnosis: As simple as Ara h 1, 2, and 3. Consultant for pediatricians. *Consult. Pediatr.* **2013**, *12*, 347–350.

27. Maleki, S.J.; Viquez, O.; Jacks, T.; Dodo, H.; Champagne, E.T.; Chung, S.-Y.; Landry, S.J. The major peanut allergen, Ara h 2, functions as a trypsin inhibitor, and roasting enhances this function. *J. Allergy Clin. Immunol.* **2003**, *112*, 190–195. [CrossRef] [PubMed]

28. Zhou, Y.; Wang, J.-S.; Yang, X.-J.; Lin, D.-H.; Gao, Y.-F.; Su, Y.-J.; Yang, S.; Zhang, Y.-J.; Zheng, J.-J. Peanut allergy, allergen composition, and methods of reducing allergenicity: A review. *Int. J. Food Sci.* **2013**, *2013*. [CrossRef] [PubMed]

chemosensors

Article

Electrochemical Immunosensor for Detection of IgY in Food and Food Supplements

Chiara Gaetani, Emmanuele Ambrosi, Paolo Ugo and Ligia M. Moretto *

Department of Molecular Sciences and Nanosystems, University Ca' Foscari of Venice, 30174 Venice, Italy;
chiara.gaetani@unive.it (C.G.); emmanuele.kizito@gmail.com (E.A.); ugo@unive.it (P.U.)
* Correspondence: moretto@unive.it; Tel.: +39-041-234-8585

Academic Editor: Igor Medintz
Received: 16 December 2016; Accepted: 24 February 2017; Published: 2 March 2017

Abstract: Immunoglobulin Y is a water-soluble protein present in high concentration in hen serum and egg yolk. IgY has applications in many fields, e.g., from food stuff to the mass production of antibodies. In this work, we have implemented an electrochemical immunosensor for IgY based on templated nanoelectrodes ensembles. IgY is captured by the templating polycarbonate and reacted with anti-IgY labeled with horseradish peroxidase. In the presence of H_2O_2 and methylene blue as the redox mediator, an electrocatalytic current is generated which scales with IgY concentration in the sample. After optimizing the extracting procedure, the immunosensor was applied for analysis of fresh eggs and food integrators. The data obtained with the biosensor were validated by SDS-PAGE and Western blot measurements.

Keywords: immunoglobulin Y; egg yolk; nanoelectrodes ensembles; electrochemical immunosensor; food supplements; SDS-PAGE

1. Introduction

Immunoglobulin Y (IgY) is a water-soluble livetin, present in egg yolk. There are three different kinds of livetins and each of them can be compared with a protein of hen serum: α (~to albumin), β (~alfa-2-macroglobulin) and γ (~gamma globulins). γ-livetins are represented in yolk by immunoglobulin Y [1]. Its function is similar to mammalian IgG, providing immunity to the chicks. Concerning the structure, IgY is more similar to mammalian IgE, since the Hinge region is not present in its molecular structure [2]. IgY is composed of two heavy chains (M_w = 68 kDa) and two light chains (M_w = 27 kDa), resulting in a molecular weight of around 190 kDa [3]. The concentration of this antibody in hen blood serum is 5–15 mg/mL, in egg yolk is 10–25 mg/mL; for this high concentration and for its complete absence in egg white, IgY is well representative of egg yolk [4].

Recently, it has been demonstrated that IgY has useful applications in many fields. Among all of them, the most promising use of IgY is in immunotherapy, as shown by the many reviews and articles recently published covering this subject [2–5]; for example, IgY can protect humans and mammalians against animal diseases caused by virus or bacteria attacks. For these purposes, IgY can be added to many foods, both for cattle and humans (kids and adults), to improve the immunological properties of aliments. Moreover, it can be easily purified from egg yolk and its production is low cost and fast. IgY from hen egg yolk is a great resource in the field of primary and secondary antibodies research and development, especially because the cross-reactivity with mammalian complement system is very low or totally absent [2].

It is therefore important to provide reliable analytical methods to monitor the content of IgY in foods and related products. Different analytical methods have been proposed to identify and quantify IgY from egg yolk: ELISA; HPLC [6–8]; a biosensor based on surface plasmon resonance [9];

a nanostructured sensor based on electrochemical detection [10]; an electrochemical immunoassay based on electrodes arrays [11]; fluorescence switch assay [12]; or resonance light scattering [13].

Analytical methods should be fast, simple, reliable, highly sensitive and specific, and if possible, not expensive: electrochemical immunosensors satisfy all these requirements. Recently, our research group proposed an electrochemical immunosensor for IgY determination [10] based on nanoelectrode ensembles (NEEs), prepared by electroless deposition of gold using a track-etched polycarbonate (PC) membrane as the hard template. The polycarbonate part of NEEs has high affinity to proteins, therefore it is possible to promote the immobilization of the biorecognition element on the PC insulating surface of the NEE, leaving the gold nanodisks of the electrode surface free for the electron exchange [14]. The proposed IgY-NEE-based immunosensor was successfully applied to detect egg yolk in tempera paintings and other works of art [10]. NEEs present geometrical and diffusion characteristics that allow very low detection limits to be achieved that potentiate the typical high specificity and selectivity of the immunoassays. For these reasons, NEEs are suitable to be used as transducers of electrochemical biosensors, in particular in the case of immunosensors [15].

In the present work, we focus on the NEE-based IgY immunosensor previously proposed [10] whose performance was optimised with respect to the previous work by decreasing both the geometrical area of the sensor and the amount of reagents. Thanks to these improvements, we have been able to analyse a larger concentration range of yolk and IgY, using a very low amount of sample. This optimisation is especially useful when the analyte is present in low concentration or if there is a small amount of sample available. SDS-PAGE and western blot (WB) assays are carried out for the validation of the method. We demonstrate its applicability for the sensitive determination of IgY in complex matrices such as egg yolk, as well as in other food based on egg components. Among the available food and food supplements based on egg, we studied IgY determination in samples that have undergone an industrial treatment, such as freeze-drying. We investigated if this process can inflict the stability of IgY and its concentration levels in the samples.

2. Materials and Methods

2.1. Materials and Instrumentation

NEEs are prepared via template deposition of gold in a nanofilter polycarbonate (PC) membrane supplied by SPI Pore Filter (47 mm diameter, thickness 6 μm, pore diameter 30 nm, nominal pore density 6×10^8 pores/cm^2), impregnated with polyvinylpyrrolidone by the producer. The geometrical area (area of the nanoelectrodes plus insulator between them) of the NEEs used for this study is 2.27 mm^2, three times smaller than the one used in [10]. The geometrical area (area of the nanoelectrodes plus insulator between them) of the NEEs used for this study is 2.27 mm^2, three times smaller than the one used in [10]. The active area (area of the nanodisks electrodes) is approximately 0.01 mm^2. More detailed information on the deposition of gold and nanoelectrodes fabrication can be found elsewhere [16,17].

Eggs used in this work were purchased at a local market. Albumin from Bovine Serum (BSA) was from Sigma Aldrich; the secondary antibody goat anti-IgY conjugated to horseradish peroxidase (HRP) HRP-conjugated was from Immunology Consultants Laboratory (USA). Freeze-dried eggs (egg white and egg yolk, separately) were from a sport-related company specialized in the production of freeze-dried products and proteins for sportsmen.

Electrochemical measurements were carried out at room temperature with a CH 1000a instrument potentiostat using a typical three-electrodes cell equipped with a platinum wire as the counter electrode, an Ag/AgCl (KCl saturated) reference electrode and NEEs as working electrodes. The supporting electrolyte used was 0.01 M phosphate-buffered saline solution (PBS). Preliminary cyclic voltammetric experiments indicated 10 mV/s as the optimal scan rate for this application.

2.2. Samples Preparation

To identify IgY, it is necessary to separate the water-soluble proteins (including IgY) from the water-insoluble lipid and lipoprotein part of egg yolk. In the literature, several methods are proposed for the extraction and purification of IgY [18–20]; we followed the extraction proposed in [20] that employs 0.2 M acetate buffer at pH 5. Briefly, after mechanical separation of yolk and egg white, a known amount of yolk is dissolved in the buffer to obtain a starting concentration of 0.5 g/mL of egg yolk, then the solution is mixed with vortex and left for two hours in a conical centrifuge tube, to let the heavy and not-soluble fraction deposit. The supernatant, containing IgY and other soluble proteins, is then separated from the pellet with a micropipette. The supernatant is further diluted in acetate buffer to desired values.

IgY was determined in two different food samples: the first was fresh egg yolk and the second was freeze-dried egg commercialised as a food integrator for sportsmen. Eight samples of yolk extracted from the same fresh egg were analysed and analysed concentrations are shown in Table 1; different samples were prepared by sequential dilution. We estimated the amount of IgY immobilized on the electrode from an average concentration of 10 mg/mL of IgY in the yolk and considering a volume of sample deposited on the sensor of 5 µL. For the concentrations of egg yolk in these samples, the amount of IgY present on the surface of the electrodes increases from 4 ng to 1 µg. A sample containing egg white was analyzed as well as a negative test.

Table 1. Description of the analysed samples from fresh egg.

Sample		(Egg Yolk) g/mL
Egg white		//
	1	0.002
	2	0.005
	3	0.01
Egg Yolk	4	0.02
	5	0.05
	6	0.1
	7	0.25
	8	0.5

Freeze-dried egg samples were prepared following procedure and amounts suggested by the provider: i.e., 15–45 g of powder in 150–200 mL of water or milk. For our purposes, we extracted the proteins with acetate buffer 0.2 M at pH 5 directly from the freeze-dried product without previous dissolution in water or milk. Analyzed samples are described in Table 2: one sample of egg white and two samples of freeze-dried egg yolk at different concentrations. For each sample, 1.5 g of powder was dissolved in 25 mL of buffer, then the sample containing egg yolk was further diluted to obtain a lower concentration value. Yolk concentration was calculated considering the dilution factor. Percentage of the protein content in the samples declared by the company is reported in column 5.

Table 2. Description of the freeze-dried samples.

Sample	Powder (g)	Acetate Buffer pH: 5 (mL)	(Egg Yolk) g/mL	Protein Content * (%)
Egg White Powder	1.5	25	0	79
Egg Yolk Powder	1.5	25	0.06	35
Egg Yolk Powder	1.5	250	0.006	35

* Protein content declared by the producer.

2.3. Operation of the Immunosensor

Scheme 1 briefly illustrates the behaviour of the immunosensor: the analyte is captured on the PC portion of the NEE surface area by incubation in the sample at pH 5. Then a blocking step is performed by immersion of the electrode in a bovine serum albumin (BSA) solution. Afterwards, IgY is recognized by a specific secondary antibody labelled with the enzyme horseradish peroxidase (HRP). The label is finally detected by adding the substrate (0.5 mM hydrogen peroxide) in a solution containing a redox mediator (0.1 mM methylene blue); under these conditions, an electrocatalytic signal is generated that scales with the analyte concentration.

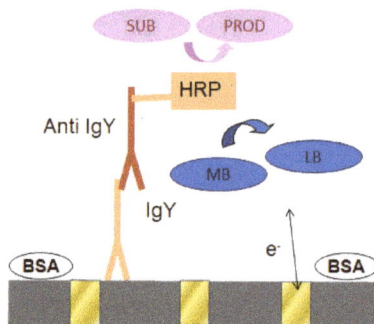

Scheme 1. Scheme of the immunosensor. IgY from the sample is captured by the PC surface. BSA is the blocking agent. The secondary antibody anti-IgY labeled with HRP recognizes IgY and reacts with the substrate (H_2O_2) and the redox mediator methylene blue (MB); leucomethylene blue (LB) is the reduced form of MB.

2.4. Capture of IgY on NEEs

The capture of IgY on NEEs was carried out in the following steps:

1. Incubation of 5 µL of the sample solution (in acetate buffer) for 30 min;
2. Four times rinsing of the NEE with 1 mL of 0.01 M PBS, the electrode is then dried by a gentle stream of air;
3. Immersion of the electrode in 1 mL of blocking solution (1% BSA in PBS) on orbital shaker plates for 30 min;
4. Four times rinsing of the electrode with 1 mL of PBS;
5. Incubation of 5 µL of 0.1 mg/mL of the secondary antibody Anti-IgY-HRP for 30 min;
6. Four times rinsing the electrode with 1 mL of PBS.

The incubation steps are performed in a wet chamber at room temperature. Also in this case, the amount of solution and reagents used to complete the immobilization procedure were three times smaller than those used in [10].

2.5. SDS-PAGE and Western Blot Analysis

To validate the results obtained by the electrochemical immunosensors, SDS-PAGE and Western blot (WB) analysis were performed. Samples were loaded on 12% polyacrylamide discontinuous gels [21] under reducing condition, and separation occurred using a constant voltage of 180 V. Proteins were stained using Commassie Blue. A broad range protein standard (Bio-Rad 161-0317) from 6.5 kDa to 200 kDa was used as the marker.

Proteins were also blotted on a polyvinylidene difluoride (PVDF) membrane with a Mini Trans-Blot® Electrophoretic Transfer Cell (instruments settings: 100 V and 350 mA, Bio-Rad, Hercules,

CA, USA). IgY detection was carried out with the same secondary antibody anti-IgY HRP-conjugated used for the immunosensor. HRP detection was done with 4-chloronaphthol and hydrogen peroxide.

3. Results and Discussion

3.1. Characterization of the Immunosensor

The capture of IgY on the NEE was monitored by cyclic voltammetry and is shown in Figure 1. The dotted-black line CV (cyclic voltammogram) was recorded at a bare NEE in 0.1 mM MB. It shows the quasi-reversible voltammetric behavior of MB, with $E_{1/2} = 0.2$ V. The dashed-blue line CV was recorded in the same solution at the NEE after the complete functionalization (described in Section 2.3—i.e., incubation with IgY, blocking with BSA and incubation with HRP-secondary antibody). It does not show any significant difference from the one recorded at the bare electrode, just a little decrease of the peak current, probably due to almost negligible adsorption on the nanoelectrode surface during the functionalization procedure. On the contrary, dramatic changes are detected at the functionalized NEE after addition of 0.5 mM hydrogen peroxide (Figure 1 solid-red line), that can be summarized as follows:

1. The CV shape moves from the peak shaped to a sigmoidal pattern, with overall current increasing with IgY amount;
2. The anodic peak disappears.

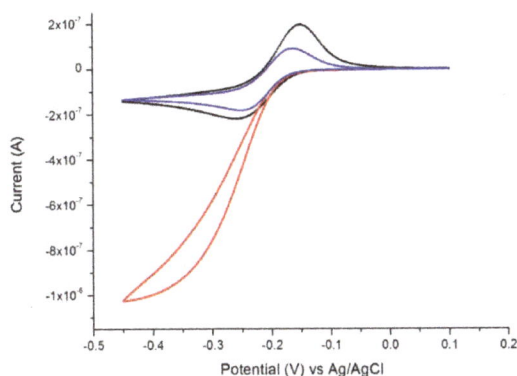

Figure 1. CVs of 0.1 mM MB in 0.01 M PBS before the functionalization (dotted-black line), after the complete functionalization of the NEE (dashed-blue line) and after the addition of 0.5 mM of the enzyme substrate (H_2O_2) (solid-red line). Scan rate 10 mV/s.

Indeed, in the presence of the enzyme, the substrate and the redox mediator, the following electrocatalytic cycle is operative:

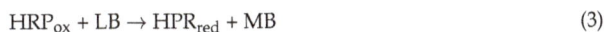

$$HRP_{red} + H_2O_2 \rightarrow HPR_{ox} + H_2O \tag{1}$$

$$MB + 2e^- + H^+ \rightarrow LB \tag{2}$$

$$HRP_{ox} + LB \rightarrow HPR_{red} + MB \tag{3}$$

where LB in the reduced Leuco form of the MB; HRP_{ox} and HRP_{red} are the oxidized and reduced form of HRP, respectively.

Note that the addition of 0.5 mM H_2O_2 to MB solution does not cause any change in the MB pattern recorded at a bare NEE [11]. To identify the appropriate concentration of substrate, tests with increasing concentration of H_2O_2 were performed: 0.5 mM, 1.0 mM and 1.5 mM H_2O_2 were successively added to the same solution and CVs of MB in the presence of the enzyme and its substrate were recorded (see Figure S1 in the Supplementary Materials).

Figure 2 presents the behavior of the immunosensor when the sample incubated contains rabbit glue (i.e., collagen) instead of egg yolk (i.e., IgY). All the following steps of the procedure are performed as described in 2.3. The dotted-black line CV corresponds to the MB recorded at the bare electrode; the dashed-blue line CV was recorded after the complete functionalization procedure (carried out with rabbit glue) and the solid-red line CV was recorded after the addition of the substrate in the same MB solution. In the case of incubation with rabbit glue, some minor changes appear in the CV pattern, however they do not match at all with the features expected for the electrocatalytic process (1)–(3). This evidence confirms the specificity of the NEE-based method for IgY detection.

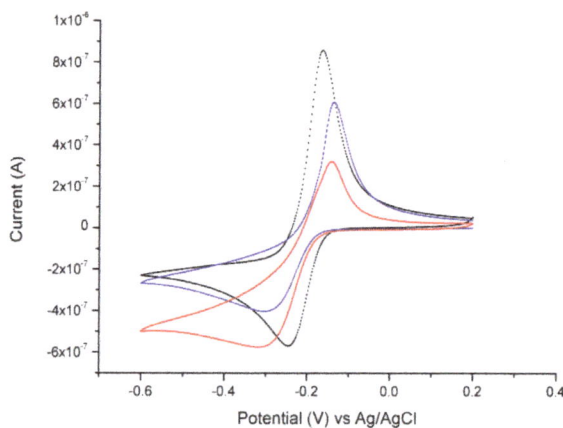

Figure 2. CVs of 0.1 mM MB in 0.01 M PBS before the functionalization (dotted-black line), after the complete functionalization of the NEE with rabbit glue as the analyte (dashed-blue line) and after addition of 0.5 mM H_2O_2 (solid-red line). Scan rate: 10 mV/s.

3.2. Determination of IgY in Egg

The immunosensor was applied to analyse samples of fresh eggs, investigating the correlation between the electrocatalytic current and the egg yolk concentration and amount of IgY. The samples described in Table 1 were analyzed and representative CVs are shown in Figure 3. The solid lines CVs, from yellow to red, were recorded in the presence of H_2O_2 at NEEs incubated in samples containing an increasing amount of IgY. A different NEE was used for each sample and the relative standard deviation of three measurements was 5%. The dotted line CV corresponds to the typical pattern of the mediator MB (at a NEE treated with egg yolk and HRP-anti-IgY) before addition of H_2O_2.

It is possible to observe that the electrocatalytic current increases with an increasing amount of IgY. Data in Figure 4 show that the dependence of the catalytic current is a function of the yolk content over the entire concentration range examined. It is important to note that the reported current values represent I_{net}, calculated by Equation (4):

$$I_{net} = I_{Ecat} - I_{cat} \tag{4}$$

where I_{cat} is the cathodic peak of MB in the absence of the substrate, I_{Ecat} is the electrocatalytic current recorded at -0.40 V, where the current achieved a plateau.

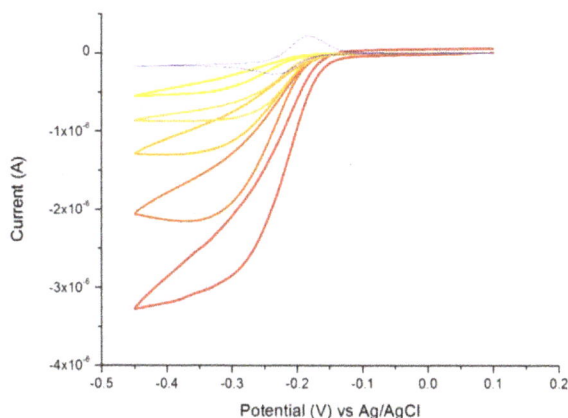

Figure 3. CVs of 0.1 mM MB in 0.01 M PBS in the absence (dashed-blue line) and in the presence of 0.5 mM H_2O_2 at NEEs incubated with egg yolk at a concentration from 0.002 g/mL to 0.5 g/mL of yolk (solid-lines from light yellow to red). Scan rate 10 mV/s.

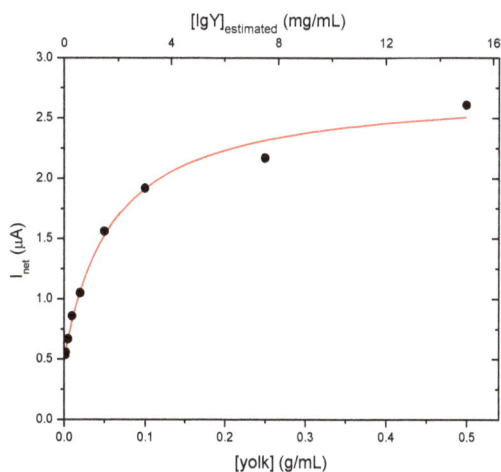

Figure 4. Electrocatalytic current vs. the concentration of yolk (lower *x*-axis) and versus the relevant estimated IgY concentration (upper *x*-axis), and best fitting curve. The curve can be used to quantify the IgY concentration.

The pattern of the electrocatalytic current increment vs the amount of yolk fits a saturation-like curve, following Equation (5):

$$y = \frac{ax}{b + x} + c \qquad (5)$$

where the fitting parameters are $a = 2.25$ µA; $b = 0.06$ g/mL; $c = 0.49$ µA, with an R^2 value of 0.991. From this curve, it is also possible to define a linear range from 0.002 to 0.05 g/mL of yolk. This trend is in agreement with the one previously observed in Figure S1 of Reference [10]. To define the amount of IgY related to the yolk concentration, semi-quantitative comparison between preliminary SDS-PAGE assays carried out on pure IgY and the electrochemical immunosensor was performed (results not shown). This test suggests that for a yolk concentration of 0.02 g/mL (sample 4 in Table 1—the sample is included in the linear range in Figure 4, lower *x*-axis) the sensor is able to detect an IgY concentration of 0.06 mg/mL, indicating that almost 3% content of egg yolk can be attributed to IgY and recognized by the immunosensor. Therefore, the concentration of IgY in the linear range considered for the yolk concentration is between 0.06 mg/mL and 1.5 mg/mL. These data are shown also in the upper axis of Figure 4.

3.3. Determination of IgY in Food Integrators

The immunosensor was applied to determine IgY in matrices of industrialized food, by following the same procedure as above, but on freeze-dried food integrators containing only egg yolk or only egg white (see Table 2).

The cyclic voltammograms recorded after the complete procedure performed on the three samples are shown in Figure 5, namely egg white 0.06 g/mL, egg yolk 0.006 g/mL and 0.06 g/mL. In all the cases, an electrocatalytic current is observed. The catalytic current recorded in the presence of egg white (Figure 5 solid-yellow line) is very low and can be ascribed to aspecific adsorption. On the other hand, the same parameter increases dramatically for egg yolk powder, scaling with the amount of food supplement analysed.

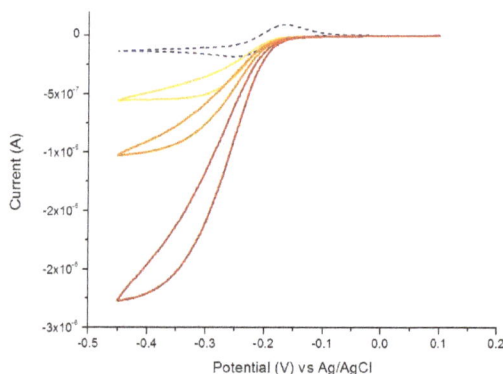

Figure 5. CVs of 0.1 mM MB in 0.01 M PBS recorded at 10 mV/s before (dashed-blue lines) and after addition (solid lines) of 0.5 mM H_2O_2 for the following samples: 0.06 g/mL egg white powder (full-yellow line), 0.006 g/mL egg yolk powder (full-orange line) and 0.06 g/mL egg yolk powder (full-red line).

By combining electrophoresis (see below) and voltammetric data (see Figure 4), it has been possible to estimate the amount of IgY detected by the immunosensor. Calculations carried out on the most diluted samples suggest a concentration of IgY close to 0.4 mg/mL, correlated to 0.005 g/mL of yolk, while the experimental concentration of the analysed sample is 0.006 g/mL of yolk. The industrial treatments of the eggs probably cause a little damage to the protein structure that involves 20% of the material.

3.4. SDS-PAGE and WB Analysis

The validation of the results obtained with the electrochemical immunosensor was carried out with SDS-PAGE and WB assays. Figure 6a,b shows the SDS-PAGE and WB results, respectively, for some of the samples described in Tables 1 and 2, containing fresh egg yolk and freeze-died egg yolk. Electrophoresis carried out in reducing conditions should provide two bands that can be correlated with IgY, one corresponding to the heavy chain (HC) and the other to the light chain (LC). Molecular weight of HC and LC of IgY according to literature are 68 and 27 kDa respectively [3]. Figure 6a, lanes 1–4, shows many protein bands, both in the samples containing fresh egg yolk and freeze-dried egg yolk. These bands represent the soluble proteins of the yolk, and considering the complexity of the matrices and the high numbers of proteins present in it, it is difficult to correlate and identify IgY bands. In order to obtain specific results, a WB assay was performed. Secondary antibody anti-IgY identified three IgY bands, as shown in Figure 6b, lanes 1–4. The first band is relevant to the heavy IgY chains and the third one is relevant to the light chains. The band that is intermediate to the first and third ones has not been associated to any fragment or chain of IgY, but is previously observed in the literature [21–24]. Verifying the correlation between WB and SDS-PAGE bands, it is possible to ascribe a molecular weight to these bands: the HC bands in WB correspond to the 67 kDa band in SDS-PAGE, the LC band corresponds to 28.6 kDa, in good agreement with literature data. The intermediate band corresponds to a M_w of 44 kDa, previously observed but not completely explained in the literature [24].

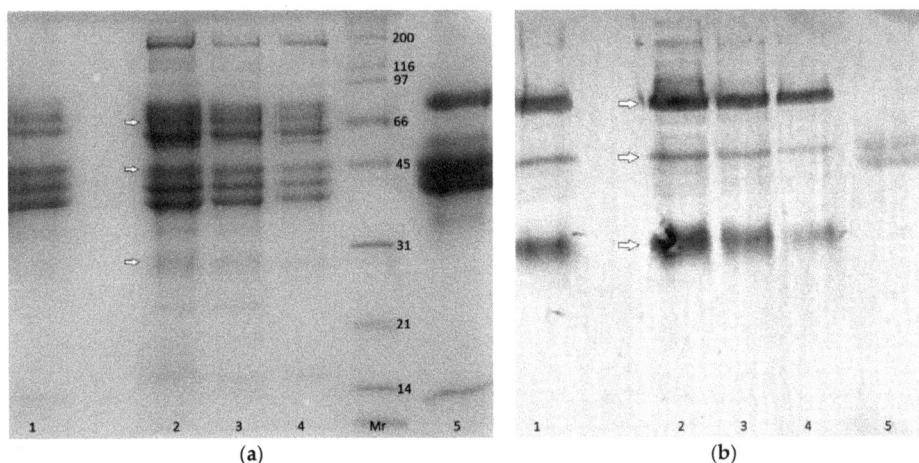

Figure 6. (a) SDS-PAGE results for the following samples: (1) freeze-dried yolk, 0.006 g/mL; (2–4) egg yolk samples 0.02 mg/mL, 0.01 mg/mL, 0.005 mg/mL, respectively; (5) egg white 0.02 mg/mL; Mr = marker. The values indicated in the marker lane correspond to the MW expressed in kDa; (b) Results of the WB, with the samples in the same order as in (a). The arrows indicate the lines considered.

A sample of egg white was analysed by SDS-PAGE and WB as well. The results are shown in lane 5 in Figure 6a,b. The SDS-PAGE presents at least five bands (Figure 6a), but none of these can be correlated to IgY (since the molecular weight of the bands does not correspond with the ones observed in lanes 1–4), indicating the absence of IgY in egg white. Despite this, the secondary antibody recognises two bands, at M_w around 45 kDa (Figure 6b, lane 5), that can be ascribed to an aspecific interaction between the antibody and an egg white protein present in high concentration in the samples. It is worth noting that this aspecific interaction was observed also with the immunosensor (see Figure 3 in Section 3.2).

It is worth commenting some advantages of the electrochemical proposed method with respect to different methods present in the literature. The combination of NEEs' high sensitivity and selectivity of the enzyme–substrate interaction allows the detection of IgY in low concentration both in simple and complex matrices. In fact, we have been able to detect 60 µg/mL of IgY in the egg yolk sample with just a simple one step extraction treatment and 0.4 mg/mL of IgY in the sample that has undergone industrial treatment such as freeze-drying. ELISA commercial kits offer the same sensibility (ng/mL), while SDS-PAGE is a bit less sensitive. Other advantages of the proposed method concern the time-consumption and costs of the techniques: the electrochemical procedure is, in fact, shorter than ELISA and definitely shorter than SDS-PAGE and WB procedures. Moreover, thanks to the low amount of chemicals required, the method has a very low cost. Other immunosensors [11,12] based on different principles present higher sensitivity, but are much more complex in the preparation and with high cost, and are not applicable to complex matrices.

4. Conclusions

The proposed electrochemical immunosensor based on gold nanoelectrodes ensemble is suitable to identify immunoglobulin Y in different samples such as fresh eggs and food supplements based on egg components. The reduction of the electrode size with respect to our previous work allowed us to exploit a wide range of analyte concentrations using small amounts of reagents. We observed that the electrocatalytic current generated by the reaction of the enzyme with its substrate follows a saturation-like trend, for a yolk concentration interval of 0.002 to 0.5 g/mL with the possibility of quantitative determination up to approximately 0.1 g/mL. The sensor is also able to identify IgY in complex matrices that have undergone an industrial treatment such as freeze-drying. The method was validated with SDS-PAGE and WB, all the results obtained with the immunosensor were verified, confirming its high sensitivity. The proposed method is fast, reliable and not expensive. Egg white analysed samples generated an almost negligible electrocatalytic current. Since these aspecific interactions were observed also with the Western blot analysis, they can be ascribed to partial aspecificity of the secondary antibody. The large concentration range analysed is comparable to those of other analytical techniques. Coupling the electrochemical immunosensor with SDS-PAGE improved the information obtained in a previous work, demonstrating the good performance of the EC immunosensor and the versatility of NEEs as a platform for different biosensors.

Supplementary Materials: The following are available online at www.mdpi.com/2227-9040/5/1/10/s1.

Author Contributions: C.G. performed all the electrochemical measurements, prepared and functionalized the nanoelectrodes ensembles, wrote the paper draft; E.A. performed the electrophoretic measurements; P.U. proposed the initial idea, discussed the results and corrected the paper draft; L.M.M. supervised the work, discussed the results and wrote the finale version of the paper.

Conflicts of Interest: The authors declare no conflict of interest.

References

1. Ulrichs, T.; Drotleff, A.M.; Ternes, W. Determination of heat-induced changes in the protein secondary structure of reconstituted livetins (water-soluble proteins from hen's egg yolk) by FTIR. *Food Chem.* **2015**, *172*, 909–920. [CrossRef] [PubMed]
2. Rahman, S.; Nguyen, S.V.; Icatlo, F.C., Jr.; Umeda, K.; Kodama, Y. Oral passive IgY-based immunotherapeutics. *Hum. Vaccines Immunother.* **2013**, *9*, 1039–1048. [CrossRef] [PubMed]
3. Santos, F.N.; Brum, B.C.; Cruz, P.B.; Molinaro, C.M.; Silva, V.L.; Chaves, S.A.M. Production and Characterization of IgY against Canine IgG: Prospect of a New Tool for the Immunodiagnostic of Canine Diseases. *Braz. Arch. Biol. Technol.* **2014**, *57*, 523–531. [CrossRef]
4. Spillner, E.; Braren, I.; Greunke, K.; Seismann, H.; Blank, S.; du Plessis, D. Avian IgY antibodies and their recombinant equivalents in research, diagnostics and therapy. *Biologicals* **2012**, *40*, 313–322. [CrossRef] [PubMed]
5. Kovacs-Nolan, J.; Mine, Y. Egg yolk Antibodies for Passive Immunity. *Annu. Rev. Food Sci. Technol.* **2012**, *3*, 163–182. [CrossRef] [PubMed]
6. Murai, A.; Kakiuchi, M.; Hmano, T.; Kobayashi, M.; Tsudzuki, M.; Nakano, M.; Matsuda, Y.; Horio, F. An ELISA for quantifying quail IgY and characterizing maternal IgY transfer to egg yolk in several quail strains. *Vet. Immunol. Immunop.* **2016**, *175*, 16–23. [CrossRef] [PubMed]
7. Fassbinder-Orth, C.A.; Wilcoxen, T.E.; Tran, T.; Boughton, R.K.; Fai, J.M.; Hofmeister, E.K.; Grindstaff, J.L.; Owen, G.C. Immunoglobulin detection in wild birds: Effectiveness of three secondary anti-avian IgY antibodies in direct ELISAs in 41 avian species. *Methods Ecol. Evol.* **2016**, *7*, 1174–1181. [CrossRef] [PubMed]
8. Potenza, M.; Sabatino, G.; Giambi, F.; Rosi, L.; Papini, A.M.; Dei, L. Analysis of egg-based model wall painting by use of an innovative combined dot-ELISA and UPLC-based approach. *Anal. Bioanal. Chem.* **2013**, *405*, 691–701. [CrossRef] [PubMed]

9. Scarano, S.; Carretti, E.; Dei, L.; Baglioni, P.; Minunni, M. Coupling non invasive and fast sampling of proteins from work of art surfaces to surface pasmon resonance biosensing: Differential and simultaneous detection of egg components for cultural heritage diagnosis and conservation. *Biosens. Bioelectron.* **2016**, *85*, 83–89. [CrossRef] [PubMed]

10. Bottari, F.; Oliveri, P.; Ugo, P. Electrochemical immunosensor based on ensemble of nanoelectrodes for immunoglobulin IgY detection: Application to identify hen's egg yolk in tempera paintings. *Biosens. Bioelectron.* **2014**, *52*, 403–410. [CrossRef] [PubMed]

11. Wilson, M.S.; Nie, W. Electrochemical Multianalyte Immunoassays Using an Array-Based Sensor. *Anal. Chem.* **2006**, *78*, 2507–2513. [CrossRef] [PubMed]

12. Wang, Q.; Fu, X.; Huang, X.; Wu, F.; Ma, M.; Cai, Z. A rapid triple-mode fluorescence switch assay for immunoglobulin detection by using quantum dots-gold nanoparticles nanocomposites. *Sens. Actuators B* **2016**, *231*, 779–786. [CrossRef]

13. Wang, Q.; Wu, S.; Ma, M.; Cai, Z. Determination of Egg Yolk Immunoglobulin by Resonance Light Scattering of Affinity-Labeled Au Nanoparticles. *Food Anal. Methods* **2016**, *9*, 2052–2059. [CrossRef]

14. Mucelli, S.P.; Zamuner, M.; Tormen, M.; Stanta, G.; Ugo, P. Nanoelectrode ensembles as recognition platform for electrochemical immunosensors. *Biosens. Bioelectron.* **2008**, *23*, 1900–1903. [CrossRef] [PubMed]

15. Ongaro, M.; Ugo, P. Bioelectroanalysis with nanoelectrode ensembles and arrays. *Anal. Bioanal. Chem.* **2013**, *405*, 3715–3729. [CrossRef] [PubMed]

16. Menon, V.P.; Martin, C.R. Fabrication and evaluation of nanoelectrode ensembles. *Anal. Chem.* **1995**, *67*, 1920–1928. [CrossRef]

17. De Leo, M.; Pereira, F.C.; Moretto, L.M.; Scopece, P.; Polizzi, S.; Ugo, P. Towards a better understanding of gold electroless deposition in track-etched templates. *Chem. Mater.* **2007**, *19*, 5955–5964. [CrossRef]

18. Gee, S.C.; Bate, I.M.; Thomas, T.M.; Rylatt, D.B. The purification of IgY from chicken egg yolk by preparative electrophoresis. *Protein Expr. Purif.* **2003**, *30*, 151–155. [CrossRef]

19. Chang, H.M.; Lu, T.C.; Chen, C.C.; Tu, Y.Y.; Hwang, J.Y. Isolation of Immunoglobulin from Egg Yolk by Anionic Polysaccharides. *J. Agric. Food Chem.* **2000**, *48*, 995–999. [CrossRef] [PubMed]

20. Sugita-Konishi, Y.; Shibata, K.; Yun, S.S.; Hara-Kudo, Y.; Yamaguchi, K.; Kumagai, S. Immune Functions of Immunoglobulin Y Isolated from Egg Yolk of Hens Immunized with Various Infectious Bacteria. *Biosci. Biotechnol. Biochem.* **1996**, *60*, 886–888. [CrossRef] [PubMed]

21. Laemmli, U.K. Cleavage of structural proteins during the assembly of the head of bacteriophage T4. *Nature* **1970**, *227*, 680–685. [CrossRef] [PubMed]

22. Hodek, P.; Trefil, P.; Simunek, J.; Hudecek, J.; Stiborova, M. Optimized Protocol of Chicken Antibody (IgY) PurificationProviding Electrophoretically Homogenous Preparations. *Int. J. Electrochem. Sci.* **2013**, *8*, 113–124.

23. Borhani, K.; Mobarez, A.M.; Khabiri, A.R.; Behmanesh, M.; Khoramabadi, N. Production of specific IgY Helicobacter pylori recombinant OipA protein and assessment of its inhibitory effects towards attachment of H. pylori to AGS cell line. *Clin. Exp. Vaccine Res.* **2015**, *4*, 177–183. [CrossRef] [PubMed]

24. Pauly, D.; Chacana, P.A.; Calzado, E.G.; Brembs, B.; Schade, R. IgY Technology: Extraction of Chicken Antibodies from Egg Yolk by Polyethylene Glycol (PEG) Precipitation. *J. Vis. Exp.* **2011**. [CrossRef] [PubMed]

chemosensors

MDPI

Article

Design of an Affibody-Based Recognition Strategy for Human Epidermal Growth Factor Receptor 2 (HER2) Detection by Electrochemical Biosensors

Hoda Ilkhani, Andrea Ravalli and Giovanna Marrazza *

Department of Chemistry "Ugo Schiff", University of Florence, Via della Lastruccia 3,
50019 Sesto Fiorentino (FI), Italy; jeyranh@yahoo.com (H.I.); andrea.ravalli@unifi.it (A.R.)
* Correspondence: giovanna.marrazza@unifi.it; Tel.: +39-055-457-3320

Academic Editors: Paolo Ugo and Ligia Moretto
Received: 26 July 2016; Accepted: 30 November 2016; Published: 2 December 2016

Abstract: In this study, we have designed and realized three simple electrochemical bioassays for the detection of the human epidermal growth factor receptor 2 (HER2) cancer biomarker using magnetic beads coupling screen-printed arrays. The different approaches were based on a sandwich format in which affibody (Af) or antibody (Ab) molecules were coupled respectively to streptavidin or protein A-modified magnetic beads. The bioreceptor-modified beads were used to capture the HER2 protein from the sample and sandwich assay was performed by adding the labeled secondary affibody or the antibody. An enzyme-amplified detection scheme based on the coupling of secondary biotinylated bioreceptor with streptavidin-alkaline phosphatase enzyme conjugate was then applied. The enzyme catalyzed the hydrolysis of the electro-inactive 1-naphthyl-phosphate to the electro-active 1-naphthol, which was detected by means of differential pulse voltammetry (DPV). Each developed assay has been studied and optimized. Furthermore, a thorough comparison of the analytical performances of developed assays was performed. Finally, preliminary experiments using serum samples spiked with HER2 protein were also carried out.

Keywords: affibody; immunosensor; magnetic beads; electrochemical detection; cancer biomarker

1. Introduction

One of the biggest factors associated with successful treatment outcome is the early detection of cancer [1]. Unfortunately, for many types of cancers, the first outward symptoms appear late in disease progression; therefore, cancer biomarker detection in biological fluids including serum, sputum and urine has an important role in early cancer detection [2,3]. The human epidermal growth factor receptor 2 (HER2) protein is a member of the epidermal growth factor receptor (EGFR or ErbB) family and is a trans-membrane tyrosine kinase receptor [4]. The level of HER2 in serum has a direct relationship with the risk of diseases such as ovarian, lung, gastric and oral cancers [5,6]. The extracellular domain (ECD) of cleaved HER2 protein enters into the serum, serving as an indicator of increased HER2 expression [4,7]. The HER2 concentration in the serum of breast cancer patients is 15–75 ng/mL, which is elevated when compared to that of normal individuals (2–15 ng/mL) [8]. Monitoring the level of HER2 protein could also be a good indicator of antitumor treatment efficiency [9]. Fluorescence in situ hybridization (FISH) and immunohistochemistry (IHC) are the most commonly used methods for HER2 analysis. Both procedures are complex, involve time-consuming steps and require specially trained personnel to carry them out. Therefore, several new methods are reported for HER2 detection in biological fluids [10–13].

Immunosensors are important analytical tools designed to detect the binding event between the antibody and antigen without the need for separation and washing steps [4]. The most common types of immunosensors are found in electrochemical [7,14–21], optical [22,23], and gravimetric [24] sensors. It is well known that electrochemical detection methods have been able to sensitively and quickly detect biomolecule targets with high selectivity [25–27].

Recently, instead of antibodies (Abs), affibodies (Afs) have been used as new bioreceptors in novel immunosensors for improving the selectivity and sensitivity of the assay [28,29]. Affibody molecules are engineered small proteins with 58 amino acid residues ($\approx$7 kDa) based on a single polypeptide and 3 R-helical bundle structure (the smallest and fastest known cooperatively folding structural domain), as derived from one of the Immunoglobulin G (IgG) binding domains of Staphylococcal protein A [30]. They have high affinity and selectivity for a wide variety of applications such as detection reagents [31] and inhibit receptor interactions [32].

Recently, there has been a focus on how apply the magnetic beads (MBs) coated with proteins, polymers or other molecules in different fields of biochemical science [33,34]. The protein-coated MBs have shown a variety of applications in immunosensors because of their high, specific affinity to biomolecules, possibility for solution-phase bio-recognition reaction, and easy washing and collection [35,36]. The protein A-coated magnetic beads can be used to immuno-precipitate target proteins from crude cell lysates using selected primary Abs. In addition, specific Abs can be chemically cross-linked to the protein A-coated surface to create reusable immuno-precipitation beads, thus avoiding the co-elution of antibodies with target antigens [37]. The streptavidin-coated magnetic beads (Strept-MBs) provide a fast and convenient method for manual or automated immuno-precipitation, protein interaction studies, DNA-protein pull downs and the purification of biotinylated proteins and nucleic acids. They use a recombinant form of streptavidin with a mass of 53 kDa and a neutral isoelectric point [38]. Streptavidin is covalently coupled to the surface of the MBs. For each streptavidin molecule on the bead surface, there are four biotin-binding available sites. Unlike avidin, streptavidin has no carbohydrate groups, resulting in low nonspecific binding. Furthermore, the MBs exhibit low nonspecific binding in the presence of complex biological samples such as blood serum and whole cells [39].

In this study, different protocols using antibody (Ab) and affibody (Af) as capture and signaling bioreceptor were applied to disposable electrochemical immunosensors based on the sandwich assay for HER2 detection. The first protocol relies on immobilization of the antibody on the magnetic beads coated with protein A (Prot A-MBs) as capture bioreceptor and the use of the biotinylated affibody as a signaling bioreceptor. The second protocol relies of the use of biotinylated affibody as a capture bioreceptor and biotinylated antibody as a signaling bioreceptor. The last protocol relies on immobilization of the biotinylated affibody on the magnetic beads coated with streptavidin (Strept-MBs) as capture bioreceptors and the use of secondary biotinylated affibody as signaling bioreceptor as well. An enzyme-amplified detection scheme based on the coupling of secondary biotinylated bioreceptors with streptavidin-alkaline phosphatase conjugates was then applied. The enzyme catalyzed the hydrolysis of the electroinactive 1-naphthyl-phosphate to 1-naphthol; this product is electroactive and is detected by means of differential pulse voltammetry (DPV). Moreover, we have evaluated potential application of the bioassays for serum sample analysis. In all approaches, eight screen-printed electrochemical cells are used as transducers.

2. Materials and Methods

2.1. Chemicals and Reagents

Dynabeads® paramagnetic beads, coated with protein A (ProtA-MBs) and with streptavidin (Strept-MBs) were provided by Invitrogen (Milan, Italy). The monoclonal anti-human HER2 antibody (Ab1), the biotinylated anti-human HER2 antibody (Biot-Ab2) and the HER2 protein (R&D Systems) were obtained from Space SRL (Milan, Italy). Biotinylated anti-HER2 affibody molecules were purchased from Abcam (Cambridge, UK). Sodium phosphate dibasic dihydrate and sodium phosphate monobasic monohydrate were purchased from Merck (Milan, Italy). Sodium chloride, Trizma base, diethanolamine, magnesium chloride, potassium chloride, polyoxyethylene sorbitan monolaurate

(Tween 20), 1-naphtyl phosphate, bovine serum albumin (BSA), streptavidin-alkaline phosphatase (S-AP) and human serum sample were obtained from Sigma (Milan, Italy). All solutions were prepared using water from a Milli-Q Water Purification System (Millipore, UK).

All the buffers used in this study are as follows:

- Buffer A: phosphate-buffered saline (PBS, 0.10 M, pH = 5.0), containing 140 mM NaCl (with and without 0.05% Tween 20);
- Buffer B: phosphate-buffered saline (PBS, 0.10 M, pH = 7.4), containing 0.10 M KCl (with and without 0.05% Tween 20);
- Buffer C: diethanolamine buffer (DEA, 0.10 M, pH = 9.6), containing 1.0 mM MgCl$_2$ and 100 mM KCl (with and without 0.05% Tween 20 and 0.1% BSA);
- Buffer D: Tris buffer (20 mM, pH = 7.4), containing 150 mM NaCl (with and without 0.05% Tween 20).

2.2. Apparatus

Electrochemical measurements were performed using Palmsens Electrochemical Interface system (Palm Instruments BV, Houten, The Netherlands). The transducer was composed of eight screen-printed electrochemical cells, each one composed of a graphite working electrode (diameter = 2.0 mm), a silver pseudo-reference electrode and a graphite counter electrode. The arrays were produced in house by a DEK 248 screen-printing machine (DEK, Weymouth, UK). In order to use sensor array in combination with the magnetic beads, each array was placed on a suitable holding block mounting eight magnet bars of 1.5 mm diameter. The eight sensors strips, coupled with a specially designed methacrylate well box, were compatible with a standard 8-channel multi pipette. A sample mixer with a 12-tube mixing wheel and the magnet rack were purchased from Dynal Biotech (Milan, Italy).

2.3. Development of Affibody-Based Assay

The schematic representation of dual affibody sandwich assay for HER2 detection is reported in Scheme 1.

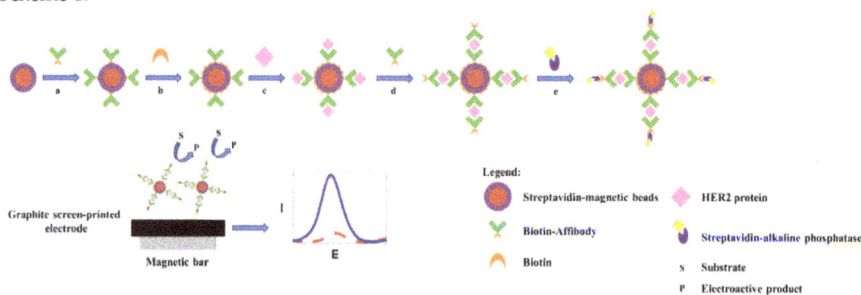

Scheme 1. Schematic representation affibody-based assay for human epidermal growth factor receptor 2 (HER2) detection: (**a**) functionalization of streptavidin-modified magnetic beads with the biotinylated affibody; (**b**) blocking step with biotin; (**c**) affinity reaction with HER2 protein; (**d**) incubation with the secondary biotinylated affibody; (**e**) addition of streptavidin-alkaline phosphatase enzyme. Electrochemical measurements were then performed in accordance with Section 2.3.5.

2.3.1. Immobilization of the Biotinylated Affibody

The streptavidin-modified magnetic beads (Strept-MB) were firstly washed with buffer A (added with 0.005% Tween 20) three times and, after the removal of the supernatant with the help of a magnetic rack, re-suspended in 100 μL of buffer A added with Tween 20. Then, 400 μL of 5 μg/mL of the biotinylated affibody (Biot-Af) in buffer A was added to the beads suspension and left to incubate overnight.

2.3.2. Blocking Step

The affibody-modified beads (Strept-MB/Af) were washed and streptavidin free-sites were blocked by the addition of 1 mM biotin (prepared in buffer B containing 1% w/v BSA) for 90 min. After three washing steps in buffer D (containing Tween 20), the beads were stocked in buffer D at 4 °C for at least one week.

2.3.3. Reaction with HER2 Protein

To perform the calibration curve, 50 μL of the modified beads were incubated with 200 μL of HER2 protein at different concentrations prepared in buffer D for 20 min at room temperature. Three washing steps in buffer B-added Tween 20 were then performed.

In order to evaluate the selectivity of the developed assays, 5, 10 and 20 ng/mL of vascular endothelial growth factor (VEGF) protein, prepared in buffer D, were incubated instead of HER2 with the modified beads.

2.3.4. Binding with the Biotinylated Affibody and Streptavidin-Alkaline Phosphatase Enzyme

Firstly, 250 μL of 5 μg/mL biotinylated affibody, as signaling bioreceptor, prepared in buffer B were incubated with the HER2-modified beads for 45 min at room temperature. Subsequently, three washing steps in buffer C added with Tween 20 were performed.

Finally, the beads were incubated with 500 μL of streptavidin-alkaline phosphatase (3.9 U/mL) prepared in buffer C added with 0.1% w/v BSA for 10 min at room temperature, followed by three washing steps with buffer C added with Tween 20.

After the removal of the supernatant, the beads were re-suspended in 50 μL of buffer C.

2.3.5. Electrochemical Measurements

For the electrochemical measurements, 4 μL of the MB suspension was placed onto each working electrode of the eight screen-printed electrochemical cells and fixed in position through the help of a home-made magnet-holding block. Each well of the arrays was then filled with 60 μL of a solution containing 1 mg/mL of 1-naphthyl phosphate enzyme substrate prepared in buffer C.

After 6 min of incubation, DPV measurements were performed at room temperature using the following parameters: potential range from −0.05 to 0.6 V, step potential 7 mV, modulation amplitude 70 mV, interval time of 0.1 s. The current peak height was taken as the analytical signal.

Each measurement was repeated at least 10 times using different screen-printed arrays. Percentage Relative Standard Deviation (%RSD) values were calculated as measure of inter-assay reproducibility.

2.4. Development of Antibody/Affibody Based Assay

The scheme of antibody/affibody-based bioassay is illustrated in the Supplementary Materials (Figure S6).

2.4.1. Immobilization of the Antibody

Firstly, the protein A-modified magnetic beads (ProtA-MB) were washed with buffer A (added with 0.005% Tween 20) three times and, after the removal of the supernatant with the help of a magnetic rack, re-suspended in 100 μL of buffer A added with Tween 20. Then, 400 μL of 50 μg/mL of the antibody (Ab) in buffer A were added to the beads suspension and left to incubate for 45 min.

2.4.2. Blocking Step

Subsequently, the Ab1-modified beads (ProtA-MB/Ab) were washed and the protein A-free site blocking step was performed using the addition of 5% w/v casein prepared in buffer A for 30 min. After three washing steps in buffer D (containing Tween 20), the beads were stocked in buffer D at 4 °C for at least one week.

2.4.3. Reaction with HER2 Protein and Biotinylated Affibody, Labeling with Streptavidin-Alkaline Phosphatase and Electrochemical Measurements

The following steps were carried out as reported in the previous sections. Reaction with HER2 protein and evaluation of non-specific interaction with VEGF were carried out in accordance with Section 2.3.3.

The reaction with biotinylated affibody and labeling with streptavidin-alkaline phosphatase were carried out in accordance with Section 2.3.4.

Electrochemical measurements were then performed as reported in Section 2.3.5.

2.5. Development of Affibody/Antibody-Based Assay

The scheme of antibody/affibody based bioassay is illustrated in supplementary information (Figure S7).

Streptavidin-modified magnetic beads were functionalized with the biotinylated affibody and blocked with biotin in accordance with Sections 2.3.1 and 2.3.2. Furthermore, HER2 affinity reaction and non-specific test with VEGF protein were performed in accordance with Section 2.3.3.

Binding with Biotinylated Antibody and Streptavidin-Alkaline Phosphatase Enzyme

Firstly, 250 µL of 1 µg/mL secondary biotinylated antibody (Biot-Ab2), as signaling bioreceptor, prepared in buffer B, were incubated with the HER2-modified beads for 60 min at room temperature. Subsequently, three washing steps in buffer C added with Tween 20 were performed.

Finally, the beads were incubated with 500 µL of streptavidin-alkaline phosphatase enzyme (3.9 U/mL) prepared in buffer C added with 0.1% w/v BSA for 10 min at room temperature, followed by three washing steps with buffer C added with Tween 20.

Electrochemical measurements were then performed in accordance with Section 2.3.5.

2.6. Analysis of Serum Samples

The human serum samples were filtered (Filtropur S, diameter of filter pores 0.2 µM), diluted 1:2 with buffer D and spiked with HER2 protein solution (range of concentration 0–20 ng/mL). Then, 50 µL of the affibody-modified bead suspension was incubated with 200 µL of HER2 serum samples and the experiments were carried out as reported in Sections 2.3.4 and 2.3.5.

3. Results and Discussion

As mentioned before, in this study, different protocols using antibody (Ab) and affibody (Af) as capture and signaling bioreceptors were both applied to disposable electrochemical immunosensors based on the sandwich assay for HER2 detection. To achieve the best conditions, key parameters that affect the read-out response of each assay were studied and optimized. The optimization parameters of the developed assays are reported in the Supplementary Materials. In the following section, the studies of affibody-based assay are focused on and reported.

3.1. Optimization of Experimental Parameters

The optimization of the experimental parameters in the case of dual affibody sandwich assay was performed in order to find the best conditions for HER2 binding and detection. The suitable experimental conditions were chosen in accordance with the current difference value (ΔI) obtained in the presence of 10 ng/mL HER2 (I_{HER2}) and the blank (0 ng/mL HER2, I_{Blank}), and the percentage Relative Standard Deviation (%RSD) values (Table 1).

Firstly, the concentration and the incubation time of the biotinylated affibody (Biot-Af) on the surface of the streptavidin-modified magnetic beads (Strept-MBs) were optimized (Table 1, assay step a). In particular, 1, 5 and 10 µg/mL of biotinylated-affibody solutions (prepared in buffer A) were incubated with the Strept-MB, followed by blocking step with biotin, affinity reaction with HER2 protein and incubation with secondary biotinylated affibody labeled with streptavidin-AP.

The Biot-Af concentration of 1.0 μg/mL was not sufficient to bind the HER2, while a similar current difference was observed using an affibody concentration of 5.0 and 10 μg/mL ($\Delta I_{Af. 5\ \mu g/mL}$ = 3.3 μA; $\Delta I_{Af. 10\ \mu g/mL}$ = 3.2 μA). Thus, the concentration of Biot-Af of 5.0 μg/mL was selected for the further experiments.

Table 1. Experimental parameters optimization for affibody-based sandwich assay. Current difference ($\Delta I = I_{HER2} - I_{Blank}$) represents the difference between the current obtained using 10 ng/mL (I_{HER2}) and 0 ng/mL (I_{Blank}) HER2 buffered solutions. The letters of assay step column are in accordance with Scheme 1. Percentage Relative Standard Deviation (%RSD) values were calculated using at least 10 measurements obtained by different screen-printed arrays.

Assay Step	Parameter		Current Difference (μA) ($\Delta I = I_{HER2} - I_{Blank}$)	%RSD
a	Biot-Af concentration	1 μg/mL	1.5	8
		5 μg/mL	3.3	7
		10 μg/mL	3.2	10
	Biot-Af incubation time	120 min	1.7	7
		240 min	2.0	10
		o.n.	3.3	7
b	Biotin incubation time	30 min	1.2	8
		60 min	2.2	9
		90 min	3.3	7
c	HER2 incubation time	10 min	1.3	6
		20 min	3.3	7
		60 min	3.0	9
d	Biot-Af concentration	1 μg/mL	1.2	8
		5 μg/mL	3.3	7
		10 μg/mL	3.5	10
	Biot-Af incubation time	30 min	1.6	7
		45 min	3.3	7
		60 min	2.9	9

Biot-Af: biotinylated affibody; o.n.: overnight.

The incubation time of the primary Biot-Af with the streptavidin was also evaluated. The Biot-Af was left to incubate for 120, 240 min and overnight (o.n.) with the Strept-MBs. The best incubation time was found to be overnight probably due to the complete coverage of the surface of magnetic beads which ensures a higher recognition of the target protein (Table 1, assay step a). After the functionalization of the Strept-MBs with the Biot-Af, the conjugates were blocked with various blocking agents (1.0 mM Biotin solution containing 1% w/v BSA, BSA 1% and milk powder 5% for different incubation times: 30, 60 and 90 min). The best blocking step in terms of sensitivity and reproducibility (n = 10) was performed using 1 mM biotin containing 1% w/v BSA for 90 min (in accordance with ref. [40]) as reported in Table 1, assay step b and in Supplementary Materials Section 1.

The incubation time of the affinity reaction with the HER2 protein was also optimized. As can be observed in Table 1 (assay step c), 10 min as incubation time seems insufficient to bind the whole amount of the protein, while similar values in current difference were obtained using 20 or 60 min ($\Delta I_{45\ min.}$ = 3.3 μA; $\Delta I_{60\ min.}$ = 3.0 μA). The best results, in terms of signal-to-noise ratio and reproducibility (%RSD = 7, n = 10), were obtained for an incubation time of 20 min which was selected as optimal. Finally, the concentration and the incubation time of the Biot-Af, used as secondary biotinylated bioreceptor, was evaluated (Table 1, assay step d). Particularly, the current difference using a concentration of 5.0 μg/mL is considerably higher than that at 1.0 μg/mL, whereas similar behavior was obtained using an affibody concentration of 10 µg/mL. With respect to the Biot-Af incubation time, 45 min was found as optimal both in terms of current difference and of %RSD.

3.2. Sensitivity and Reproducibility

Using the optimized conditions, the calibration curves for the quantification of HER2 in the case of all assays were obtained (Figure 1A) and the related DPV scans were plotted in Figure 1B. A linear response in the range of HER2 0–20 ng/mL was obtained for affibody/affibody (Af/Af: y = 0.33x, coefficient of determination (R^2) = 0.997), affibody/antibody (Af/Ab: y = 0.23x, R^2 = 0.993) and for antibody/affibody (Ab/Af: y = 0.21x, R^2 = 0.98) sandwich assays with limit of detection (LOD), calculated as 3 times standard deviation of blank divided the slope of calibration curve ($3S_{Blank}/Slope$) of 1.8, 2.6 and 3.4 ng/mL respectively. Reproducibility of the proposed assays were also evaluated using at least 10 measurements performed on different screen-printed arrays. Results showed a mean %RSD value of 7 for Af/Af assay, 10 for Af/Ab assay and 11 for Ab/Af assay. Taking in consideration both the LOD and the %RSD, the Af/Af assay showed better performance in comparison with Af/Ab and Ab/Af mixed sandwich assays (and also with respect to the previous dual antibody-based sandwich assay reported in literature [19]) for HER2 determination (whose cut-off in serum sample was set to 15 ng/mL).

Furthermore, the selectivity of the developed assays was also verified using 5, 10 and 20 ng/mL of the vascular endothelial growth factor (VEGF) protein, involved in the metastatic process of breast cancer [41], as a nonspecific molecule. No significant variation respect to the signal of the blank was found, confirming the high specificity of the affibody bioreceptors for the detection of the HER2 cancer protein (Figure 1A, inset).

The affibody-based assay was thus selected to evaluate the possibility of detecting the HER2 protein in serum samples.

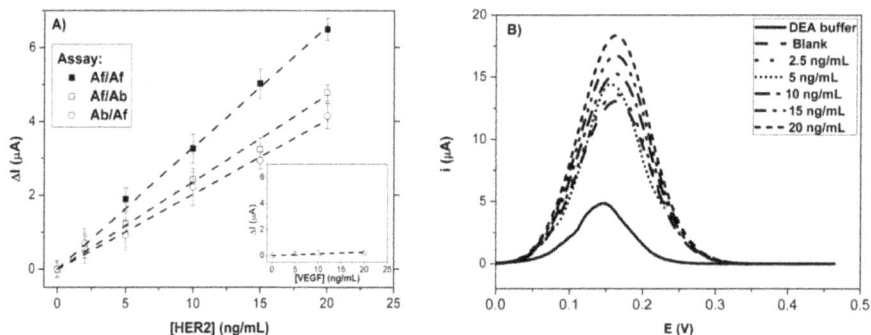

Figure 1. (**A**) Calibration curves for HER2 protein obtained using affibody/affibody (Af/Af), affibody/antibody (Af/Ab) and antibody/affibody (Ab/Af)-based assays. Inset: nonspecific interaction evaluation with vascular endothelial growth factor (VEGF) protein; (**B**) Differential pulse voltammetry (DPV) scans of Af/Af assay. Blank signal was subtracted from each measurement. Electrochemical measurements were performed in accordance with Section 2.3.5. Each measurement was repeated at least 10 times using different screen-printed sensors.

3.3. Analysis of Serum Samples

Once the suitability of the dual affibody-based assay to detection of HER2 standard solutions was verified, experiments with HER2 fortified serum samples were carried. Batches of a non-pathologic serum AB group from females were filtered, diluted 1:2 in buffer D and spiked with HER2 protein in order to have four different final concentrations in the range of 5–20 ng/mL. Results are summarized in Table 2.

Table 2. Recovery, Bias and %RSD for HER2 determination in fortified serum samples.

HER2 Spiked (ng/mL)	HER2 Found (ng/mL)	Recovery (%)	Bias (%)	%RSD
5	5.5	110	10	12
10	11	108	8	11
15	15	97	−3	14
20	19	95	−5	13

Good recovery and difference between the average of measurements made on the same sample and its true value (bias values) (respectively in the range between 95%–110% and −5%–10%) were obtained. Furthermore, the mean %RSD, calculated using at least 10 measurements obtained using different screen-printed arrays, was 13%. These results confirm the suitability of the use of the proposed magnetic beads-based affibody assay for real sample analysis.

4. Conclusions

The potential of affibodies as bioreceptors in sandwich assay has been investigated by the optimization and the application of three strategies for the detection of HER2 protein. The best results were obtained by the use of affibody as both a capture and signaling bioreceptor. This biosensor showed the best sensitivity and detection limit and a good linear range in HER2-buffered solutions and serum samples. With this comparison, the importance of a deep study on the different analytical approaches for HER2 detection to obtain the performance of the best assay configurations has been demonstrated. Our results open up the way for the development of a new generation of biosensors for highly sensitive detection in a variety of analyses.

Supplementary Materials: The following are available online at www.mdpi.com/2227-9040/4/4/23/s1, Section 1: Response of graphite screen-printed arrays to the 1-naphthol; Section 2: Optimization of experimental parameters of affibody/affibody assay; Section 3: Optimization of experimental parameters of antibody/affibody assay; Section 4: Optimization of experimental parameters of affibody/antibody assay. Figure S1: Calibration curve of 1-naphthol; Figure S2: optimization of Af concentration as signaling bioreceptor; Figure S3: Optimization of the incubation time of the Af with HER2 as a signaling bioreceptor; Figure S4: Blocking agent optimization: BSA 1% (w/v), dried milk powder 5% (w/v) and IgG 10 mg/mL; Figure S5: Incubation time optimization for dried milk powder 5% (w/v) as blocking agent; Figure S6: Schematic representation of antibody/affibody-based assay for HER2 detection; Figure S7: Schematic representation of affibody/antibody assay for HER2 detection. Table S1: Experimental parameters optimization for antibody/affibody assay; Table S2: Experimental parameters optimization for affibody/antibody assay.

Author Contributions: Experiment details were conceived and designed by H.I. and G.M.; Data analysis was carried out by H.I., A.R. and G.M.; G.M. contributed reagents/materials/analysis tools; all contributors wrote the paper. Authorship must be limited to those who have contributed substantially to the work reported.

Conflicts of Interest: The authors declare no conflict of interest.

References

1. Ankerst, D.P.; Liss, M.; Zapata, D.; Hoefler, J.; Thompson, I.M.; Leach, R.J. A case control study of sarcosine as an early prostate cancer detection biomarker. *BMC Urol.* **2015**, *15*, 99. [CrossRef] [PubMed]
2. Payne, R.C.; Allard, J.W.; Anderson-Mauser, L.; Humphreys, J.D.; Tenney, D.Y.; Morris, D.L. Automated assay for HER-2/neu in serum. *Clin. Chem.* **2000**, *46*, 175–182. [PubMed]
3. Ravalli, A.; Marrazza, G. Gold and magnetic nanoparticles-based electrochemical biosensors for cancer biomarker determination. *J. Nanosci. Nanotechnol.* **2015**, *15*, 3307–3319. [CrossRef] [PubMed]
4. Diaconu, I.; Cristea, C.; Hârceagă, V.; Marrazza, G.; Berindan-Neagoe, I.; Săndulescu, R. Electrochemical immunosensors in breast and ovarian cancer. *Clin. Chim. Acta* **2013**, *425*, 128–138. [CrossRef] [PubMed]
5. Slamon, D.J.; Godolphin, W.; Jones, L.A.; Holt, J.A.; Wong, S.G.; Keith, D.E.; Levin, W.J.; Stuart, S.G.; Udove, J.; Ullrich, A. Studies of the HER-2/neu proto-oncogene in human breast and ovarian cancer. *Science* **1989**, *244*, 707–712. [CrossRef] [PubMed]

6. Omenn, G.S.; Guan, Y.; Menon, R. A new class of protein cancer biomarker candidates: Differentially expressed splice variants of ERBB2 (HER2/neu) and ERBB1 (EGFR) in breast cancer cell lines. *J. Proteom.* **2014**, *107*, 103–112. [CrossRef] [PubMed]
7. Qureshi, A.; Gurbuz, Y.; Niazi, J.H. Label-free capacitance based aptasensor platform for the detection of HER2/ErbB2 cancer biomarker in serum. *Sens. Actuators B Chem.* **2015**, *220*, 1145–1151. [CrossRef]
8. Takahashi, Y.; Miyamoto, T.; Shiku, H.; Asano, R.; Yasukawa, T.; Kumagai, I.; Matsue, T. Electrochemical detection of epidermal growth factor receptors on a single living cell surface by scanning electrochemical microscopy. *Anal. Chem.* **2009**, *81*, 2785–2790. [CrossRef] [PubMed]
9. Patris, S.; De Pauw, P.; Vandeput, M.; Huet, J.; Van Antwerpen, P.; Muyldermans, S.; Kauffmann, J.M. Nanoimmunoassay onto a screen printed electrode for HER2 breast cancer biomarker determination. *Talanta* **2014**, *130*, 164–170. [CrossRef] [PubMed]
10. Arkan, E.; Saber, R.; Karimi, Z.; Shamsipur, M. A novel antibody–antigen based impedimetric immunosensor for low level detection of HER2 in serum samples of breast cancer patients via modification of a gold nanoparticles decorated multiwall carbon nanotube-ionic liquid electrode. *Anal. Chim. Acta* **2015**, *874*, 66–74. [CrossRef] [PubMed]
11. Engfeldt, T.; Orlova, A.; Tran, T.; Bruskin, A.; Widström, C.; Karlström, A.E.; Tolmachev, V. Imaging of HER2-expressing tumours using a synthetic Affibody molecule containing the 99mTc-chelating mercaptoacetyl-glycyl-glycyl-glycyl (MAG3) sequence. *Eur. J. Nucl. Med. Mol. Imaging* **2007**, *34*, 722–733. [CrossRef] [PubMed]
12. Emami, M.; Shamsipur, M.; Saber, R.; Irajirad, R. An electrochemical immunosensor for detection of a breast cancer biomarker based on antiHER2-iron oxide nanoparticle bioconjugates. *Analyst* **2014**, *139*, 2858–2866. [CrossRef] [PubMed]
13. Gohring, J.T.; Dale, P.S.; Fan, X. Detection of HER2 breast cancer biomarker using the opto-fluidic ring resonator biosensor. *Sens. Actuators B Chem.* **2010**, *146*, 226–230. [CrossRef]
14. Chen, H.; Tang, D.; Zhang, B.; Liu, B.; Cui, Y.; Chen, G. Electrochemical immunosensor for carcinoembryonic antigen based on nanosilver-coated magnetic beads and gold-graphene nanolabels. *Talanta* **2012**, *91*, 95–102. [CrossRef] [PubMed]
15. Zahmatkeshan, M.; Ilkhani, H.; Paknejad, M.; Adel, M.; Sarkar, S.; Saber, R. Analytical Characterization of Label-Free Immunosensor Subsystems Based on Multi-Walled Carbon Nanotube Array-Modified Gold Interface. *Comb. Chem. High Throughput Screen.* **2015**, *18*, 83–88. [CrossRef] [PubMed]
16. Ilkhani, H.; Sarparast, M.; Noori, A.; Zahra Bathaie, S.; Mousavi, M.F. Electrochemical aptamer/antibody based sandwich immunosensor for the detection of EGFR, a cancer biomarker, using gold nanoparticles as a signaling probe. *Biosens. Bioelectron.* **2015**, *74*, 491–497. [CrossRef] [PubMed]
17. Ravalli, A.; Pilon Dos Santos, G.; Ferroni, M.; Faglia, G.; Yamanaka, H.; Marrazza, G. New label free CA125 detection based on gold nanostructured screen-printed electrode. *Sens. Actuators B Chem.* **2013**, *179*, 194–200. [CrossRef]
18. Taleat, Z.; Ravalli, A.; Mazloum-Ardakani, M.; Marrazza, G. CA125 Immunosensor Based on Poly-Anthranilic Acid Modified Screen-Printed Electrodes. *Electroanalysis* **2013**, *25*, 269–277. [CrossRef]
19. Al-Khafaji, Q.A.M.; Harris, M.; Tombelli, S.; Laschi, S.; Turner, A.P.F.; Mascini, M.; Marrazza, G. An Electrochemical Immunoassay for HER2 Detection. *Electroanalysis* **2012**, *24*, 735–742. [CrossRef]
20. Mucelli, S.P.; Zamuner, M.; Tormen, M.; Stanta, G.; Ugo, P. Nanoelectrode ensembles as recognition platform for electrochemical immunosensors. *Biosens. Bioelectron.* **2008**, *23*, 1900–1903. [CrossRef] [PubMed]
21. Eletxigerra, U.; Martinez-Perdiguero, J.; Merino, S.; Barderas, R.; Torrente-Rodríguez, R.M.; Villalonga, R.; Pingarrón, J.M.; Campuzano, S. Amperometric magnetoimmunosensor for ErbB2 breast cancer biomarker determination in human serum, cell lysates and intact breast cancer cells. *Biosens. Bioelectron.* **2015**, *70*, 34–41. [CrossRef] [PubMed]
22. Liu, X.; Liu, R.; Tang, Y.; Zhang, L.; Hou, X.; Lv, Y. Antibody-biotemplated HgS nanoparticles: Extremely sensitive labels for atomic fluorescence spectrometric immunoassay. *Analyst* **2012**, *137*, 1473. [CrossRef] [PubMed]
23. Eletxigerra, U.; Martinez-Perdiguero, J.; Barderas, R.; Pingarrón, J.M.; Campuzano, S.; Merino, S. Surface plasmon resonance immunosensor for ErbB2 breast cancer biomarker determination in human serum and raw cancer cell lysates. *Anal. Chim. Acta* **2016**, *905*, 156–162. [CrossRef] [PubMed]
24. Miller, S.A.; Hiatt, L.A.; Keil, R.G.; Wright, D.W.; Cliffel, D.E. Multifunctional nanoparticles as simulants for a gravimetric immunoassay. *Anal. Bioanal. Chem.* **2011**, *399*, 1021–1029. [CrossRef] [PubMed]

25. Ilkhani, H.; Arvand, M.; Ganjali, M.R.; Marrazza, G.; Mascini, M. Nanostructured Screen Printed Graphite Electrode for the Development of a Novel Electrochemical Genosensor. *Electroanalysis* **2013**, *25*, 507–514. [CrossRef]

26. Xu, X.W.; Weng, X.H.; Wang, C.L.; Lin, W.W.; Liu, A.L.; Chen, W.; Lin, X.H. Detection EGFR exon 19 status of lung cancer patients by DNA electrochemical biosensor. *Biosens. Bioelectron.* **2016**, *80*, 411–417. [CrossRef] [PubMed]

27. Torati, S.R.; Reddy, V.; Yoon, S.S.; Kim, C. Electrochemical biosensor for Mycobacterium tuberculosis DNA detection based on gold nanotubes array electrode platform. *Biosens. Bioelectron.* **2016**, *78*, 483–488. [CrossRef] [PubMed]

28. Justino, C.I.L.; Freitas, A.C.; Pereira, R.; Duarte, A.C.; Rocha Santos, T.A.P. Recent developments in recognition elements for chemical sensors and biosensors. *TrAC Trend. Anal. Chem.* **2015**, *68*, 2–17. [CrossRef]

29. Ravalli, A.; da Rocha, C.G.; Yamanaka, H.; Marrazza, G. A label-free electrochemical affisensor for cancer marker detection: The case of HER2. *Bioelecrochemistry* **2015**, *106*, 268–275. [CrossRef] [PubMed]

30. Orlova, A.; Magnusson, M.; Eriksson, T.L.J.; Nilsson, M.; Larsson, B.; Höiden-Guthenberg, I.; Widström, C.; Carlsson, J.; Tolmachev, V.; Ståhl, S.; et al. Tumor imaging using a picomolar affinity HER2 binding Affibody molecule. *Cancer Res.* **2006**, *66*, 4339–4348. [CrossRef] [PubMed]

31. Hansson, M.; Ringdahl, J.; Robert, A.; Power, U.; Goetsch, L.; Nguyen, T.N.; Uhlén, M.; Ståhl, S.; Nygren, P. An in vitro selected binding protein (affibody) shows conformation-dependent recognition of the respiratory syncytial virus (RSV) G protein. *Immunotechnology* **1999**, *4*, 237–252. [CrossRef]

32. Ekerljung, L.; Lindborg, M.; Gedda, L.; Frejd, F.Y.; Carlsson, J.; Lennartsson, J. Dimeric HER2-specific affibody molecules inhibit proliferation of the SKBR-3 breast cancer cell line. *Biochem. Biophys. Res. Commun.* **2008**, *377*, 489–494. [CrossRef] [PubMed]

33. Samadi-Maybodi, A.; Nejad-Darzi, S.K.H.; Ilkhani, H. A new sensor for determination of paracetamol, phenylephrine hydrochloride and chlorpheniramine maleate in pharmaceutical samples using nickel phosphate nanoparticles modified carbon past electrode. *Anal. Bioanal. Electrochem.* **2011**, *3*, 134–145.

34. Sakudo, A.; Onodera, T. Virus capture using anionic polymer-coated magnetic beads (review). *Int. J. Mol. Med.* **2012**, *30*, 3. [CrossRef] [PubMed]

35. Marszałł, M.P.; Buciński, A. A protein-coated magnetic beads as a tool for the rapid drug-protein binding study. *J. Pharm. Biomed. Anal.* **2010**, *52*, 420–424. [CrossRef] [PubMed]

36. Laschi, S.; Miranda-Castro, R.; Gonzalez-Fernandez, E.; Palchetti, I.; Reymond, F.; Rossier, J.S.; Marrazza, G. A new gravity-driven microfluidic-based electrochemical assay coupled to magnetic beads for nucleic acid detection. *Electrophoresis* **2010**, *31*, 3727–3736. [CrossRef] [PubMed]

37. Barizuddin, S.; Balakrishnan, B.; Stringer, R.C.; Dweik, M. Highly specific and rapid immuno-fluorescent visualization and detection of E. coli O104: H4 with protein-A coated magnetic beads based LST-MUG assay. *J. Microbiol. Methods* **2015**, *115*, 27–33. [CrossRef] [PubMed]

38. Kay, B.K.; Thai, S.; Volgina, V.V. High-throughput biotinylation of proteins. *Methods Mol. Biol.* **2009**, *498*, 185–198. [PubMed]

39. Heineman, W.R.; Halsall, H.B. Strategies for electrochemical immunoassay. *Anal. Chem.* **1985**, *57*, 1321A–1331A. [CrossRef] [PubMed]

40. Florea, A.; Ravalli, A.; Cristea, C.; Sandulescu, R.; Marrazza, G. An Optimized Bioassay for Mucin1 Detection in Serum Samples. *Electroanalysis* **2015**, *27*, 1594–1601. [CrossRef]

41. Ferrara, N.; Gerber, H.-P.; LeCouter, J. The biology of VEGF and its receptors. *Nat. Med.* **2003**, *9*, 669–676. [CrossRef] [PubMed]

chemosensors

MDPI

Article

Miniaturized Aptamer-Based Assays for Protein Detection

Alessandro Bosco [1,†,‡], Elena Ambrosetti [1,2,3,†], Jan Mavri [4], Pietro Capaldo [1,2,3] and Loredana Casalis [1,*]

1   Elettra-Sincrotone S.C.p.A., SS 14 km 163,5 in AREA Science Park, Basovizza, Trieste 34149, Italy; alessandro.bosco@ki.se (A.B.); elena.ambrosetti@elettra.eu (E.A.); pietrocapaldo@gmail.com (P.C.)
2   Department of Physics, University of Trieste, Via Valerio 9, Trieste 34127, Italy
3   INSTM-ST Unit, SS 14 km 163,5 in AREA Science Park, Basovizza, Trieste 34149, Italy
4   Centre of Excellence for Biosensors, Instrumentation and Process Control, Tovarniška 26, Ajdovščina SI-5270, Slovenia; janmavr@gmail.com
*   Correspondence: loredana.casalis@elettra.eu; Tel.: +39-040-375-8291; Fax: +39-040-938-0902
†   These authors contributed equally to this work.
‡   Present address: Department of Medical Biochemistry and Biophysics, Karolinska Institutet, Scheeles väg 2, SE-17177 Stockholm, Sweden.

Academic Editors: Paolo Ugo and Ligia Moretto
Received: 11 July 2016; Accepted: 29 August 2016; Published: 2 September 2016

Abstract: The availability of devices for cancer biomarker detection at early stages of the disease is one of the most critical issues in biomedicine. Towards this goal, to increase the assay sensitivity, device miniaturization strategies empowered by the employment of high affinity protein binders constitute a valuable approach. In this work we propose two different surface-based miniaturized platforms for biomarker detection in body fluids: the first platform is an atomic force microscopy (AFM)-based nanoarray, where AFM is used to generate functional nanoscale areas and to detect biorecognition through careful topographic measurements; the second platform consists of a miniaturized electrochemical cell to detect biomarkers through electrochemical impedance spectroscopy (EIS) analysis. Both devices rely on robust and highly-specific protein binders as aptamers, and were tested for thrombin detection. An active layer of DNA-aptamer conjugates was immobilized via DNA directed immobilization on complementary single-stranded DNA self-assembled monolayers confined on a nano/micro area of a gold surface. Results obtained with these devices were compared with the output of surface plasmon resonance (SPR) assays used as reference. We succeeded in capturing antigens in concentrations as low as a few nM. We put forward ideas to push the sensitivity further to the pM range, assuring low biosample volume (μL range) assay conditions.

Keywords: biosensors; aptamers; AFM; nanoarray; EIS

1. Introduction

The rapid and reliable detection of multiple biomarkers simultaneously in small sample volumes is increasingly requested in current clinical practice and represents a fundamental step towards personalized medicine [1]. Nowadays, most of the available diagnostic devices are solid-state based analytical assays, where a functionalized surface works as the active element for biorecognition. Site-specific immobilization of multiple active elements on the same surface to probe low biosample volumes requires miniaturization and is generally technologically demanding, since it entails successive steps specifically tailored to each biomarker probe.

One promising strategy for multiplexing probe immobilization is DDI (DNA-directed immobilization) [2–5], where different DNA-conjugated antibodies targeting different biomarkers are immobilized via Watson-Crick base pairing on surface-tethered complementary DNA sequences. However the synthesis of DNA-protein conjugates is quite challenging, requires click-chemistry kits and careful optimization in order to assure a final construct that is both suitable for immobilization and preserves the original affinity.

Nucleic acid aptamers can offer a valid alternative to antibodies. Aptamers are oligosequences selected in vitro to bind a target with high affinity. In particular, they show highly specific binding activity since aptamer-target interaction is based on three-dimensional folding patterns, resulting in dissociation constants in the picomolar range. In addition, aptamers, if compared to standard proteins, show higher stability, and ease of chemical modification. Aptamer production is fully automated, highly reproducible, and low-cost. Moreover, aptamers are naturally integrable in the context of DDI: DNA-aptamer constructs are produced simply by adding a surface binding sequence to the aptamer during oligo-synthesis.

In this paper, we exploited aptamers as the active recognition elements of miniaturized DNA-based biosensors. We modified aptamers with a DNA tag (cDNA), meant to hybridize to micron-sized surface-grafted complementary DNA monolayers, to create the desired functional areas (Figure 1). In particular, we carefully optimized the immobilization strategy to implement DNA-modified aptamers onto two innovative, DNA-based miniaturized sensor platforms: (i) atomic force microscopy (AFM)-based nanoarrays, and (ii) electrochemical impedance spectroscopy (EIS)-based microelectrodes. Both of these platforms have been designed and developed to allow for the sensitive detection of biomolecules in small sample volumes.

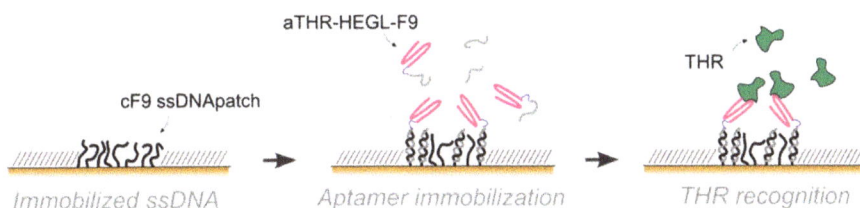

Figure 1. Cartoon representing the DDI strategy to immobilize an aptamer on the micron-sized DNA-based biosensor.

In the case of nanoarrays, we capitalize here on the exploitation of AFM-based lithography, nanografting [6–8], and AFM topographic imaging in physiological environment successfully demonstrated in the work of Bano et al. for the detection of multiple proteins in standardized human serum [9], and successively by Bosco and Ganau et al., for the ultimate integration with a device for cell sorting, to measure the secretome of few selected cells [10]. By AFM nanografting we confined nanospots of thiol-modified single-stranded DNA monolayers inside a bio-repellent, self-assembled monolayer (SAM) of oligoethyleneglycol-terminated alkanethiols. The density of the DNA-confined SAM can be tuned via nanografting, and adjusted to the steric requirements of the biorecognition elements. On such active DNA nanospots, the aptamer-cDNA conjugate is then loaded through Watson-Crick base-pairing, in a process known as DNA-directed immobilization [9]. On the other side, an electrochemical impedance-based biosensor, with a three-electrode design developed in our group for the recognition of nucleic acids [11,12], are tested towards protein detection through modified aptamers. The working micro-electrode is fully covered by a functional, thiolated, single-stranded DNA SAM, on which the aptamer-cDNA conjugate is then loaded via DDI.

As a proof of principle, in this study we used an extensively-investigated aptamer for thrombin (THR) [13] that shows a well-characterized structure and binding properties (K_D = 50 nM), confirmed by several studies [14,15]. Two constructs were proposed in this regard: a simple design consisting of the aptamer region on the 5' side, extending with the immobilization region complementary to the DNA grafted monolayer on the 3' side (Supplementary Materials, Figure S1a); a second design similar to the first one, but containing a hexaethilenglycol-spacer (HEGL) between the two regions to improve the functionality of aptamer binding site (Supplementary Materials, Figure S1b). As stated in several works, the use of polyethyleneglycol groups can significantly improve biorecognition sensitivity by reducing nonspecific interactions and steric hindrance effects [16,17]. In both cases the surface linking oligo sequence (with an alkanethiol linker to bind to the gold surface), and the complementary one on the aptamer side, were carefully selected to have minimal influence on the protein binding—aptamer region in order to avoid possible interferences during aptamer immobilization and then on the aptamer binding interaction with THR.

2. Materials and Methods

2.1. Preliminary Affinity Characterization of the DNA-Aptamer Constructs

Affinity characterization of the DNA-aptamer constructs was first carried out in silico using the web service UNAFold, developed by Zucker and his coworkers [18]. The software application [19] for hybridization of two different oligonucleotide strands was used to generate temperature-dependent concentration plots of the tested sequence pairs in buffer, considering an ionic composition of 150 mM Na$^+$, 2 mM Mg^{2+}, similar to the binding buffer used in further wet experiments. The calculated concentrations from the plot data for hybridized and free sequences at 25 °C were further used to determine the theoretical dissociation constants (K_D). In this regard theoretical K_D values were calculated and compared for the thiolated linker sequence cF9 (5′-CTTCACGATTGCCACTTTCCAC-3′) vs. the protein binding aptamer sequence (5′-GGTTGGTGTGGTTGG-3′) and the complementary hybridization region F9 (5′-GTGGAAAGTGGCAATCGTGAAG-3′) used in the aptamer constructs. The designed aptamer constructs were further tested in bulk conditions for their functionality, testing the binding affinity for human thrombin (HT) in a qualitative electrophoretic mobility shift assay (EMSA) (Supplementary Materials). Two different aptamer constructs were prepared: the one in which the F9 sequence is directly extending from the aptamer sequence (5′-GGTTGGTGTGGTTGGGT GGAAAGTGGCAATCGTGAAG-3′), named F9aTHR and with hexaethyleneglycol (HEGL) linker in between, (5′-GGTTGGTGTGGTTGG-HEGL-GTGGAAAGTGGCAATCGTGAAG-3′), named F9-HEGL-aTHR. All the reagents in the binding experiments including buffers, human thrombin and materials used for synthesis of aptamer constructs F9aTHR and F9-HEGL-aTHR, were ordered from Sigma-Aldrich Corp. (Saint Louis, MO, USA).

To test aptamer affinity on surfaces, a Biacore X100 Surface Plasmon Resonance (GE Healthcare, Little Chalfont, Buckinghamshire, UK) instrument was used at a constant temperature of 25 °C. A continuous flow (5 μL/min) of PBS buffer (running buffer) was maintained during all the experiments. First, a biotinilated cF9 sequence (cF9-biotin, 2 μM in PBS buffer) was immobilized over the Biacore SA gold chip surface. The immobilization through streptavidin-biotin binding was stopped after reaching a binding level of ~1200 RU, corresponding to an amount of ssDNA on the surface that ensure an efficient attachment of the molecules in the following steps of the experiment; then the surface was rinsed with two 1 min pulses of 50 mM NaOH solution, in order to remove unbound cF9-biotin [20]. The hybridization was carried out by incubation with F9-HEGL-aTHR at 30 μM in TE buffer with 1 M NaCl until reaching a binding level of ~500 RU, to form an active layer suitable to detect a binding signal also with the lowest concentrations of analyte. The immobilization procedure was followed by a flow of running buffer for 2 h in order to remove aptamers non-specifically bound to the surface and to stabilize the baseline. After this procedure the signal remained constant without any baseline drifting. Binding affinity tests were performed injecting different dilutions of thrombin (0, 0.2, 0.8, 3.1, 12.5, 50, and 200 nM) in running buffer at a flow rate of 30 μL/min for 3 min (association phase) and afterwards flushing with running buffer for 5 min (dissociation phase). For the regeneration of the surface 1 min pulses of a 50 mM NaOH solution were used, followed by a stabilization time of 5 min [21]. Binding affinity parameters were calculated using the BIAevaluation 3.1 software (Biacore GE Healthcare, Little Chalfont, Buckinghamshire, UK).

2.2. AFM-Based Nanoarrays

All AFM experiments were carried out on a XE-100 Park Instrument (Park System Corp., KANC 4F, Suwon, Korea) with a customized liquid cell. A tip-assisted AFM-based nanolithography technique has been used to fabricate DNA nanoarrays with high surface density (1–2 × 10^13 molecules/cm^2): using Si cantilevers (NSC36B Mikromasch (Mikromasch, Innovative Solutions Bulgaria Ltd., Sofia, Bulgaria), spring constant: 0.6 N/m) multiple nanografting assembled monolayers (NAM) of thiol-modified single-stranded DNA (ssDNA), named cF9, were prepared by serial AFM-based nanografting inside a self-assembled monolayer (SAM) of a top oligo ethylene glycol terminated alkanethiol, TOEG ((1-mercaptoundec-11-yl)hexa(ethyleneglycol), HS-(CH2)11-(OCH2CH2)6-OH from Sigma Aldrich) on ultraflat gold surfaces [22] following standard protocols reported earlier [9,23]. The DNA patches were obtained promoting the replacement of the TOEG molecules with the oligonucleotides by the AFM tip scanning an area of 1 μm × 1 μm or less at high force (about 100 nN) in the presence of a solution of thiolated ssDNA sequences (5 μM in TE buffer 1 M NaCl) at a scan rate of 2 Hz. Several patches (6–8 for each experiment, to guarantee good statistics) of cF9 ssDNA were produced.

Aptamer immobilization was performed via DDI, incubating the ssDNA SAM with 1:1 mix of F9-HEGL-aTHR and F9 at 2 μM in TE buffer with 1 M NaCl in order to avoid steric hindrance by reducing the aptamer surface density, hence, preserving the activity.

Aptamer-thrombin binding was promoted through the incubation of aptamer nanopatches for one hour with a solution (volume of about 100 μL) containing thrombin at different concentrations (THR buffer: 20 mM Tris pH 7.4, 140 mM NaCl, 5 mM KCl, 1 mM MgCl$_2$, 1 mM CaCl$_2$); topographic height variations over the NAMs were measured with AFM in gentle contact with standard silicon cantilevers (CSC38 Mikromasch, spring constant: 0.06 N/m) at a 1 Hz scan rate, applying a force of 0.1 nN to detect and quantify binding affinity.

2.3. EIS-Based Devices

Electrochemical impedance spectroscopy (EIS)-based devices were fabricated using optical lithography techniques. They consist of a three-electrode electrochemical cell with microfabricated working (WE) and the counter (CE) gold electrodes and a classical mm-sized Ag/AgCl reference electrode (RE). As in standard electrochemical setup, the potential (AC, 10 mV rms) is applied across WE and RE, whereas the current is measured across WE and CE. The differential capacitance (C$_d$) defines the charge density (σ$_M$) change at the metal surface for a small variation of the applied potential (φ):

$$C_d = \frac{\partial \sigma_M}{\partial \varphi}$$

is obtained by fitting the current response of the device upon application of the AC voltage at four frequencies: 100 Hz, 200 Hz, 250 Hz, and 400 Hz, by using the HEKA PG340 USB bipotentiostat (HEKA Elektronik, Dr. Schulze, GmbH, Lambrecht, Germany). As already explained in [11,12], at each frequency we collect 200 complete periods from which we compute the root mean squared value of the measured current, $I_{rms} = 2\pi f V_{rms} C_d$, and the relative uncertainties using error propagation analysis. We also proved that the device is reusable up to several times and the data are reproducible on different sensors with a standard deviation of only a few percent.

C$_d$ variations, recorded upon exposing the electrochemical cell to aptamer-DNA conjugate/analyte containing solutions, were quantitatively connected to the number of biorecognition events. In particular, C$_d$ is mostly affected by molecular layer height changes, the replacement of water molecules in the biological layer, and by electrical charge redistribution upon biorecognition.

Electrodes were first patterned on clean microscope slides using MEGAPOSIT^{TM} SPR^{TM} 220 1.2 (Series Photo-Resist) (The Dow Chemical Company, Michigan, USA) as the optical positive resist. The slides were then metalized in an e-beam evaporator, with a 20-nm Ti layer, in order to promote the adhesion, followed by 80-nm Au layer deposition and then kept in an acetone bath overnight in order to perform the lift-off process. After lift-off, the electrodes were coated again with an insulating layer of NANO^{TM}SU8-2002 (MicroChem Corp., Westborough, MA, USA), shaped to expose only the circular part of WE and CE. SU8 was used as insulating material to reduce the active surface area for biorecognition. The circular WE exposed to the solution, the arc of the CE and the profile of the SU8 layer, which covers the rest of the metal electrodes, are visible in Figure 2.

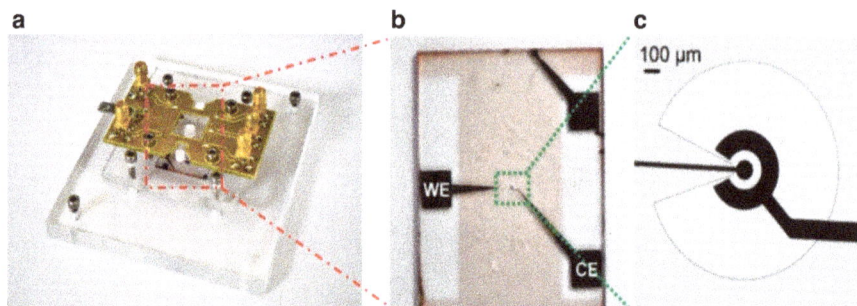

Figure 2. (a) Sample holder equipped with an electronic card to bring the signal to the HEKA-bipotentiostat; (b) microfabricated working and counter gold electrodes (black) on the patterned insulation layer (SPR 220 1.2) (pink); and (c) zoom-in of the central part of the electrodes. The diameter of the WE in contact with the solution is 100 μm. The patterned resist used to electrically insulate the electrodes is clearly recognizable as the darker-gray outer area of the image.

The insulating SU8-layer had a thickness of about 1.5 μm, measured using a 3D surface profilometer. Gold electrode functionalization with ssDNA molecules was carried out using the well-established procedures for creating DNA SAMs on gold [24–26]. Initially the electrodes were wetted for 15 min with a drop of a high-ionic-strength buffer, TE NaCl 1 M, containing cF9-SH to create a low density DNA SAM (2.1×10^{12} molecules/cm^2) [12]. Low density monolayers were chosen to avoid steric hindrance limitations to the hybridization efficiency. In this regime DNA hybridization follows Langmuir-like kinetics [12,27], while electrostatic charges are largely screened allowing for fast hybridization kinetics and reducing the limit of detection of the device. After each functionalization step the devices were rinsed with the proper DNA/protein buffer solution, prior to capacitance measurements. SAM hybridization was performed with 1 μM DNA-aptamer construct in about 100 μL solution (same volume as for the nanoarray measurements) containing 100 mM KCl. To follow hybridization kinetics, we initially measured at a rate of four measurements/min for 15 min, then we slowed down to one measurement/min until the C_d differential variations between successive points were lower than 6%, which we considered to be the "steady-state" of our measure.

3. Results and Discussion

3.1. Preliminary Affinity Characterization of the DNA-Aptamer Constructs

All of the experiments were carried out using a buffer (THR buffer) that folds the aptamer in the functional conformation for the recognition of the antigen. To this aim, preliminary tests were carried out to check that both the affinity of the DNA-aptamer construct for the surface immobilized complementary strand and the one for the ligand were preserved. In this regard an estimation of hybridization affinity between the thiolated linker and aptamer construct was first performed in silico. According to UNAFold analysis, the calculated K_D value of cF9-F9aTHR binding in bulk conditions resulted to be in the order of magnitude of attomolar, much lower than the K_D value obtained for F9 binding to the aptamer region alone (6.5 μM), therefore, suggesting that the immobilization of the DNA-aptamer constructs is occurring essentially through the cF9-F9 pairing. Using EMSA, we also verified that both DNA-aptamer constructs were able to bind thrombin (Supplementary Materials,

Figure S2). However, F9-HEGL-aTHR was observed to bind more efficiently to THR compared to F9aTHR, so in the following experiments we decided to focus on the first construct.

Binding affinity analysis was then performed on surface immobilized aptamers via SPR, following the DNA-aptamer immobilization procedure described in the previous session. A wide THR concentration range (0.8–200 nM) was screened with Biacore. The signal response is proportional to the THR concentration (Figure 3a); the mostly straight binding lines in the association phase obtained with analyte injections up to 12.5 nM show a possible mass transfer contribution. These curves do not display a sufficient curvature to perform kinetic analysis. Although an equilibrium state has not been reached for all the concentrations tested, we recorded the binding level at the end of the association phase. These values, plotted against THR concentration, were fitted with a single site interaction model. This "non-steady-state" analysis implies an underestimation of the binding affinity [28,29]. Therefore, we can conclude from the SPR data that the K_D is in the range 10–100 nM, in agreement with the value reported in the literature for the aptamer-THR binding (K_D = 50 nM) (Figure 3b).

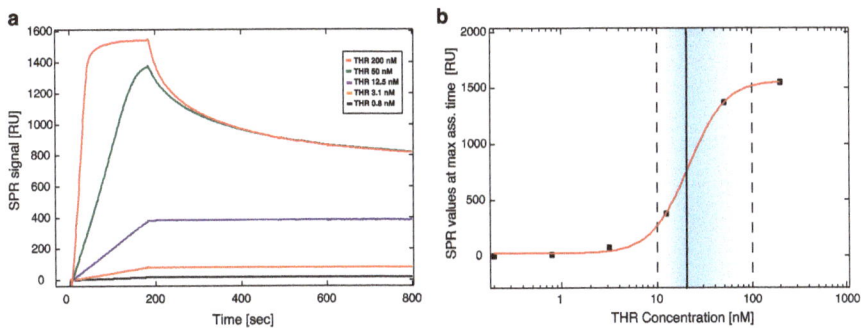

Figure 3. SPR characterization: (**a**) sensograms of different THR concentrations (0.8–200 nM); and (**b**) binding affinity analysis; SPR responses at the end of the association phase are plotted against THR concentration and fitted (red line) to a single site interaction model ([RU] = $RU_{max} - (1/(1 + K_D/[A]))$) the black vertical line indicates the value of K_D found from the RU model; the light-blue area is the range of possible K_D values (see text).

3.2. AFM Mechanical Sensing on Nanostructured Assay

The AFM nanoarray platform was built starting from cF9 DNA nanografting, on which we loaded via DDI the F9-HEGL-aTHR aptamer construct (Figure 4 left, top image).

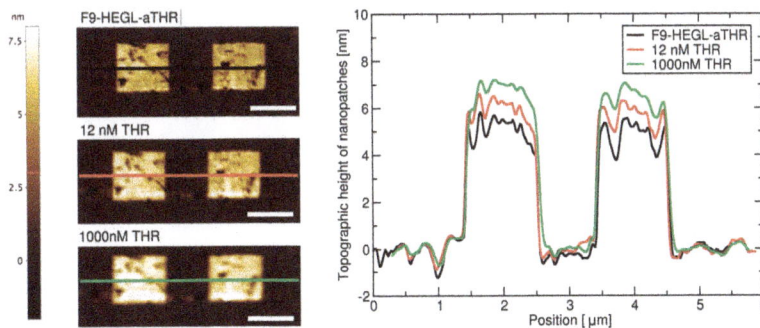

Figure 4. Images of the nanopatches on the gold surface (**left**, bar 1 μm) and their topographic profiles (**right**). Black: F9-HEGL-aTHR aptamer immobilization; red: THR 12 nM incubation; and green: THR 1000 nM incubation.

By using AFM topographic imaging, we monitored height profile changes over DNA-aptamers loaded by DDI on cF9 nanografted arrays (Figure 4: left, upper patches; right, black profile) vs. different THR concentrations in the binding buffer (Figure 4 left, mid and lower images; Figure 4 right, red and green profiles correspond to 12 nM and 1000 nM THR, respectively), due to the conformation change of the aptamer upon binding the ligand. In all these steps the TOEG SAM embedding the patches is used as a reference level to monitor topographic height changes [9]. Differential height changes are shown in Figure 5a. By plotting height change vs. concentration, we extracted an effective dissociation constant for this system in the range 10–100 nM, a value that is in good agreement both with literature and with the SPR results obtained using the same molecular construct. As one can clearly see in Figure 5a, the data are affected by significantly large errors that could be attributed to the fact that when the ligand binds to the aptamer, it folds to a more compact shape.

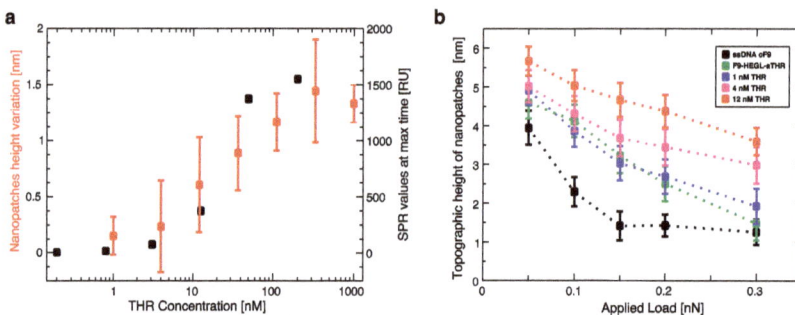

Figure 5. (**a**) Comparison of binding curves obtained from the height variation measured with the AFM on DNA nanopatches (left) and SPR data (right) vs. THR concentration; and (**b**) evaluation of the static compressibility on DNA nanopatches at low concentrations of THR.

This results in a reduced AFM differential height, since the height increase due to protein uptake is counterbalanced by the height decrease due to a change in binder conformation. Additionally, the softness of the nanopatches contributes to such large errors: the coupling of the aptamer to the nucleic acid linker, in fact, increases the compressibility of the entire nanopatch, giving more variability while measuring AFM height variation [30]. We then performed a so-called "compressibility analysis": we monitored the softness of the patches, exposed to different THR concentrations, by imaging progressively the patches at increased tip load. The variation of (absolute) patch height vs. tip load for the bare DNA patches, the patches loaded with the aptamer constructs and then the last ones exposed to different THR concentrations are shown in Figure 5b. As shown in Figure 5b, while applying only 200 pN, nanopatches can be compressed up to 40% of the height measured at minimum force (about 50 pN below which the AFM tip loss its contact with the surface). However, from such analysis we can conclude that the higher the THR concentration, the stiffer the patch towards tip compression. The increase of nanopatch stiffness upon THR immobilization is due to the high compactness of globular proteins (as THR), much higher than for nucleic acids at the ionic strength used for the experiments. A stiffness change of about 30%–35% upon 4 nM THR binding can, in fact, be estimated from the compressibility experiments, at variance with the patch height that stays almost constant.

3.3. EIS on Self Assembled Monolayers

Finally we performed an EIS analysis of the DNA-aptamer construct binding. First we checked the stability of the DNA-aptamer SAM-covered electrode in KCl buffer solution (Figure 6a, in orange); then, we changed the buffer with the one for THR, which contains divalent ions. The capacitance-change kinetics measured in-situ upon changing of the buffer solution (Figure 6a, in cyan) has to be attributed to the electrode interface rearrangement caused by the conformational reorganization of the nucleic acids upon the interaction with the divalent salt solution. At this point, after stabilization, we challenged the device for thrombin detection. In particular, we monitored the capacitance response to three different THR concentrations, on the same device, regenerating it (using a solution with 2 M NaCl) after each test. The capacitance variation upon binding of 4 nM THR is shown in green. After regeneration, we tested again the stability of the aptamer-functionalized electrode (in black) and incubated the sensor with a solution with a 40 nM thrombin (in blue). We compared the maximum capacitance variation between the two equilibrium states at different concentration (4, 10 and 40 nM) by the EIS sensor obtaining 12.3% ± 3.9%, 21.1% ± 3.3%, and 44.9% ± 2.7% respectively, in very good agreement with SPR data (Figure 6b).

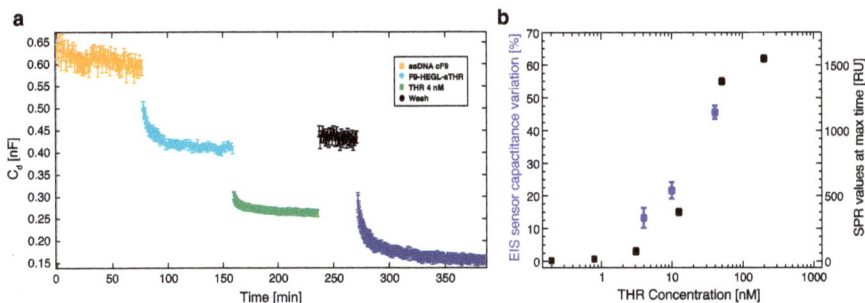

Figure 6. (**a**) Capacitance measured with the EIS setup at different steps of the immobilization; and (**b**) comparison of data obtained from the capacitance variation measured with the EIS setup (left) and SPR data (right) vs. THR concentration.

4. Conclusions

Within this work we proved that short oligo-labelled aptamers can be used for the detection of biomolecules in miniaturized, surface-based devices, as nanoarrays and electrochemical microfabricated cells, in a scheme which allows for multiplexing analysis. First, we demonstrated in bulk that by linking a short oligo via a polyethileneglycol linker to an aptamer designed and optimized for thrombin recognition, the binding affinity was preserved. Then, by immobilizing the DNA-aptamer construct on a SPR chips via DNA-directed immobilization and performing binding affinity measurements via SPR, we obtained a value of K_D in the range of 10–100 nM, in good agreement with the available literature for in bulk studies, demonstrating that the aptamer functionality was retained upon surface immobilization. Successful thrombin detection was achieved by tuning aptamer surface density on the active area to low values in order to allow aptamers conformational rearrangement necessary to bind the ligands. Additionally, we demonstrated that carefully optimized miniaturized devices could be used in a quantitative manner to determine thrombin concentration in solution, with the given, ultimate advantage with respect to other assays to be used with low volume biosamples. We used two miniaturized platforms, one based on AFM lithography and topography readout, and one on electrochemical impedance spectroscopy measurements. In both cases the measured binding affinity curves overlapped with SPR data impressively well. In particular, although these measurements are preliminary, we observed that at low thrombin concentrations (<5 nM), the EIS sensor is more sensitive than SPR, and that could be pushed to reach lower detection limits (we expect, in the pM range). On the other hand, DNA nanoarrays also allowed for an estimation of the binding affinity which resulted in good agreement with SPR data. Here, the change of conformation of aptamer upon THR binding towards a more compact structure worked reducing the expected height variation due to immobilization of THR on the device. Although still detectable, the error associated to the measurements makes this device less reliable than SPR chips to measure binding affinity. However, mechanical compressibility measurements performed on the nanopatches helped to validate the result.

The height change measurements on the nanoarrays, moreover, can be combined with the monitoring of roughness variations on the patch and outside it to infer the occurrence of specific binding only on the patch, and exclude the presence of aspecific binding outside it, even without the use of (expensive) sandwich schemes.

We put forward the idea to complement our data with AFM-based force spectroscopy measurements to investigate the variation of the mechanical properties of the DNA-aptamer conjugate structures upon ligand binding. This additional strategy might be particularly profitable given the peculiar properties of aptamers, which undergo a global conformational change upon binding. Furthermore, combining the output of these two techniques, a structural-mechanical model for the interface may be derived to be, in turn, used to describe the details of capacitance changes at the gold electrode.

Supplementary Materials: The following are available online at www.mdpi.com/2227-9040/4/3/18/s1.

Acknowledgments: This work has been supported by a Project financed by Cross-Border Cooperation Programme Italy-Slovenia 2007–2013 (Project PROTEO, Code N. CB166) (to E.A., J.M., L.C.) and by a grant from Associazione Italiana per la Ricerca sul Cancro (AIRC) (AIRC 5 per mille 2011, No. 12214) (to A.B., P.C., L.C.). We would like to thank Pietro Parisse and Paola Storici for stimulating discussions.

Author Contributions: A.B. designed the experiments, performed the AFM measurements, and helped with EIS measurements; E.A. designed the experiments, performed the SPR measurements and part of the AFM experiments; J.M. designed and provided all the DNA-aptamer constructs; P.C. performed all the EIS experiments; L.C. designed the experiments and coordinated the work. All the authors have contributed to write the manuscript and have given approval to the final version of it.

Conflicts of Interest: The authors declare no conflict of interest.

References

1. Stegh, A.H. Toward personalized cancer nanomedicine—Past, present, and future. *Integr. Biol.* **2013**, *5*, 48–65. [CrossRef] [PubMed]

2. Niemeyer, C.M.; Sano, T.; Smith, C.L.; Cantor, C.R. Oligonucleotide-directed self-assembly of proteins: Semisynthetic DNA-streptavidin hybrid molecules as connectors for the generation of macroscopic arrays and the construction of supramolecular bioconjugates. *Nucleic Acids Res.* **1994**, *22*, 5530–5539. [CrossRef] [PubMed]

3. Niemeyer, C.M.; Boldt, L.; Ceyhan, B.; Blohm, D. DNA-directed immobilization: Efficient, reversible, and site-selective surface binding of proteins by means of covalent DNA-streptavidin conjugates. *Anal. Biochem.* **1999**, *268*, 54–63. [CrossRef] [PubMed]

4. Niemeyer, C.M. The developments of semisynthetic DNA-protein conjugates. *Trends Biotechnol.* **2002**, *20*, 395–401. [CrossRef]

5. Niemeyer, C.M. Semisynthetic DNA-protein conjugates for biosensing and nanofabrication. *Angew. Chem. Int.* **2010**, *49*, 1200–1216. [CrossRef] [PubMed]

6. Liu, M.; Amro, N.A.; Chow, C.S.; Liu, G.Y. Production of nanostructures of DNA on surfaces. *Nano Lett.* **2002**, *2*, 863–867. [CrossRef]

7. Liu, M.; Amro, N.A.; Liu, G.Y. Nanografting for surface physical chemistry. *Annu. Rev. Phys. Chem.* **2008**, *59*, 367–86. [CrossRef] [PubMed]

8. Mirmomtaz, E.; Castronovo, M.; Grunwald, C.; Bano, F.; Scaini, D.; Ensafi, A.A.; Scoles, G.; Casalis, L. Quantitative study of the effect of coverage on the hybridization efficiency of surface-bound DNA nanostructures. *Nano Lett.* **2008**, *8*, 4134–4139. [CrossRef] [PubMed]

9. Bano, F.; Fruk, L.; Sanavio, B.; Glettenberg, M.; Casalis, L.; Niemeyer, C.M.; Scoles, G. Toward multiprotein nanoarrays using nanografting and DNA directed immobilization of proteins. *Nano Lett.* **2009**, *9*, 2614–2618. [CrossRef] [PubMed]

10. Ganau, M.; Bosco, A.; Palma, A.; Corvaglia, S.; Parisse, P.; Fruk, L.; Beltrami, A.P.; Cesselli, D.; Casalis, L.; Scoles, G. A DNA-based nano-immunoassay for the label-free detection of glial fibrillary acidic protein in multicell lysates. *Nanomedicine* **2015**, *11*, 293–300. [CrossRef] [PubMed]

11. Ianeselli, L.; Grenci, G.; Callegari, C.; Tormen, M.; Casalis, L. Development of stable and reproducible biosensors based on electrochemical impedance spectroscopy: Three-electrode vs. two-electrode setup. *Biosens. Bioelectron.* **2014**, *15*, 1–6. [CrossRef] [PubMed]

12. Capaldo, P.; Alfarano, S.R.; Ianeselli, L.; Dal Zilio, S.; Bosco, A.; Parisse, P.; Casalis, L. Circulating disease biomarkers detection in complex matrices: Real-time, in-situ measurements of DNA/miRNA hybridization via electrochemical impedance spectroscopy. *ACS Sens.* **2016**, *8*, 1003–1010. [CrossRef]

13. Bock, L.C.; Griffin, L.C.; Latham, J.A.; Vermaas, E.H.; Toole, J.J. Selection of single-stranded DNA molecules that bind and inhibit human thrombin. *Nature* **1992**, *355*, 564–566. [CrossRef] [PubMed]

14. Padmanabhan, K.; Padmanabhan, K.P.; Ferrara, J.D.; Sadler, J.E.; Tulinsky, A. The structure of alpha-thrombin inhibited by a 15-mer single-stranded DNA aptamer. *J. Biol. Chem.* **1993**, *268*, 17651–17654. [PubMed]

15. Padmanabhan, K.; Tulinsky, A. An ambiguous structure of a DNA 15-mer thrombin complex. *Acta Cryst.* **1996**, *52*, D272–D282. [CrossRef] [PubMed]

16. Charles, P.T.; Stubbs, V.R.; Soto, C.M.; Martin, B.D.; White, B.J.; Taitt, C.R. Reduction of non-specific protein adsorption using Poly(ethylene)glycol (PEG) modified polyacrylate hydrogels in immunoassays for staphylococcal enterotoxin b detection. *Sensors* **2009**, *9*, 645–655. [CrossRef] [PubMed]

17. Arya, S.K.; Solanki, P.R.; Datta, M.; Malhotra, B.D. Recent advances in self-assembled monolayers based biomolecular electronic devices. *Biosens. Bioelectron.* **2009**, *24*, 2810–2817. [CrossRef] [PubMed]

18. Markham, N.R.; Zuker, M. UNAFold: Software for nucleic acid folding and hybridization. In *Bioinformatics: Structure, Function and Applications—Methods in Molecular Biology*; Keith, J.M., Ed.; Humana Press: Totowa, NJ, USA, 2008; pp. 3–31.

19. The DINAMelt Web Server. Available online: http://www.mfold.rna.albany.edu/?q=DINAMelt/software (accessed on 1 July 2016).

20. Zhang, D.; Yan, Y.; Li, Q.; Yu, T.; Cheng, W.; Wang, L.; Ju, H.; Ding, S. Label-free and high-sensitive detection of salmonella using a surface plasmon resonance DNA-based biosensor. *J. Biotechnol.* **2012**, *160*, 123–128. [CrossRef] [PubMed]

21. Pasternak, A.; Hernandez, F.; Rasmussen, L.; Vester, B.; Wengel, J. Improved thrombin binding aptamer by incorporation of a single unlocked nucleic acid monomer. *Nucleic Acids Res.* **2011**, *39*, 1155–1164. [CrossRef] [PubMed]

22. Gupta, P.; Loos, K.; Korniakov, A.; Spagnoli, C.; Cowman, M.; Ulman, A. Facile route to ultraflat SAM-protected gold surfaces by "amphiphile splitting". *Angew. Chem.* **2004**, *43*, 520–523. [CrossRef] [PubMed]

23. Castronovo, M.; Scaini, D. The atomic force microscopy as a lithographic tool: Nanografting of DNA nanostructures for biosensing applications. *Methods Mol. Biol.* **2011**, *749*, 209–221. [PubMed]

24. Steel, A.B.; Herne, T.M.; Tarlov, M.J. Electrochemical quantitation of DNA immobilized on gold. *Anal. Chem.* **1998**, *70*, 4670–4677. [CrossRef] [PubMed]

25. Levicky, R.; Herne, T.M.; Tarlov, M.J.; Satija, S.K. Using self-assembly to control the structure of DNA monolayers on gold: A neutron reflectivity study. *J. Am. Chem. Soc.* **1998**, *120*, 9787–9792. [CrossRef]

26. Peterson, A.W.; Wolf, L.K.; Georgiadis, R.M. Hybridization of mismatched or partially matched DNA at surfaces. *J. Am. Chem. Soc.* **2002**, *682*, 14601–14607. [CrossRef]

27. Peterson, A.W.; Heaton, R.J.; Georgiadis, R.M. The effect of surface probe density on DNA hybridization. *Nucleic Acids Res.* **2001**, *29*, 5163–5168. [CrossRef] [PubMed]

28. Hulme, E.C.; Trevethick, A. Ligand binding assays at equilibrium: Validation and interpretation. *Br. J. Pharmacol.* **2010**, *161*, 1219–1237. [CrossRef] [PubMed]

29. Basic Theory of Affinity. Available online: https://www.biacore.com/lifesciences/help/basic_theory_of_affinity/ (accessed on 1 July 2016).

30. Bosco, A.; Bano, F.; Parisse, P.; Casalis, L.; DeSimone, A.; Micheletti, C. Hybridization in nanostructured DNA monolayers probed by AFM: Theory vs. experiment. *Nanoscale* **2012**, *4*, 1734–1741. [CrossRef] [PubMed]

chemosensors

MDPI

Article

Unique Properties of Core Shell Ag@Au Nanoparticles for the Aptasensing of Bacterial Cells

Ezat Hamidi-Asl [1,*], Freddy Dardenne [2], Sanaz Pilehvar [1], Ronny Blust [2] and Karolien De Wael [1]

[1]   AXES Research Group, Department of Chemistry, University of Antwerp, Groenenborgerlaan 171,
      B-2020 Antwerp, Belgium; sanaz.pilehvar@uantwerpen.be (S.P.); karolien.dewael@uantwerpen.be (K.D.W.)
[2]   Sphere Research Group, Department of Biology, University of Antwerp, Groenenborgerlaan 171,
      B-2020 Antwerp, Belgium; freddy.dardenne@uantwerpen.be (F.D.); ronny.blust@uantwerpen.be (R.B.)
*   Correspondence: ehamidiasl@yahoo.com; Tel.: +98-11-3233-7926

Academic Editors: Paolo Ugo and Ligia Moretto
Received: 18 April 2016; Accepted: 17 August 2016; Published: 29 August 2016

Abstract: In this article, it is shown that the efficiency of an electrochemical aptasensing device is influenced by the use of different nanoparticles (NPs) such as gold nanoparticles (Au), silver nanoparticles (Ag), hollow gold nanospheres (HGN), hollow silver nanospheres (HSN), silver–gold core shell (Ag@Au), gold–silver core shell (Au@Ag), and silver–gold alloy nanoparticles (Ag/Au). Among these nanomaterials, Ag@Au core shell NPs are advantageous for aptasensing applications because the core improves the physical properties and the shell provides chemical stability and biocompatibility for the immobilization of aptamers. Self-assembly of the NPs on a cysteamine film at the surface of a carbon paste electrode is followed by the immobilization of thiolated aptamers at these nanoframes. The nanostructured (Ag@Au) aptadevice for *Escherichia coli* as a target shows four times better performance in comparison to the response obtained at an aptamer modified planar gold electrode. A comparison with other (core shell) NPs is performed by cyclic voltammetry and differential pulse voltammetry. Also, the selectivity of the aptasensor is investigated using other kinds of bacteria. The synthesized NPs and the morphology of the modified electrode are characterized by UV-Vis absorption spectroscopy, scanning electron microscopy, energy dispersive X-ray analysis, and electrochemical impedance spectroscopy.

Keywords: aptasensor; *Escherichia coli*; nanoparticles; electrochemistry; core shell nanostructures

1. Introduction

It is well known that the properties of detection systems composed of nano-dimensional elements are different from those of common bulky ones [1]. Among the various nanoparticles (NPs), noble metal nanomaterials such as Ag, Au, Pt, and Pd have attracted a lot of attention for their unique physical and chemical properties like tunable surface plasmonics, high-efficiency electrochemical sensing, enhanced fluorescence, and quantum conductance [2]. Of particular interest are binary metallic nanostructures showing multiple characteristics [3]. In this case, core/shell nanoparticle architectures, in which a layer of metal surrounds another NP core, have shown specific properties different from those of their monometallic counterparts and even alloys [4]. In this structure, the stability and surface chemistry of the shell nanoparticles can be improved along with accessing the physicochemical nature of the core layer. This synergy between two metals can be coordinated by shape, size, and composition. The core/shell nanostructures might exhibit favorable electrocatalytic activity, taking place on the shell of the NPs while the core metal dramatically affects the performance of the whole NPs [5]. Due to their particular electronic and catalytic impacts as well as their good stability, convenience of electron transfer, and biocompatibility [6], gold and silver are ideal choices for the construction of core/shell nanostructures in biosensors. Tang et al. reported a silver–gold core shell (Ag@Au) label-free amperometric immune biosensor [7]. Li et al. introduced a hydrogen peroxide sensor based on Au@Ag@C core-double shell nanocomposites [8]. Eksi et al. developed an electrochemical immunosensor for the determination of E. coli using Ag@Au bioconjugates and anti-E. coli modified PS-microwells [9]. Here we report, for the first time, on the effect of using Ag@Au core shell NPs for the electrochemical aptasensing of gram-negative bacterium E. coli.

E. coli is commonly found in the intestinal track of humans and other warm-blooded animals. It can be transmitted to humans through the consumption of contaminated food or water and is often used as a biomarker to identify fecal contamination [10]. During recent years, a lot of attention has been given to the design of biosensors for the recognition of single cells, viruses, and bacteria [11]. Different techniques have been developed and improved for the monitoring of live targets and their viability such as super resolution fluorescence microscopy, [12] scanning electrochemical microscopy, [13] capillary electrophoresis electrospray ionization mass spectrometry [14], laser ablation inductively coupled plasma mass spectroscopy [15], and chemical patterning-based single cell trapping [16,17]. Among them, electrochemical biosensors are of particular interest due to their remarkable advantages such as low cost, good sensitivity, and fast response. Some examples include a graphene-based potentiometric biosensor for the immediate detection of *Staphylococcus aureus* [18], the detection of *Salmonella Typhimurium* using a carbon nanotubes based aptasensor [19] and the real-time detection of cytokines released from immune cells after mitogenic activation [20].

The immobilization of a biorecognition element on the working electrode is a key factor in the development of a biosensor. Artificial oligo nucleotides (aptamers) are potentially well suited for targeting motile objects through the binding of different components on the cell surface such as proteins, polysaccharides, or flagella [21], given the examples in literature such as an aptamer-based Au NP biosensor for the detection of flu viruses [22], aptamer-conjugated NPs for the delivery of paclitaxel to MUC1-positive tumor cells [23] and targeting prostate cancer cells within PLA-PEG-COOH NPs aptamer bioconjugates [24].

This work reports on the joint action of aptamers and Ag@Au core shell NPs on a carbon paste electrode (CPE) for the detection of E. coli. Up to now, only a mixture of nanomaterials instead of core shell NPs have been used for the immobilization of aptamers [25–27]. Our electrochemical aptasensor is based on CPE modified with different NPs such as gold nanoparticles (Au), silver nanoparticles (Ag), hollow gold nanospheres (HGN), hollow silver nanospheres (HSN), a silver–gold core shell (Ag@Au), a gold–silver core shell (Au@Ag), and silver/gold alloy nanoparticles (Ag/Au). A comparison of the responses of these different NP-based aptasensors is provided. On the other hand, we report on the development of a label-free aptasensor via direct monitoring of the electrochemical signal of E. coli bacteria.

2. Experimental Section

Experimental details can be found in the supporting information including materials, sequence of aptamers, bacterial strains, culture conditions, procedure for the synthesis of nanoparticles, and preparation of the aptasensor. For the sensing of E. coli, we used an aptamer mixture (aptamer cocktail) composed of three different DNA aptamers, as reported by Kim et al. [28,29].

3. Results and Discussion

3.1. Electrochemical Response of E. coli at Bare Electrode

Figure 1A shows the cyclic voltammograms of a bare CPE in different concentrations of E. coli cells (CFU/mL). In microbiology, CFU (colony forming units) designates the number of viable bacteria in a sample. No oxidation process is observed for cell concentrations lower than 10^7 CFU/mL. A weak peak appears at ca +0.80 V (vs. Ag/AgCl) from a concentration of 5×10^7 CFU/mL (curve d), increasing with increasing concentrations of E. coli (curve e and f). This irreversible oxidation peak can possibly be attributed to the oxidation of guanine in the bacterial cell cytoplasm to 8-oxo-guanine [30,31]. To obtain more sensitive signals, differential pulse voltammograms (DPV) were recorded (see inset of Figure 1A).

Figure 1B illustrates the cyclic voltammograms of bare CPE in 5×10^7 CFU/mL of different bacteria: (a) *Pseudomonas aeruginosa*; (b) *Staphylococcus aureus*; and (c) *Escherichia coli*. A similar process for both other types of bacteria is observed, approximately at the same potential [31,32]. The inset of this figure shows the DPV of a bare CPE in different bacteria solution. Accordingly, the peak around +0.80 V in DPV was selected as a characteristic signal for monitoring of *E. coli* and other bacteria.

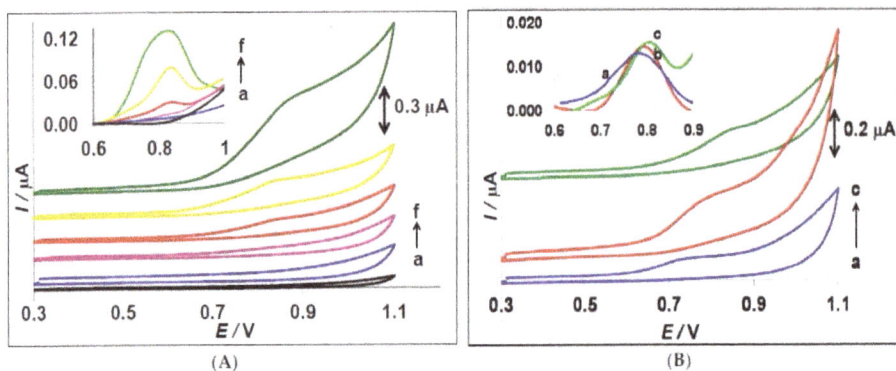

Figure 1. (**A**) Cyclic voltammograms of bare CPE in (a) 0.0, (b) 10^6, (c) 10^7, (d) 5×10^7, (e) 10^8, and (f) 10^9 CFU/mL *E. coli* in PBS 0.1 M pH 7.0, scan rate 100 mV·s^{-1}; *Inset:* Differential pulse voltammograms of bare CPE in different concentrations of *E. coli*, pulse height: 0.05 V, scan rate: 20 mV·s^{-1}. (**B**) Cyclic voltammograms of bare CPE in 5×10^7 CFU/mL of different bacteria in PBS 0.1 M pH 7.0: (a) *Pseudomonas aeruginosa*, (b) *Staphylococcus aureus*, and (c) *Escherichia coli*. Inset: Differential pulse voltammograms of bare CPE in different bacteria.

3.2. Characterization of the Nanoparticles

Figure S1 provides the overview of the different synthesized nanoparticles. They are red, deep yellow, pale pink or yellowish red, dark reddish pink, pale yellow, blue, and dark orange for gold, silver, a silver/gold alloy, the silver–gold core shell, the gold-silver core shell, hollow gold nanospheres, and hollow silver nanospheres, respectively. The corresponding UV-Vis absorption spectroscopic data are shown in Figure S2. Generally, the absorption peaks of the metallic nanoparticles are linked to the surface plasmon resonances (SPR) absorption. The SPR effect in metallic compounds is of interest for a variety of applications because of the large electromagnetic field enhancement that occurs in the vicinity of the metal surface and the resonance wavelength depends on the size, shape, and local dielectric environment of nanoparticles [33]. The absorbance wavelengths are as follows: λ_{max} = 406, 420, 450, 525, 535, 565, and 640 nm for Ag, Au@Ag, HSN, Ag/Au, Ag@Au, Au, and HGN, respectively. This is in agreement with the reported colors and wavelengths [6]. Usually, checking the color and λ_{max} is the most straightforward way to confirm the quality of the synthesized nanoparticles [34]. Figures S3 and S4 represent the SEM images and EDX analyses for all synthesized nanoparticles. As can be seen in Figure S3, the size distribution of nanoparticles ranges from 30 to 100 nm. In the solutions containing silver NPs, there is a higher tendency for agglomeration compared to solutions containing gold NPs [35,36], clearly illustrated in Figure S3B, C, and E; hollow silver nanospheres, especially, form clusters (Figure S3G). The energy dispersive X-ray analysis is used to provide elemental identification (Figure S4) [37]. The indicative peaks for both Au and Ag elements in Figures S4C–E prove the existence of binary nanostructures at the surface of the working electrode [8].

3.3. Characterization of the Modified CPE

A cysteamine film is formed at the surface of CPE to capture the nanoparticles in a next step. During successive CV scanning, the cysteamine is chemically adsorbed at the CPE via amine oxidation, while the thiol end group is free to interact with neighboring molecules [38]. After immersing the cysteamine-modified CPE in the nanoparticles solution, a strong interaction between either Au or Ag nanoparticles and the thiol group present at the electrode will appear because of the inherent affinity between sulfur and some metals like gold, silver, palladium, and copper [39].

Cyclic voltammetry was selected to determine the surface coverage of the working electrode. The cysteamine-modified CPE, after overnight immersing in a nanoparticle solution, was washed and CV was performed in a phosphate buffer solution (PBS) 0.1 M pH 7.0. Five successive cyclic voltammograms were recorded without a change in the current potential readout, confirming the stability of fixed nanoparticles at the cysteamine film during potential cycling. Hence, it acts as a strong template for the immobilization of aptamers. The amperometric responses of a bare CPE and a cysteamine modified CPE are shown in Figure S5.

Figure 2A illustrates the current potential behavior of a gold nanoparticles modified CPE in PBS solution. An oxidation and reduction peak for gold can be observed at 0.96 V and 0.53 V, respectively [40]. Also, both the CV of a bare CPE (red CV) and Cys-CPE (green CV) are shown in this figure, acting as background curves. Figure 2B represents the current potential behavior of a silver nanoparticles modified CPE, showing a characteristic peak for the oxidation and reduction of silver [41]. The cyclic voltammogram of an Ag/Au alloy nanoparticles modified CPE indicates, in Figure 2C, both the existence of Ag and Au at the surface. However, the oxidation of Ag in the alloy starts at a less positive potential in comparison with pure Ag. Indeed, silver alloys are more active than pure Ag and they act as an intermediate bridge to accelerate the electron transport [42].

The current potential behavior of Ag@Au-coated CPE is shown in Figure 2D. In this cyclic voltammogram, a quasi-reversible peak (aa') and irreversible peak (bb') appear for silver and gold. The peaks a ($E_{pa} = 0.2$ V) and a' ($E_{pc} = 0.1$V) belong to the oxidation and reduction of silver nanoparticles ($\Delta E_p = 90$ mV), respectively [41,43]. This sharp characteristic peak proves the existence of silver at the surface of the working electrode. The oxidation and reduction of Au (bb') takes place at more positive potentials. A series of anodic peaks ($b_{(I)}$, $b_{(II)}$, and $b_{(III)}$) correspond to intermediate steps leading to the formation of an AuO/OH layer [44]. The cathodic peak at 0.45 V on the return scan (b) represents the reduction of the AuO/OH layer [45]. The comparison between Ag/Au and Ag@Au-modified CPE leads to two key observations: first, there is almost no difference in the peak potentials for the oxidation and the reduction of gold and silver; secondly, the peak currents of the alloy modified CPE are higher than the silver–gold core shell modified CPE. This can be explained by the position of Ag in these two binary nanostructures. In the core shell structure, Ag is surrounded by Au as a shell, while in the alloy structure the metal of Ag is as accessible as Au. Additionally, there is a higher tendency for aggregation of the nanoparticles whenever silver is accessible [36]. Therefore, the amount of stabilized particles would be increased due to silver agglomeration. The higher peak current somehow indicates the bigger portion of these metals at the surface of the working electrode. The SEM images are in good agreement with these observations.

Figure 2E shows the cyclic voltammogram of an Au@Ag-modified CPE. In this structure, Au particles are covered by an Ag shell. Therefore, Ag is the outer element. As can be seen, there is no clear peak, except the huge charging current that belongs to poly(vinylpyrrolidinone) (PVP) [46], which is necessary to synthesize this type of core shell. Therefore, this nanostructure is not useful for our purposes, i.e., the immobilization of aptamers. Figure 2F,G depict the CV of HGN and HSN, respectively. The positions of the oxidation and reduction peaks are similar to Au and Ag in Figure 2A,B.

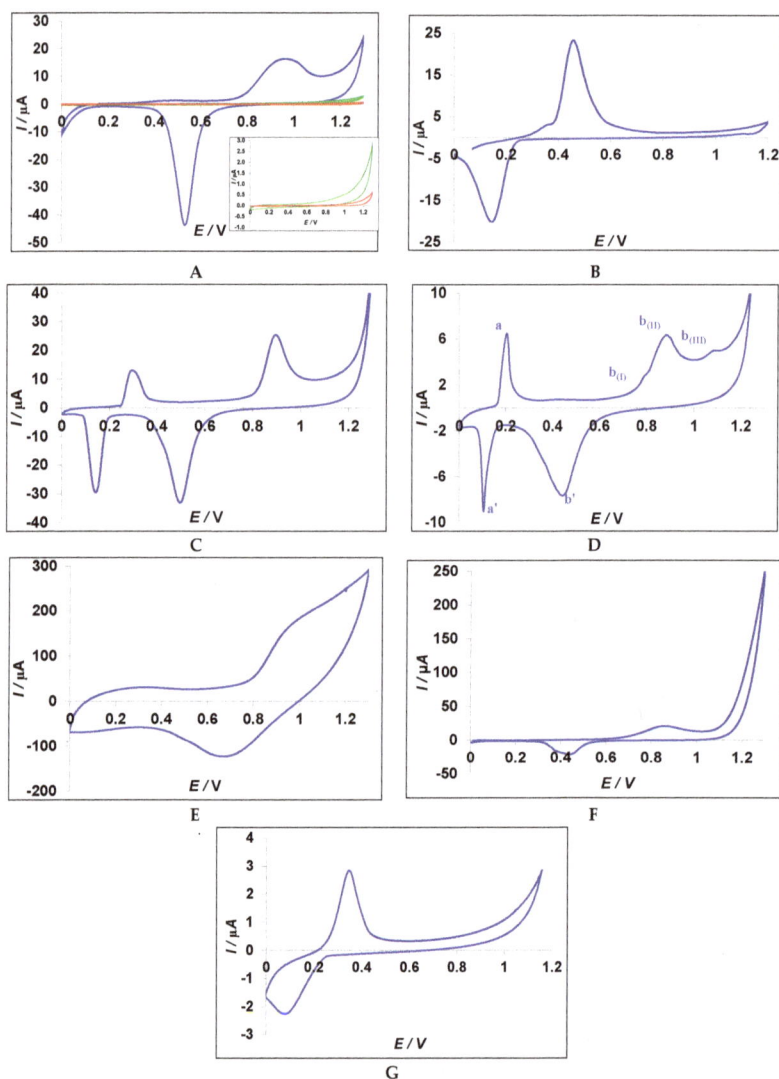

Figure 2. Cyclic voltammograms of modified CPE with different synthesized nanoparticles in PBS 0.1 M pH 7.0, scan rate 100 mV·s^{-1}. In Figure 2A, both the CV of a bare CPE and Cys-CPE are plotted in red and green, respectively (inset: zoom of the curves). (**A**) Gold nanoparticles (Au); (**B**) silver nanoparticles (Ag); (**C**) silver–gold alloy (Ag/Au); (**D**) silver–gold core shell (Ag@Au); (**E**) gold–silver core shell (Au@Ag); (**F**) hollow gold nanospheres (HGN); (**G**) hollow silver nanospheres (HSN).

3.4. Electrochemical Detection of E. coli and the Role of Ag@Au Nanoparticles

The whole procedure to fabricate the aptasensor is presented in Figure 3 and explained in the experimental section of the supporting information. Briefly, after preparation of CPE, a film of cysteamine was formed at the surface of CPE by performing cyclic voltammetry [38]. Then, it was immersed in the nanoparticles solution overnight to form a nanotemplate at the surface of the working electrode. For the immobilization of the aptamer, a modified CPE was soaked in the aptamer solution

(2.5 μM, overnight). Next, the aptasensor was transferred to suspensions with different concentrations of living bacteria (30 min) to allow binding to *E. coli*. Subsequently, it was gently washed and moved to 0.1 M PBS pH 7.0 for voltammetric measurements.

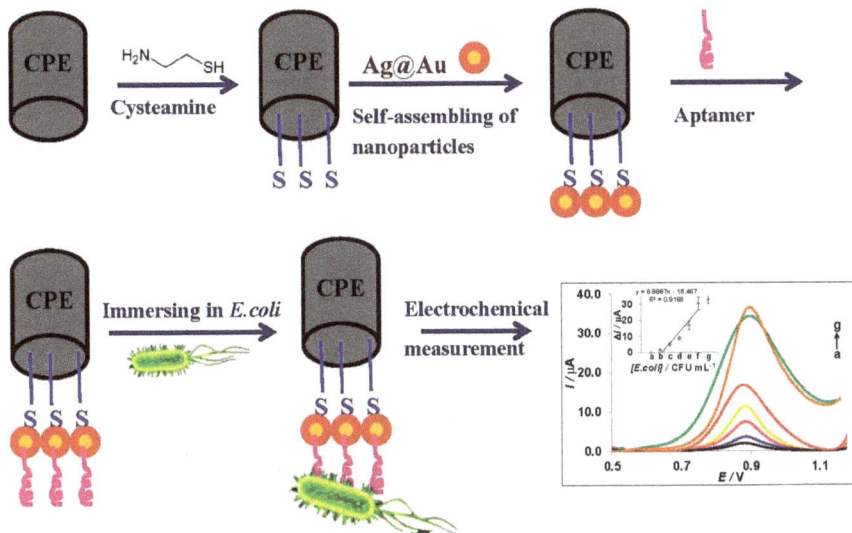

Figure 3. Schematic diagram of the procedure.

As observed in the inset of Figure 1B, the oxidation of the bacteria solution shows a representative signal around 0.80 V, using DPV at the surface of a bare CPE. Figure 4A shows the DPVs of the Ag@Au aptasensor in different concentrations of *E. coli*. There is an enhanced signal around 0.88 V when increasing the bacteria concentration from 0.0 to 10^5 CFU/mL. The oxidation signal of *E. coli* at the surface of a bare CPE could only be observed at concentrations higher than 5×10^5 CFU/mL (see Figure 1A). However, the bacteria oxidation process at bare CPE occurs at less positive potential in comparison to the aptamer-modified CPE. This small shift in the position of the potential can be attributed to the coverage of CPE. On the other hand, the oxidation of gold happens approximately at this potential (see Figure 2A,C,D). Therefore, we assume an overlap between these two oxidation signals, guanine oxidation in *E. coli*, and gold oxidation of the nanostructures, playing a synergetic effect in its detection. The stability of gold at the surface of the working electrode is confirmed by successive CV scanning, as shown in Figure 2.

Figure 4B illustrates the calibration curve expressing the variation of ΔI (I_1–I_0)/μA vs. the concentration of *E. coli*. I_0 and I_1 belong to the DPV current of the aptasensor in PBS and *E. coli*, respectively. The best performance is attributed to the Ag@Au-aptasensor and there is a linear relationship between peak current and the concentration of *E. coli* to 10^4 CFU/mL. For higher concentrations, the current no longer changes significantly because of the saturation of the surface [29]. The calculated LOD for this aptasensor is around 90 CFU/mL, while the reported LOD in the literature for this aptamer cocktail is 370 CFU/mL at the surface of a gold electrode [28]. Hence, the silver–gold core shell binary nanoparticles improve the efficiency of the aptasensor by four times.

The obtained LOD values for the alloy and the Au NPs modified aptasensor are 245 and 300 CFU/mL, respectively. So, the silver/gold alloy leads to a higher sensitivity than obtained at the Au NPs modified electrode. A better performance for Ag@Au in comparison with the Ag/Au alloy can be explained by the position of silver in these binary nanostructures. In the first one, silver is covered by the gold nanoparticles, so that agglomeration of the Ag NPs cannot be initiated. In contrast, in the alloy structure, the silver particles are easily accessible and tend to aggregate. This phenomenon, observed in the SEM images, limits the performance of the nanoframes. In comparison with other nanoparticles, hollow gold nanospheres showed less effect on the efficiency of this electrochemical aptasensor, probably due to their narrow wall [6].

In general, a better diagnostic performance of the core shell nanoparticles modified sensor can be explained by the high surface-to-volume ratio of the assembled nanoparticles enhancing the density of the immobilized aptamers [24]. Also, the layer of nanoparticles may work as an intervening "spacer" matrix to keep the immobilized aptamers away from the substrate matrix in the mobile phase, resulting in more accessible binding sites for the target [47,48]. Therefore, these nanomaterials can significantly affect the aptasensing of *E. coli*.

Silver nanoparticles did not show any promising response in the aptasensor. A lower tendency of Ag to interact with thiols, compared to Au-thiol, might cause this difference [49,50]. For this reason, a lower amount of Ag NPs can self-assemble on the cysteamine-modified CPE and consequently fewer aptamers are immobilized at this nanoframe. Also, Ag NPs did not oxidize in the potential range of the bacteria. Therefore, these nanoparticles alone could not have any added value in the detection strategy. The experiments with HSN and Au@Ag, with the silver as a cover for gold NPs, were performed and similar results as with Ag NPs were observed. The difficulty with Au@Ag was the presence of the surfactant in the solution of these nanoparticles. The addition of PVP and CTAB was necessary for its synthesis, but problematic for electrochemical detection as a self-assembly monolayer is formed at the surface of the working electrode which disturbs the signals. As shown in Figure 2E, it was not possible to obtain a sharp signal similar to that observed for the other nanoparticles.

As a conclusion of Figure 4B, an increment in the efficiency of aptasensing is observed in the following order: Ag < HGN < Au < Ag/Au < Ag@Au. The excellent performance of the core shells in the series can be explained by the fact that Au NPs and Ag NPs have a different surface plasmon band [51] and the extinction coefficient of the surface plasmon band for the Ag NPs is nearly four times larger than for Au NPs of the same size [52]. Therefore, in the composition of Ag@Au core shells, the physical properties of Ag nanoparticles combined with the surface chemistry of Au allows for functionalization with an aptamer. Additionally, Ag NPs in the core decrease the resistance of electron transfer, enhancing the sensitivity [7].

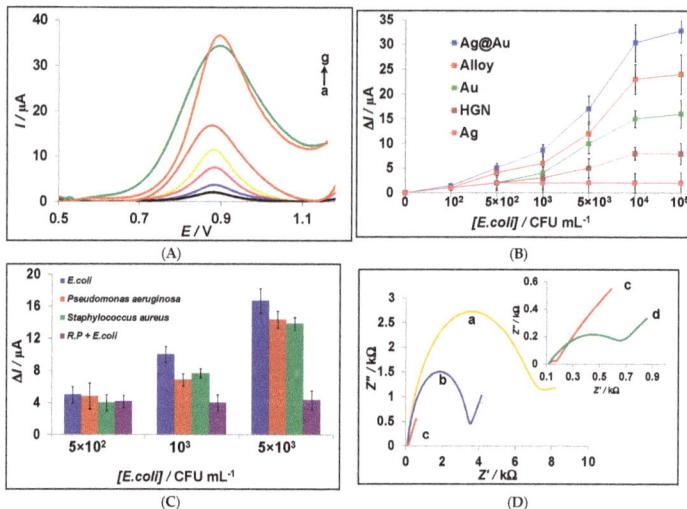

Figure 4. *Cont.*

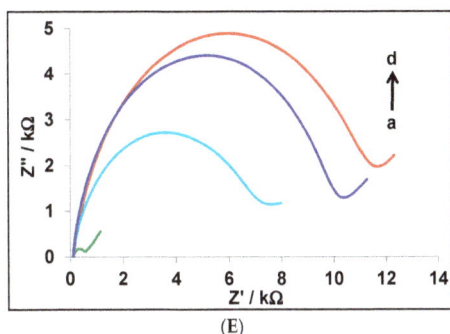

(E)

Figure 4. (**A**) Differential pulse voltammograms of Ag@Au aptasensor in (a) 0.0, (b) 10^2, (c) 5×10^2, (d) 10^3, (e) 5×10^3, (f) 10^4, and (g) 10^5 CFU/mL *E. coli* in PBS 0.1 M pH 7.0; (**B**) calibration curve for variation of ΔI (µA) vs. concentration of *E. coli* using modified CPE with different synthesized nanoparticles. (**C**) Histogram of selectivity study for different bacteria: *Escherichia coli, Pseudomonas aeruginosa, Staphylococcus aureus*, and random primer (R.P.); (**D**) electrochemical impedance spectra for (a) bare CPE, (b) Cys-CPE, (c) Ag@Au-Cys-CPE, and (d) Apt-Ag@Au-Cys-CPE; (**E**) electrochemical impedance spectra at Apt-Ag@Au-Cys-CPE in different concentrations of *E. coli* (a) 0.0, (b) 10^3, (c) 10^4, and (d) 10^5 CFU/mL in the presence of 5 mM $[Fe(CN)_6]^{3-/4-}$, 0.1 M NaCl, 0.1 M PBS pH 7.0; the concentration of aptamer is 2.5 µM.

The specificity of the aptasensor was examined for other bacterial species. Figure 4C shows the histogram of a selectivity study for *Escherichia coli, Pseudomonas aeruginosa*, and *Staphylococcus aureus* in three concentrations using the Ag@Au aptasensor. Also, in another set of experiments, a random primer (RP) was used instead of the specific aptamer cocktail for *E. coli*. As seen in lower concentration ranges (5×10^2 CFU/mL), there was no recognition between different species. However, the detection ability improved when increasing the concentration of bacteria. The random primer could not detect any kind of targets, indicating the good performance of the aptamer cocktail.

Electrochemical impedance spectroscopy was selected to characterize the electrode surface in each step of the preparation of the Ag@Au aptasensor and the detection of *E. coli* (Figure 4D). The EIS measurements were performed in the presence of $[Fe(CN)_6]^{3-/4-}$, as electroactive marker ions. As EIS is a method of measuring the impedance value of the electrode surface, it can recognize the events happening at the surface of the working electrode. The diameter of the semicircle in EIS represents the electron transfer resistance, R_{et}; therefore, the further impeding in the electron transfer, the higher the R_{et}. The values of R_{et} for bare CPE and cysteamine-modified CPE are 7.07 kΩ (curve a) and 3.54 kΩ (curve b), respectively. The decrease in R_{et} after modification of CPE by cysteamine is due to the oxidation of CPE surface during pre-treatment. Consequently, there are some positive sites at CPE that can accelerate the negative marker ions at the surface of working electrode. After immersing the modified CPE in the core shell solution, the R_{et} changes (curve c = 0.10 kΩ), indicating that Ag@Au nanoparticles are chemically fixed to the cysteamine film at the surface of CPE. The R_{et} has decreased due to the enhancement of the conductivity of the surface of the working electrode [38]. After immobilization of the aptamer at this nanostructure template, the R_{et} has increased (curve d = 0.75 kΩ) because of the electrostatic repulsion between negatively charged DNA in the aptamer structure and the marker ions. When *E. coli* bacteria (10^3 CFU/mL) are captured by the aptamer, the R_{et} dramatically increased (Figure 4E, curve b = 8.85 kΩ) because of the bulky size of the bacteria in comparison to the aptamer, which prevents the marker ions from reaching the surface of the working electrode [53]. Curves c and d in Figure 4E show the EIS data for different concentrations of *E. coli*. As can be seen, the diameter of the semicircles increases with higher concentrations of the target.

4. Conclusions

A qualitative investigation of NPs with different structures on the efficiency of an aptasensor for *E. coli* was performed. Various NPs were synthesized and their performances were compared. In this study, for the first time, NPs were used as a template for the immobilization of an aptamer cocktail at the surface of a carbon paste electrode to produce an electrochemical aptasensor for *E. coli*. The results showed that silver–gold core shell NPs improved the efficiency of the aptasensing of *E. coli*. In these binary NPs, the silver is positioned inside the nanostructure, and surrounded by gold particles as a shell. Therefore, the outside layer of the NPs provides a support for the covalent attachment of aptamers, functionalized with a thiol group. The conductive silver metal as a core increases the efficiency of the transduction of the signals. The label-free aptasensing strategy uses the intrinsic oxidation peak of bacterial cells with an increased sensitivity for the detection of *E. coli* compared to the sensitivity of the aptasensors constructed with other nanomaterials.

Supplementary Materials: Supplementary materials are available online at http://www.mdpi.com/2227-9040/4/3/16/s1.

Acknowledgments: Ezat Hamidi-Asl was financially supported by Belspo (University of Antwerp). The authors are thankful to Femke De Croock for her technical support and to Stanislav Trashin for his worthwhile comments on the manuscript.

Author Contributions: E.H.-A. and S.P. performed the measurements and wrote the manuscript. F.D., R.B. and K.D.W. coordinated the project and commented on the manuscript.

Conflicts of Interest: The authors declare no conflict of interest.

References

1. Yin, Y.; Rioux, R.M.; Erdonmez, C.K.; Hughes, S.; Somorjai, G.A.; Alivisatos, A.P. Formation of hollow nanocrystals through the nanoscale kirkendall effect. *Science* **2004**, *304*, 711–714. [CrossRef] [PubMed]
2. Campion, A.; Kambhampati, P. Surface-enhanced raman scattering. *Chem. Soc. Rev.* **1998**, *27*, 241–250. [CrossRef]
3. Milliron, D.J.; Hughes, S.M.; Cui, Y.; Manna, L.; Li, J.; Wang, L.-W.; Paul Alivisatos, A. Colloidal nanocrystal heterostructures with linear and branched topology. *Nature* **2004**, *430*, 190–195. [CrossRef] [PubMed]
4. Toshima, N.; Harada, M.; Yonezawa, T.; Kushihashi, K.; Asakura, K. Structural analysis of polymer-protected palladium/platinum bimetallic clusters as dispersed catalysts by using extended X-ray absorption fine structure spectroscopy. *J. Phys. Chem.* **1991**, *95*, 7448–7453. [CrossRef]
5. Ferrando, R.; Jellinek, J.; Johnston, R.L. Nanoalloys: From theory to applications of alloy clusters and nanoparticles. *Chem. Rev.* **2008**, *108*, 845–910. [CrossRef] [PubMed]
6. Khlebtsov, N.G.; Dykman, L.A. Optical properties and biomedical applications of plasmonic nanoparticles. *J. Quant. Spectrosc. Radiat. Transf.* **2010**, *111*, 1–35. [CrossRef]
7. Tang, D.; Yuan, R.; Chai, Y. Ligand-functionalized core/shell Ag@Au nanoparticles label-free amperometric immun-biosensor. *Biotechnol. Bioeng.* **2006**, *94*, 996–1004. [CrossRef] [PubMed]
8. Li, Y.; Zhang, Y.; Zhong, Y.; Li, S. Enzyme-free hydrogen peroxide sensor based on Au@Ag@C core-double shell nanocomposites. *Appl. Surf. Sci.* **2015**, *347*, 428–434. [CrossRef]
9. Eksi, H.; Güzel, R.; Güven, B.; Boyaci, I.H.; Solak, A.O. Fabrication of an electrochemical *E. coli* biosensor in biowells using bimetallic nanoparticle-labelled antibodies. *Electroanalysis* **2015**, *27*, 343–352. [CrossRef]
10. Belluco, S.; Barco, L.; Roccato, A.; Ricci, A. *Escherichia coli* and enterobacteriaceae counts on poultry carcasses along the slaughterline: A systematic review and meta-analysis. *Food Control* **2016**, *60*, 269–280. [CrossRef]
11. Søndergaard, R.V.; Christensen, N.M.; Henriksen, J.R.; Kumar, E.K.P.; Almdal, K.; Andresen, T.L. Facing the design challenges of particle-based nanosensors for metabolite quantification in living cells. *Chem. Rev.* **2015**, *115*, 8344–8378. [CrossRef] [PubMed]
12. Jones, S.A.; Shim, S.-H.; He, J.; Zhuang, X. Fast, three-dimensional super-resolution imaging of live cells. *Nat. Methods* **2011**, *8*, 499–505. [CrossRef] [PubMed]
13. Bergner, S.; Wegener, J.; Matysik, F.-M. Monitoring passive transport of redox mediators across a confluent cell monolayer with single-cell resolution by means of scanning electrochemical microscopy. *Anal. Methods* **2012**, *4*, 623–629. [CrossRef]

14. Nemes, P.; Knolhoff, A.M.; Rubakhin, S.S.; Sweedler, J.V. Metabolic differentiation of neuronal phenotypes by single-cell capillary electrophoresis–electrospray ionization-mass spectrometry. *Anal. Chem.* **2011**, *83*, 6810–6817. [CrossRef] [PubMed]

15. Drescher, D.; Giesen, C.; Traub, H.; Panne, U.; Kneipp, J.; Jakubowski, N. Quantitative imaging of gold and silver nanoparticles in single eukaryotic cells by laser ablation icp-ms. *Anal. Chem.* **2012**, *84*, 9684–9688. [CrossRef] [PubMed]

16. Collins, J.M.; Lam, R.T.S.; Yang, Z.; Semsarieh, B.; Smetana, A.B.; Nettikadan, S. Targeted delivery to single cells in precisely controlled microenvironments. *Lab Chip* **2012**, *12*, 2643–2648. [CrossRef] [PubMed]

17. Trouillon, R.; Passarelli, M.K.; Wang, J.; Kurczy, M.E.; Ewing, A.G. Chemical analysis of single cells. *Anal. Chem.* **2012**, *85*, 522–542. [CrossRef] [PubMed]

18. Hernández, R.; Vallés, C.; Benito, A.M.; Maser, W.K.; Xavier Rius, F.; Riu, J. Graphene-based potentiometric biosensor for the immediate detection of living bacteria. *Biosens. Bioelectron.* **2014**, *54*, 553–557. [CrossRef] [PubMed]

19. Zelada-Guillén, G.A.; Riu, J.; Düzgün, A.; Rius, F.X. Immediate detection of living bacteria at ultralow concentrations using a carbon nanotube based potentiometric aptasensor. *Angew. Chem. Int. Ed.* **2009**, *48*, 7334–7337. [CrossRef] [PubMed]

20. Liu, Y.; Kwa, T.; Revzin, A. Simultaneous detection of cell-secreted tnf-α and ifn-γ using micropatterned aptamer-modified electrodes. *Biomaterials* **2012**, *33*, 7347–7355. [CrossRef] [PubMed]

21. Liu, J.; Morris, M.D.; Macazo, F.C.; Schoukroun-Barnes, L.R.; White, R.J. The current and future role of aptamers in electroanalysis. *J. Electrochem. Soc.* **2014**, *161*, H301–H313. [CrossRef]

22. Le, T.T.; Adamiak, B.; Benton, D.J.; Johnson, C.J.; Sharma, S.; Fenton, R.; McCauley, J.W.; Iqbal, M.; Cass, A.E.G. Aptamer-based biosensors for the rapid visual detection of flu viruses. *Chem. Commun.* **2014**, *50*, 15533–15536. [CrossRef] [PubMed]

23. Yu, C.; Hu, Y.; Duan, J.; Yuan, W.; Wang, C.; Xu, H.; Yang, X.-D. Novel aptamer-nanoparticle bioconjugates enhances delivery of anticancer drug to muc1-positive cancer cells *in vitro*. *PLoS ONE* **2011**, *6*, e24077. [CrossRef] [PubMed]

24. Farokhzad, O.C.; Jon, S.; Khademhosseini, A.; Tran, T.-N.T.; LaVan, D.A.; Langer, R. Nanoparticle-aptamer bioconjugates: A new approach for targeting prostate cancer cells. *Cancer Res.* **2004**, *64*, 7668–7672. [CrossRef] [PubMed]

25. Wang, A.Z.; Bagalkot, V.; Vasilliou, C.C.; Gu, F.; Alexis, F.; Zhang, L.; Shaikh, M.; Yuet, K.; Cima, M.J.; Langer, R.; et al. Superparamagnetic iron oxide nanoparticle–aptamer bioconjugates for combined prostate cancer imaging and therapy. *ChemMedChem* **2008**, *3*, 1311–1315. [CrossRef] [PubMed]

26. Medley, C.D.; Bamrungsap, S.; Tan, W.; Smith, J.E. Aptamer-conjugated nanoparticles for cancer cell detection. *Anal. Chem.* **2011**, *83*, 727–734. [CrossRef] [PubMed]

27. Kashefi-Kheyrabadi, L.; Mehrgardi, M.A. Aptamer-conjugated silver nanoparticles for electrochemical detection of adenosine triphosphate. *Biosens. Bioelectron.* **2012**, *37*, 94–98. [CrossRef] [PubMed]

28. Kim, Y.S.; Chung, J.; Song, M.Y.; Jurng, J.; Kim, B.C. Aptamer cocktails: Enhancement of sensing signals compared to single use of aptamers for detection of bacteria. *Biosens. Bioelectron.* **2014**, *54*, 195–198. [CrossRef] [PubMed]

29. Kim, Y.S.; Song, M.Y.; Jurng, J.; Kim, B.C. Isolation and characterization of DNA aptamers against *Escherichia coli* using a bacterial cell-systematic evolution of ligands by exponential enrichment approach. *Anal. Biochem.* **2013**, *436*, 22–28. [CrossRef] [PubMed]

30. Zhang, J.-J.; Gu, M.-M.; Zheng, T.-T.; Zhu, J.-J. Synthesis of gelatin-stabilized gold nanoparticles and assembly of carboxylic single-walled carbon nanotubes/Au composites for cytosensing and drug uptake. *Anal. Chem.* **2009**, *81*, 6641–6648. [CrossRef] [PubMed]

31. Feng, J.; Ci, Y.-X.; Lou, J.-L.; Zhang, X.-Q. Voltammetric behavior of mammalian tumor cells and bioanalytical applications in cell metabolism. *Bioelectrochem. Bioenerg.* **1999**, *48*, 217–222. [CrossRef]

32. Nonner, W.; Eisenberg, B. Electrodiffusion in ionic channels of biological membranes. *J. Mol. Liq.* **2000**, *87*, 149–162. [CrossRef]

33. Schaadt, D.M.; Feng, B.; Yu, E.T. Enhanced semiconductor optical absorption via surface plasmon excitation in metal nanoparticles. *Appl. Phys. Lett.* **2005**, *86*, 063106. [CrossRef]

34. Liu, M.; Guyot-Sionnest, P. Synthesis and optical characterization of Au/Ag core/shell nanorods. *J. Phys. Chem. B* **2004**, *108*, 5882–5888. [CrossRef]

35. Shen, X.S.; Wang, G.Z.; Hong, X.; Zhu, W. Nanospheres of silver nanoparticles: Agglomeration, surface morphology control and application as sers substrates. *Phys. Chem. Chem. Phys.* **2009**, *11*, 7450–7454. [CrossRef] [PubMed]

36. Li, J.; Lin, Y.; Zhao, B. Spontaneous agglomeration of silver nanoparticles deposited on carbon film surface. *J. Nanopart. Res.* **2002**, *4*, 345–349. [CrossRef]

37. Danscher, G.; Hansen, H.J.; Møller-Madsen, B. Energy dispersive x-ray analysis of tissue gold after silver amplification by physical development. *Histochemistry* **1984**, *81*, 283–285. [CrossRef] [PubMed]

38. Hu, G.-Z.; Zhang, D.-P.; Wu, W.-L.; Yang, Z.-S. Selective determination of dopamine in the presence of high concentration of ascorbic acid using nano-au self-assembly glassy carbon electrode. *Colloids Surf. B* **2008**, *62*, 199–205. [CrossRef] [PubMed]

39. Love, J.C.; Estroff, L.A.; Kriebel, J.K.; Nuzzo, R.G.; Whitesides, G.M. Self-assembled monolayers of thiolates on metals as a form of nanotechnology. *Chem. Rev.* **2005**, *105*, 1103–1170. [CrossRef] [PubMed]

40. Zhang, L.; Jiang, X.; Wang, E.; Dong, S. Attachment of gold nanoparticles to glassy carbon electrode and its application for the direct electrochemistry and electrocatalytic behavior of hemoglobin. *Biosens. Bioelectron.* **2005**, *21*, 337–345. [CrossRef] [PubMed]

41. Choi, Y.-J.; Luo, T.-J.M. Electrochemical properties of silver nanoparticle doped aminosilica nanocomposite. *Int. J. Electrochem.* **2011**, *2011*. [CrossRef]

42. Tang, M.H.; Hahn, C.; Klobuchar, A.J.; Ng, J.W.D.; Wellendorff, J.; Bligaard, T.; Jaramillo, T.F. Nickel-silver alloy electrocatalysts for hydrogen evolution and oxidation in an alkaline electrolyte. *Phys. Chem. Chem. Phys.* **2014**, *16*, 19250–19257. [CrossRef] [PubMed]

43. Zhang, W.; Tu, C.Q.; Chen, Y.F.; Li, W.Y.; Houlachi, G. Cyclic voltammetric studies of the behavior of lead-silver anodes in zinc electrolytes. *J. Mater. Eng. Perform.* **2013**, *22*, 1672–1679. [CrossRef]

44. Rouya, E.; Cattarin, S.; Reed, M.L.; Kelly, R.G.; Zangari, G. Electrochemical characterization of the surface area of nanoporous gold films. *J. Electrochem. Soc.* **2012**, *159*, K97–K102. [CrossRef]

45. Tremiliosi-Filho, G.; Dall'Antonia, L.H.; Jerkiewicz, G. Growth of surface oxides on gold electrodes under well-defined potential, time and temperature conditions. *J. Electroanal. Chem.* **2005**, *578*, 1–8. [CrossRef]

46. Gasparotto, L.H.S.; Gomes, J.F.; Tremiliosi-Filho, G. Cyclic-voltammetry characteristics of poly(vinyl pyrrolidone) (pvp) on single-crystal pt surfaces in aqueous h2so4. *J. Electroanal. Chem.* **2011**, *663*, 48–51. [CrossRef]

47. Ebato, H.; Gentry, C.A.; Herron, J.N.; Mueller, W.; Okahata, Y.; Ringsdorf, H.; Suci, P.A. Investigation of specific binding of antifluorescyl antibody and fab to fluorescein lipids in langmuir-blodgett deposited films using quartz crystal microbalance methodology. *Anal. Chem.* **1994**, *66*, 1683–1689. [CrossRef] [PubMed]

48. Okahata, Y.; Kawase, M.; Niikura, K.; Ohtake, F.; Furusawa, H.; Ebara, Y. Kinetic measurements of DNA hybridization on an oligonucleotide-immobilized 27-mhz quartz crystal microbalance. *Anal. Chem.* **1998**, *70*, 1288–1296. [CrossRef] [PubMed]

49. Lavrich, D.J.; Wetterer, S.M.; Bernasek, S.L.; Scoles, G. Physisorption and chemisorption of alkanethiols and alkyl sulfides on Au(111). *J. Phys. Chem. B* **1998**, *102*, 3456–3465. [CrossRef]

50. Fischer, D.; Curioni, A.; Andreoni, W. Decanethiols on gold: The structure of self-assembled monolayers unraveled with computer simulations. *Langmuir* **2003**, *19*, 3567–3571. [CrossRef]

51. Mulvaney, P. Surface plasmon spectroscopy of nanosized metal particles. *Langmuir* **1996**, *12*, 788–800. [CrossRef]

52. Link, S.; Wang, Z.L.; El-Sayed, M.A. Alloy formation of gold–silver nanoparticles and the dependence of the plasmon absorption on their composition. *J. Phys. Chem. B* **1999**, *103*, 3529–3533. [CrossRef]

53. Queirós, R.B.; de-los-Santos-Álvarez, N.; Noronha, J.P.; Sales, M.G.F. A label-free DNA aptamer-based impedance biosensor for the detection of *E. coli* outer membrane proteins. *Sens. Actuators B Chem.* **2013**, *181*, 766–772. [CrossRef]

Chapter 2:
Reviews

chemosensors

MDPI

Review

Guanine Quadruplex Electrochemical Aptasensors

Ana-Maria Chiorcea-Paquim and Ana Maria Oliveira-Brett *

Department of Chemistry, Faculty of Science and Technology, University of Coimbra,
3004-535 Coimbra, Portugal; anachior@ipn.pt
* Correspondence: brett@ci.uc.pt; Tel.: +351-239-854-487

Academic Editors: Paolo Ugo and Ligia Moretto
Received: 9 May 2016; Accepted: 26 July 2016; Published: 30 July 2016

Abstract: Guanine-rich nucleic acids are able to self-assemble into G-quadruplex four-stranded secondary structures, which are found at the level of telomeric regions of chromosomes, oncogene promoter sequences and other biologically-relevant regions of the genome. Due to their extraordinary stiffness and biological role, G-quadruples become relevant in areas ranging from structural biology to medicinal chemistry, supra-molecular chemistry, nanotechnology and biosensor technology. In addition to classical methodologies, such as circular dichroism, nuclear magnetic resonance or crystallography, electrochemical methods have been successfully used for the rapid detection of the conformational changes from single-strand to G-quadruplex. This review presents recent advances on the G-quadruplex electrochemical characterization and on the design and applications of G-quadruplex electrochemical biosensors, with special emphasis on the G-quadruplex aptasensors and hemin/G-quadruplex peroxidase-mimicking DNAzyme biosensors.

Keywords: G-quadruplex; G4; GQ; aptasensor; DNAzyme; DNA electrochemical biosensor

1. Introduction

DNA sequences rich in guanine (G) bases are able to self-assemble into four-stranded secondary structures called G-quadruplexes (G4 or GQ), (Scheme 1). The G4s are formed by G-quartet building blocks, which are planar associations of four G bases, held together by eight Hoogsteen hydrogen bonds (Scheme 1B). The G-quartets are stacked on top of each other, stabilized by π-π hydrophobic interactions and by monovalent cations, such as K^+ and Na^+, which are coordinated to the lone pairs of electrons of O6 in each G.

The G4 structures are very polymorphic, being classified according to the number of strands (monomer, dimer or tetramer, Scheme 1C), according to strand polarity (i.e., the relative arrangement of adjacent strands in parallel or antiparallel orientations), glycosidic torsion angle (anti or syn) and the orientation of the connecting loops (lateral, diagonal or both) [1–4]. Different G4 topologies have been observed by nuclear magnetic resonance (NMR) or crystallography, either as native structures or complexed with small molecules [5].

The G4 sequences are found in chromosomes' telomeric regions, oncogene promoter sequences, RNA 5′-untranslated regions (5′-UTR) and other biologically-relevant regions of the genome, where they may influence the gene metabolism process and also participate in other important biological processes, e.g., DNA replication, transcriptional regulation and genome stability [1–14]. Moreover, G4 formation has been associated with a number of diseases, such as cancer, HIV and diabetes [3,5]. Due to their extraordinary stiffness and biological role, G4s become relevant in areas ranging from structural biology to medicinal chemistry, supra-molecular chemistry, nanotechnology and biosensor technology.

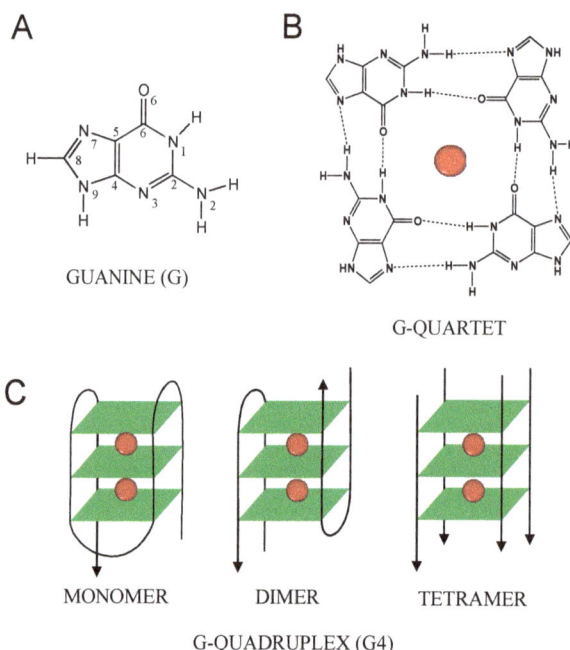

Scheme 1. (**A**) Chemical structure of the guanine (G) base; (**B**) G-quartet and (**C**) G-quadruplex (G4); the cations that stabilize the G4s are shown as red balls. Adapted from [14] with permission.

The G4 structures have emerged as a new class of cancer-specific molecular targets for anticancer drugs, since the G4 stabilization by small organic molecules can lead to telomerase inhibition and telomere dysfunction in cancer cells [2,15,16]. The G-rich oligonucleotides (ODNs) are also able to self-organize in G4-based two-dimensional networks and long nanowires, relevant for nanotechnology applications [17,18]; therefore, the assembly of G4 nanostructures and devices has been extensively revised in the literature [2,3,5,6,19].

The G4 structures were studied using different experimental techniques, such as molecular absorption, circular dichroism, molecular fluorescence, mass spectrometry, NMR, surface plasmon resonance, crystallography or atomic force microscopy (AFM) [20–23]. The electrochemical research on DNA is of great relevance to explain many biological mechanisms, and the nucleic acids redox behavior and adsorption processes have been studied for a long time [24–31], but only recently started to be used for the detection of G4 configurations [19,31,32].

Aptamers are a special class of small synthetic oligonucleotides able to form secondary and tertiary structures, larger than small molecule drugs, but smaller than antibodies, with the advantage of being highly specific in binding to small molecules, proteins, nucleic acids and even cells and tissues [33–36]. The aptamers bind to the targets by a lock and key mode, and the name "aptamer" is from the Latin word aptus, meaning "to fit" [36]. Among them, short aptamers that adopt G4 configurations received increased attention, and the electrochemical sensing devices based on G4 nucleic acid aptamers are highly selective and sensitive, fast, accurate, compact, portable and inexpensive.

In this review, recent advances on the G4 structure in nucleic acid electrochemistry and the design and applications of the G4 electrochemical biosensors that use redox labels as amplification strategies, i.e., the G4 electrochemical aptasensors and the hemin/G4 HRP-mimicking DNAzyme electrochemical biosensors, will be presented.

2. G4 Electrochemistry

The first report on the electrochemical oxidation of G4 structures concerned the investigation of two, thrombin-binding aptamer (TBA) sequences, 15-mer $d(G_2T_2G_2TGTG_2T_2G_2)$ (Scheme 2A) and 19-mer $d(G_3T_2G_3TGT_3T_2G_3)$ (Scheme 2B), using differential pulse (DP) voltammetry at a glassy carbon electrode (GCE) [32,37]. The different adsorption patterns of the TBA sequences observed by AFM onto highly oriented pyrolytic graphite (HOPG) were correlated with their voltammetric behavior in the presence/absence of K^+ ions. In Na^+-containing solutions, the oxidation of both TBA sequences showed one anodic peak corresponding to the oxidation of G residues in the TBA single strands. The G oxidation occurred at the C_8-H position, in a two-step mechanism involving the total loss of four electrons and four protons. Upon the addition of K^+, both sequences folded into G4 structures, causing the decrease of the G oxidation peak current and the occurrence of a new G4 peak at a higher potential, due to the oxidation of G residues in the G4 configuration. In the absence of K^+ ions, in only Na^+ ion-containing solutions, G4 formation also occurred, but was much slower.

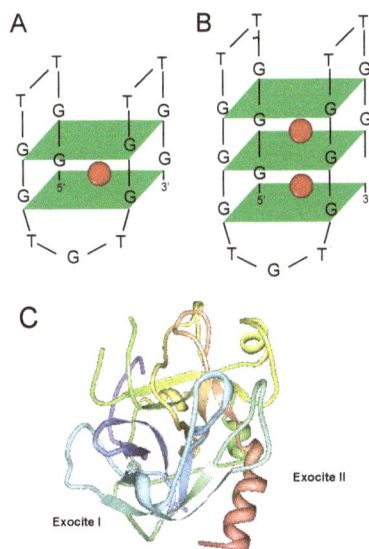

Scheme 2. Unimolecular antiparallel G4 structures formed by the thrombin binding aptamers (TBA): (**A**) $d(G_2T_2G_2TGTG_2T_2G_2)$ and (**B**) $d(G_3T_2G_3TGT_3T_2G_3)$; (**C**) thrombin tertiary structure. Adapted from [38] with permission.

The decrease of the G oxidation peak current was due to a decrease in the number of free G residues in single-stranded TBA, and the increase of the G4 oxidation peak current was due to an increased number of G4 structures that were more difficult to oxidize. The adsorption of TBA in a G4 conformation, as rod-like-shaped aggregates, was observed by AFM [32,37].

The 10-mer ODNs that contain blocks of 8–10 G residues, $d(G)_{10}$, $d(TG_9)$ and $d(TG_8T)$, form parallel tetra-molecular G4s (Scheme 1C, right) and were investigated by AFM and DP voltammetry. The influence of the ODN sequence and concentration, pH (Figure 1), the presence of monovalent cations, Na^+ vs. K^+ (Figure 2A,C), and incubation time (Figure 2B,C) was determined [19,32,39,40]. The formation of G4 structures and higher-order nanostructures, due to the presence of a long contiguous G region, and the influence of the thymine residues at the 5′ and 3′ molecular ends in $d(TG_9)$ and $d(TG_8T)$ were clarified. DP voltammetry allowed the detection of the single strands' association into G4s and G-based nanostructures, in freshly-prepared solutions, at concentrations 10-times lower than usually detected using other techniques currently employed to study the formation of G4s.

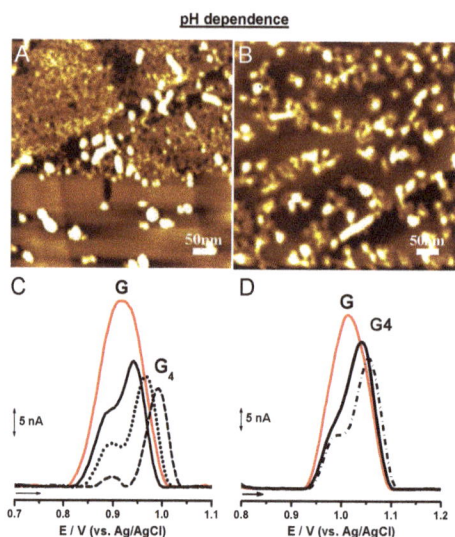

Figure 1. d(G)$_{10}$ sequence G4 formation pH dependence at (**A,C**) pH = 7.0 and (**B,D**) pH = 4.5: (A,B) AFM images of 0.3 μM d(G)10 spontaneously adsorbed onto HOPG; and (C,D) bioelectrocatalyzed voltammograms baseline corrected of 3.0 μM d(G)$_{10}$ after (—) 0 h, (—) 24 h, (E, •••) 72 h, (F, •●■) 5 days and (E, ■■■) 14 days of incubation. Adapted from [39] with permission.

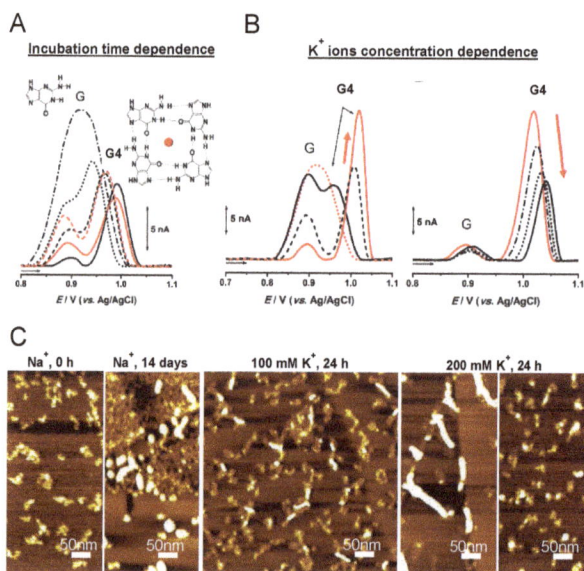

Figure 2. Incubation time and K$^+$ ion concentration dependence on the G4 formation of the d(G)$_{10}$ sequence; (**A,B**) DP voltammograms baseline corrected for d(G)$_{10}$: (**A**) in the absence of K$^+$ ions at (•●•) 0 h, (•••) 24 h, (■■■) 48 h and (—) 14 days of incubation and in the presence of 1 mM K$^+$ ions at (■■■) 0 h and (—) 24 h of incubation; (**B**) in the absence (•••) and in the presence of (—, left) 100 μM, (■■■) 5 mM, (—) 100 mM, (•●•) 200 mM, (•••) 500 mM and (—, right) 1 M K$^+$ ions, 0 h of incubation; (**C**) AFM images of d(G)$_{10}$ in the absence/presence of different K$^+$ ion concentrations and different incubation times. Adapted from [40] with permission.

Single-stranded ODNs were observed only in Na$^+$ ion solutions at short incubation times and were detected in AFM as thin polymeric structures and in DP voltammetry by the occurrence of only the G oxidation peak. The G4 structures were formed very slowly in Na$^+$ ions, after a long incubation time, faster in K$^+$ ions, after a short incubation time, and were detected by AFM as spherical aggregates and by DP voltammetry by the decrease of the G oxidation peak current and the occurrence/increase of the G4 oxidation peak current, as well as a shift to positive potentials, in a K$^+$ ion concentration- and in a time-dependent manner. The presence of K$^+$ ions strongly stabilizes and accelerates the G4 formation. For increased d(G)$_{10}$ concentrations, long G-nanowires were formed, demonstrating the potential of G-rich DNA sequences as a scaffold for nanotechnological applications [19,32,39,40].

The *Tetrahymena* telomeric repeat sequence d(TG$_4$T) forms parallel-stranded tetra-molecular G4s in the presence of Na$^+$ and K$^+$ ions and is considered to be a simple model for biologically-relevant G4s. It has also provided high resolution structural data on drug-DNA interactions. The transformation of the d(TG$_4$T) from single-stranded into G4 configurations, influenced by the Na$^+$ and K$^+$ ion concentration, was successfully detected using AFM on HOPG and DP voltammetry at GCE (Figures 3 and 4) [41]. The d(TG$_4$T) in a G4 conformation self-assembled very quickly in K$^+$ ion solutions and slowly in Na$^+$ ion solutions. The optimum K$^+$ ion concentration for the G4 structure formation of d(TG$_4$T) was similar to the intracellular K$^+$ ion concentration of healthy cells. In the presence of Na$^+$ ions, d(TG$_4$T) also formed short nanowires and nanostructured films that were never observed in K$^+$ ion-containing solution, suggesting that the rapid formation of stable G4s in the presence of K$^+$ is relevant for the good function of cells.

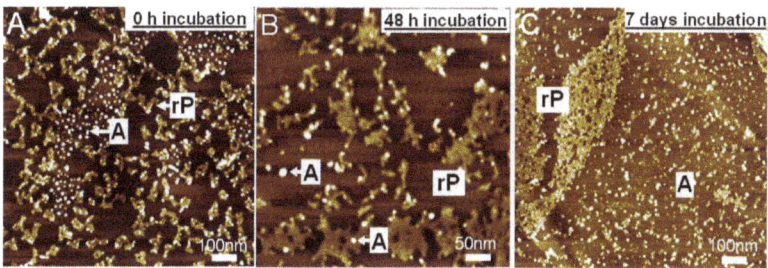

Figure 3. AFM images of d(TG$_4$T) in the presence of K$^+$ ions, after (**A**) 0 h; (**B**) 48 h and (**C**) 7 days of incubation. Adapted from [41] with permission.

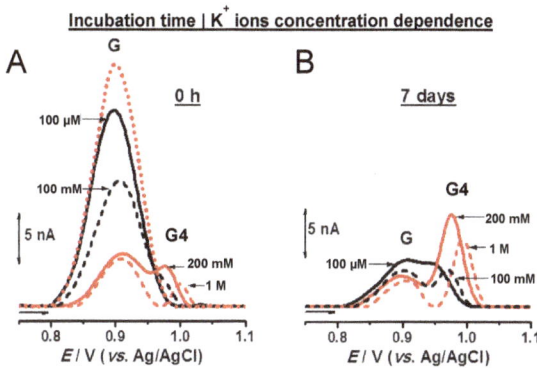

Figure 4. Incubation time and K$^+$ ion concentration dependence on the G4 formation of d(TG$_4$T). DP voltammograms baseline corrected for d(TG$_4$T), after (**A**) 0 h and (**B**) seven days of incubation; (A, •••) in the absence of K$^+$ ions and (A,B) in the presence of (—) 100 µM, (■■■) 100 mM, (—) 200 mM and (■■■) 1 M K$^+$ ions. Adapted from [41] with permission.

Synthetic polynucleotides poly(dG) and poly(G) are widely used as models to determine the interaction of drugs with G-rich segments of DNA. AFM and DP voltammetry showed that, at low incubation times, short G4 regions were formed along the poly(G) single-strands, while low adsorption large poly(G) aggregates in a G4 conformation were formed after high incubation times in the presence of either Na^+ or K^+ monovalent ions (Figure 5) [42]. The DP voltammetry in freshly-prepared poly(G) solutions showed only the G oxidation peak, due to the oxidation of G residues in the poly(G) single strand. Increasing the incubation time, the G oxidation peak current decreased; the peak disappeared; and the G4 oxidation peak in the poly(G) in a G4 conformation appeared, at a higher oxidation potential, depending on the incubation time, presenting a maximum after 10 days of incubation and reaching a steady current after ~17 days of incubation.

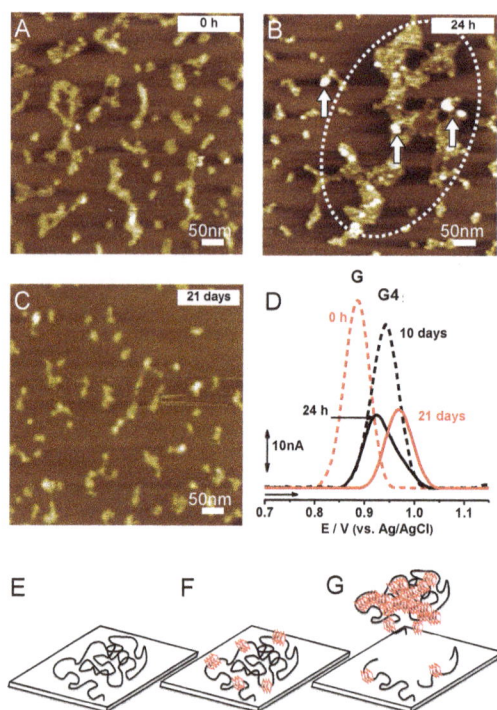

Figure 5. Poly(G) in the presence of K^+ ions: (**A–C**) AFM images, after: (A) 0 h, (B) 24 h and (C) 21 days of incubation; (**D**) DP voltammograms baseline corrected, after: (•••) 0 h, (—) 24 h, (•••) 10 days and (—) 21 days of incubation; (**E–G**) Schematic representation of the poly(G) adsorption process: (E) poly(G) single strand, (F) poly(G) single strand with short G4 regions and (G) poly(G) single strand with larger G4 regions. Adapted from [42] with permission.

The interaction between the TBA sequences $d(G_2T_2G_2TGTG_2T_2G_2)$ and $d(G_3T_2G_3TGT_3T_2G_3)$ and the serine protease thrombin (Scheme 2) was determined successfully by AFM and voltammetry, taking into account the thrombin interaction with TBA primary and secondary structures, as well as the thrombin folding in the presence of alkaline metals [32,38]. In the interaction, the TBA single strands coiled around thrombin, leading to the formation of a robust TBA-thrombin complex that maintained the thrombin symmetry and conformation, which resulted in the thrombin oxidation peaks, within the TBA-thrombin complex, occurring at more positive potentials, than in free thrombin. In the presence of K^+, the TBA sequences were folded into a G4 conformation, which facilitated the interaction with thrombin. The TBA-thrombin complexes adsorbed on the carbon electrode with the TBA in contact with the surface and the thrombin on top, far from the surface; thus, the thrombin molecule was less accessible to oxidation, also leading to the occurrence of the thrombin oxidation peaks at more positive potentials.

A large number of potent G4-binding ligands, which stabilize or promote G4 formation, has been described. Especially at the chromosomes' telomeric regions, the telomeric DNA is able to form G4 structures; therefore, the G4 ligands prevent the G4s from unwinding and opening the telomeric ends to telomerase, thus indirectly targeting the telomerase and inhibiting its catalytic activity.

A number of acridine derivatives have been specifically synthesized with the purpose of increasing binding affinity and selectivity for human telomeric G4 sequences found in chromosomes' telomeric regions, e.g., BRACO-19 [43] and RHPS4. More recently, a new series of triazole-linked acridine ligands, e.g., GL15 and GL7 [44], with enhanced selectivity for human telomeric G4s binding versus duplex DNA binding, have been designed, synthetized and evaluated.

The interactions of the GL15 triazole-acridine conjugate with the short-length *Tetrahymena* telomeric DNA repeat sequence d(TG$_4$T) and with the long chain poly(G) sequence, at the single-molecule level, by AFM and DP voltammetry, were investigated [45]. GL15 interacted with both the d(TG$_4$T) and poly(G) sequences, in a time-dependent manner, and the influence of Na$^+$ vs. K$^+$ ions was evaluated.

The G4 formation was detected in AFM, by the adsorption of small d(TG$_4$T) and poly(G) spherical aggregates, as well as large G4-based poly(G) assemblies, and the DP voltammetry showed the decrease and disappearance of the GL15 and the G oxidation peak currents and the appearance of the G4 oxidation peak (Figure 6). The GL15 strongly stabilized and accelerated the G4 formation in both Na$^+$ and K$^+$ ion-containing solutions, although only K$^+$ promoted the formation of perfectly-aligned tetra-molecular G4s. The GL15-d(TG$_4$T) complex with the G4 configuration was discrete and approximately globular, whereas the GL15-poly(G) complex with the G4 configuration was formed at a number of points along the length of the polynucleotide, analogous to beads on a string.

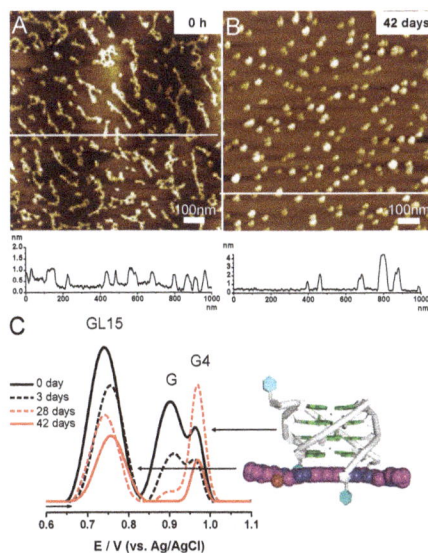

Figure 6. GL15-d(TG$_4$T) after different incubation times in the presence of K$^+$ ions: (**A,B**) AFM images and cross-section profiles through the white dotted lines and (**C**) DP voltammograms baseline corrected. Adapted from [45] with permission.

3. G4 Electrochemical Biosensors

A DNA-electrochemical biosensor is formed by an electrode (the electrochemical transducer) with a DNA probe immobilized on its surface (the biological recognition element) and is used to detect DNA-binding molecules (the analyte) that interact and induce changes in the DNA structure and electrochemical properties, which are further translated into an electrical signal [25–27,29–31,46,47]. Up to now, the G4-based electrochemical biosensors reported in the literature always used redox labels as amplification strategies. Two important types of G4 electrochemical biosensors, the G4 electrochemical aptasensors and the hemin/G4 DNAzyme electrochemical biosensors, will be revisited.

3.1. G4 Electrochemical Aptasensors

Aptamers are DNA or RNA sequences selected in vitro that present the ability to specifically bind a molecular target. Short aptamers that adopt G4 configurations can bind to a wide variety of molecular targets, mainly proteins (such as thrombin, nucleolin, signal transducer and activator of transcription STAT3, human RNase H1, protein tyrosine phosphatase Shp2, VEGF, HIV-1 integrase, HIV-1 reverse transcriptase, HIV-1 reverse transcriptase, HIV-1 nucleocapsid protein, *M. tuberculosis* polyphosphate kinase 2, sclerostin, insulin, etc.), but also some other targets (hematoporphyrin IX, hemin, ochratoxin, potassium ions, ATP) [33–35,48,49]. Many aptamers recognize specifically different positions on the analyte; for example, TBA recognizes the fibrinogen and heparin binding sites of thrombin.

The first G4 electrochemical aptasensors used TBA sequences and gold electrodes as electrochemical transducers, the aptamers' attachment being achieved using an amine or a thiol functionalization [50–52], or the affinity of biotin to avidin, streptavidin or neutravidin [53]. Depending on the assay format, two main G4 electrochemical aptasensor categories can be depicted, the sandwich-type aptasensors (also named dual-site binding) and the structure switching-based aptasensors (single-site binding) [54,55].

3.1.1. Sandwich-Type G4 Electrochemical Aptasensor

The aptamer–analyte–aptamer sandwich-type G4 electrochemical aptasensor (Scheme 3A) is composed by two aptamer layers, the first aptamer layer being immobilized on the electrode and used for capturing the analyte and the second aptamer layer being labelled and used for the electrochemical detection. The first aptamer was generally immobilized onto gold via a thiol [56–60] and, more recently, magnetic beads [61,62]. The labels of the second aptamer were either redox molecules, nanocomposites [60], nanoparticles [57,63], quantum dots [58,59] or enzymes, with catalytic activity that transformed the substrate into an electroactive product [56,64,65].

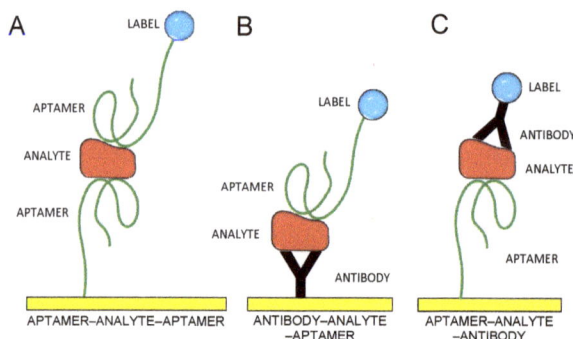

Scheme 3. Sandwich-type G4 electrochemical aptasensors: (**A**) aptamer–analyte–aptamer sandwich; the first aptamer is used for binding the analyte to the electrode, and the second labelled aptamer is used for detection; (**B**) antibody–analyte–aptamer sandwich; the analyte is bound to the surface via an antibody, and a labelled aptamer is used for detection; (**C**) aptamer–analyte–antibody sandwich; the analyte is bound to the surface via an aptamer, and a labelled antibody is used for detection. Adapted from [19] with permission.

The first G4 electrochemical aptasensor reported presented an aptamer–analyte–aptamer sandwich-type format, being developed for thrombin detection [64,65]. The sensor was built up by two aptamers, the first aptamer immobilized onto the gold electrode for capturing the thrombin onto it and the second one, a glucose dehydrogenase (GDH)-labelled anti-thrombin aptamer. The current increase generated by the electroactive product of the enzyme reaction was observed, and >10 nM thrombin were detected selectively. This approach proved for the first time that aptamers can be successfully employed in sandwich-type sensing devices, instead of and with advantages over antibodies.

Employing platinum nanoparticle labels as catalysts for the reduction of H_2O_2 to a TBA/thrombin complex allowed the amplified electrocatalytic detection of thrombin with a 1 nM detection limit [57].

In another report, gold nanoparticles' functionalization of the second aptamer improved the thrombin detection sensitivity, showing a 0.02 nM detection limit, with a 0.05–18 nM linear range [63].

The use of cadmium sulfide quantum dot labels of the secondary aptamer allowed thrombin detection with a 0.14 nM detection limit, corresponding to 28 fmol of analyte [58], while in another similar approach, thrombin determination in human serum showed a detection limit as low as 1 pM [59].

Using a more complex design, based on conductive graphene-3,4,9,10-perylenetetracarboxylic dianhydride nanocomposites as a sensor platform and PtCo nanochains–thionine–Pt–horseradish peroxidase-labelled secondary TBA for signal amplification, thrombin was detected at a linear range from 10^{-15}–10^{-9} M and a 6.5×10^{-16} M detection limit [60].

In another approach, an aptamer–analyte–aptamer sandwich-type G4 electrochemical aptasensor was based on enzymatic labelling of the second aptamer with glucose dehydrogenase (GDH), measuring the electric current generated by the glucose oxidation catalyzed by GDH and selectively detecting 1 μM of thrombin [64].

Apart from the aptamer–analyte–aptamer sandwich-type G4 electrochemical aptasensor, other design strategies were also employed. The antibody–analyte–aptamer sandwich-type aptasensor consisted of attaching the analyte to the surface via an antibody combined with a labelled aptamer that adopt the G4 configuration for detection (Scheme 3B). This sensor design was used to detect thrombin at a nanogold-chitosan composite-modified GCE, linked with the aptamer via a polyclone antibody. The electrochemical active marker used was methylene blue (MB) directly intercalated in the probing aptamer. The sensor linear response for thrombin was in the range 1–60 nM with a 0.5 nM detection limit [66].

The aptamer–analyte–antibody sandwich-type aptasensor consisted of attaching the analyte to the surface via an aptamer able to form the G4 structure, the detection being performed with a redox-labelled antibody (Scheme 3C) [67].

A sandwiched immunoassay for thrombin used a NH_2-functionalized-TBA immobilized on gold nanoparticle-doped conducting polymer nanorod electrodes and a ferrocene label bound to an antithrombin antibody [67]. The sensor used the electrocatalytic oxidation of ascorbic acid by the ferrocene moiety, presenting a wide dynamic range of 5–2000 ng·L^{-1} and a low detection limit of 5 ng·L^{-1} (0.14 pM) and was tested in a real human serum sample for the detection of spiked concentrations of thrombin.

3.1.2. Structure-Switching G4 Electrochemical Aptasensor

A different category of G4 electrochemical aptasensors is based on the aptamer structural modifications upon binding the analyte from the single-stranded to the G4 configuration, structure-switching G4 electrochemical aptasensor (Scheme 4). This strategy generally involved the direct immobilization of the aptamer on the electrode surface, while the analyte was present in solution. The electrochemical signal amplification was obtained by labelling the aptamer with a redox tag [68,69].

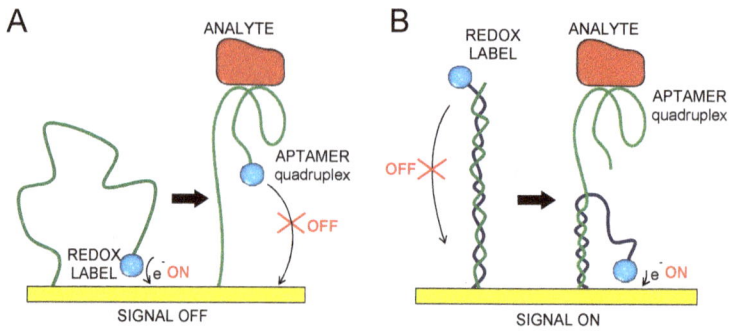

Scheme 4. Structure-switching G4 electrochemical aptasensors: the aptamer is modifying its conformation after analyte binding: (**A**) increasing the distance from the redox label to the electrode (signal off); (**B**) decreasing the distance from the redox label to the electrode (signal on). Adapted from [19] with permission.

The majority of structure-switching G4 electrochemical aptasensors were developed using gold electrodes, although, more recently, other electrochemical transducers have been employed, such as gold disk microelectrode arrays [70], modified platinum [71] or carbon electrodes [63,71–75].

The first report on a structure-switching G4 electrochemical aptasensor used a covalently-attached MB-labelled TBA on a gold electrode [68,76]. The aptasensor detection type was signal-off (Scheme 4A), i.e., in the absence of the thrombin target, the immobilized TBA remained relatively unfolded, allowing the electron transfer from the MB label to the electrode surface, while after thrombin binding, the formation of TBA in the G4 configuration was induced, which inhibited the electron transfer (Figure 7A).

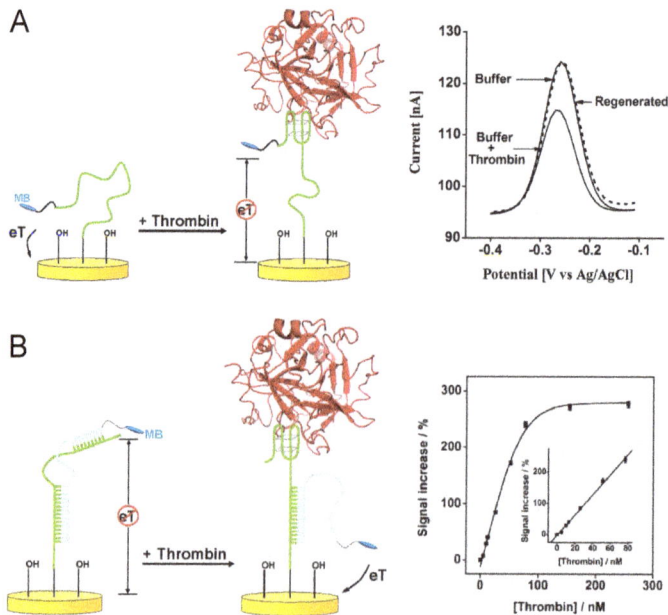

Figure 7. Structure-switching G4 electrochemical aptasensor for thrombin: (**A**) signal off: thrombin binding reduces the current from the MB redox tag; and (**B**) signal on: thrombin binding increases the current from the MB redox tag. Reproduced from [68] with permission.

The sensor was selective enough to detect thrombin directly in blood serum with a 20 nM thrombin detection limit. Similar G4 electrochemical aptasensors for thrombin were developed in parallel, using ferrocene labels [77–79]. In another design, a beacon aptamer-based biosensor for thrombin showed a linear signal between 0 and 50.8 nM of thrombin, with a 0.999 correlation factor and an 11 nM detection limit [80].

A bifunctional aptamer-based electrochemical biosensor for the detection of both thrombin and adenosine was developed [81]. The TBA was first immobilized on a gold electrode, and then, it was hybridized with an adenosine aptamer and labelled with MB. In the presence of thrombin or adenosine, the aptamer bonded to thrombin or to adenosine instead of MB, and the decrease of the MB peak current was related to the concentration of either thrombin or adenosine. The sensor showed a 3 nM thrombin and a 10 nM adenosine detection limit.

A signal-on structure-switching G4 electrochemical aptasensor (Scheme 4B) was described [82], with a short MB-tagged oligonucleotide hybridized with both the thrombin-binding portion of the TBA and the DNA sequence linking the aptamer to the electrode. The thrombin binding induced the structural modification of the TBA in a G4 configuration, liberating the 5′ end of the tagged oligonucleotide to produce a flexible, single-stranded sequence, which allowed the MB tag to react at the electrode surface, increasing the reduction peak current (Figure 7B). Comparing to the signal-off structure-switching G4 electrochemical aptasensors previously described [68,76], the signal-on aptasensor design achieved a current increase of ~ 300% with a saturated target and a 3 nM detection limit [82].

Apart from MB, other redox labels have also been employed. Among them, ferrocene [72,83–89] and $Ru(NH_3)_6]^{3+}$ [75,90] were very popular, especially for the construction of impedimetric aptasensors.

An impedimetric aptasensor for thrombin detection was described, based on different TBA sequences directly immobilized on the gold electrode and using phosphoramidite synthons for a strong thiolate anchoring of the aptamer and high flexibility [83]. In the presence of the $[Fe(CN)_6]^{3-/4-}$ redox probe, the impedimetric aptasensor exhibited high sensitivity, specificity and stability and a 3.1 ng·mL^{-1} (80 pmol/L) thrombin detection limit.

A signal-on G4 electrochemical aptasensor based on co-immobilization of MP-11 and thiol ferrocene-labeled anti-thrombin aptamer, the interaction being detected via a microperoxidase-mediated electron transfer between the ferrocene and the gold electrode surface, was described [84]. The system showed a very high sensitivity of 30 fM using electrochemical impedance spectroscopy. Another impedimetric aptasensor for thrombin, based on a layer-by-layer polyamidoamine dendrimer-modified gold electrode [91], showed in the presence of the reversible $[Fe(CN)_6]^{3-/4-}$ redox couple a linear relationship with the concentrations of thrombin in the range of 1–50 nM and a 0.01 nM detection limit.

More recently, a signal-on electrochemical aptasensor based on target-induced split aptamer fragments' conjunction was described [86]. The new design used TBA splinted into two fragments, one attached to the gold electrode and the second one modified with ferrocene, the association of thrombin inducing the association of the two fragments, thus increasing the concentration of ferrocene at the electrode surface. The signal-on electrochemical aptasensor showed a linear range of 0.8–15 nM and a 0.2 nM detection limit.

Another procedure used to improve the sensitivity of the G4 electrochemical aptasensors for the detection of thrombin used an amplification strategy based on the electrochemical active-inactive switch between monomer/dimer forms of carminic acid (CA). The CA was electroactive, while the CA dimers were electrochemically inactive [92]. With magnetic enrichment, the sensor showed a 42.4 pM detection limit.

Nanoparticle-based materials, including gold [63,73], platinum [57] and Fe_3O_4 nanoparticles [93] and quantum dot-coated silica nanospheres [94], were also used as signal amplification strategies for ultrasensitive electrochemical aptasensing. For example, an electrochemical aptasensor based on gold nanoparticles showed a linear range of 0.05–18 nM and a 0.02 nM detection limit [63], while another one based on Fe_3O_4-nanoparticles [93] showed a linear response for thrombin in the range of 1.0–75 nM and a 0.1 nM detection limit.

Based on the aptamer conformational change in the presence of K$^+$ cations, different electrochemical aptasensors have been developed for selective potassium recognition. The formation of a G4 structure in the presence of K$^+$ ions was detected by monitoring the changes on the electron transfer between redox labels and the electrode surface [95–97] or by detecting the changes on the interfacial electron transfer resistance [98]. The same strategy for specific recognition of other metal ions, such as Tb^{3+}, was used [99].

Taking advantage of the ability of thrombin to catalyze the hydrolysis of the peptide (-Ala-Gly-Arg-nitroaniline) to nitroaniline, thrombin was electrochemically detected, by quantifying the nitroaniline reaction product [56].

A different strategy for G4 electrochemical aptasensors used catalysts, such as horseradish peroxidase (HRP) [56,100]. In a simple approach, TBA was non-specifically immobilized on the electrode surface, and thrombin was detected using the HRP label, allowing a 3.5 nM detection limit, sufficient for clinical diagnostic of metastatic lung cancer, where the concentration of thrombin level detected was 5.4 nM [56].

An impedimetric biosensor based on a DNA aptamer specific to ochratoxin A (OTA) covalently immobilized onto a mixed Langmuir–Blodgett monolayer composed of polyaniline-stearic acid and deposited on indium tin oxide (ITO)-coated glass plates showed a 0.24 nM detection limit [85]. The system was further improved [87], showing a detection limit comparable to that of the HPLC method (0.12 nM), and was validated in food samples. Another design for the OTA detection proposed the use of a long polyethylene glycol spacer chain, which led to the formation of long tunnels at the surface of screen-printed carbon electrodes, with aptamers acting as gates. The aptamer changed configuration after OTA binding, and the peak current decreased [88].

In a different approach, OTA was detected at a G4 electrochemical aptasensor that used a hairpin anti-OTA aptamer and site-specific DNA cleavage of TaqaI restriction endonuclease, as well as a streptavidin-HRP tag, being able to detect as low as 0.4 pg/mL OTA with ultrahigh selectivity [100].

3.2. Hemin/G4 DNAzyme Electrochemical Biosensor

Hemin/G4 DNAzyme is an artificial catalytically-active DNA molecule (DNAzyme) that is composed of DNA in the G4 configuration with intercalated hemin molecules. Hemin is an iron-containing porphyrin, whose peroxidase activity increases in the presence of DNA, facilitating the redox reaction between H$_2$O$_2$ and a target molecule (the substrate, e.g., 3,3′,5,5′-tetramethylbenzidine, hydroquinone or ferrocene methyl alcohol), which results in the appearance of an oxidized target molecule (the electroactive product), that is electrochemically detected (Scheme 5).

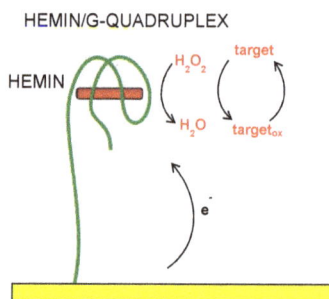

Scheme 5. Hemin/G4 peroxidase-mimicking DNAzyme electrochemical biosensor. Adapted from [19] with permission.

Hemin/G4 DNAzyme electrochemical biosensors represent nowadays one of the most popular building assays of G4-based electrochemical biosensors [101]. The most common strategy consists of the modification of the electrode by a hairpin nucleic acid oligonucleotide that contains two sequences, a sequence capable of forming a G4 structure that binds the hemin, used as the amplification strategy, and an aptamer able to specifically bind the analyte, which might form or not a G4 structure. In the presence of the analyte and hemin, the hairpin structures are opened, the hemin/G4 structures are formed on the electrode surface, while the analyte binds to the aptamer.

Since the first report on using hemin/G4 DNAzyme as the electrocatalytic label for amplifying sensing events [102], this approach attracted increasing interest in biosensor [103,104] and biofuel cell technologies [105]. In comparison with protein peroxidases, the hemin/G4 peroxidase-mimicking DNAzymes have several advantages, such as high chemical stability, low cost and simple synthesis. Hemin/G4 DNAzyme electrochemical biosensors were successfully used for the detection of cells [106,107], proteins [108–110] or low molecular weight molecules, such as adenosine monophosphate (AMP) [102,111,112], anticancer drugs [113], gaseous ligands [114], toxins [115,116], pollutant agents [117,118] or metal ions [119,120].

Later on, more complicated amplification strategies were developed to improve the sensitivity of the hemin/G4 DNAzyme HRP-mimicking activity, such as dual-amplification [121], background noise reduction [122] or autocatalytic target recycling strategies [123].

The glucose oxidase activity was followed by a hemin/G4 DNAzyme electrochemical biosensor, by attaching the enzyme to the electrode surface through the nucleic acid sequence able to form G4s in the presence of hemin [102,124]. Then, the glucose oxidase mediated the glucose oxidation to gluconic acid and H_2O_2, and the resulting H_2O_2 was analyzed through its electrocatalyzed reduction by the DNAzyme.

Another electrochemical sensing strategy, based on the G4 DNAzyme for the detection of both adenosines and hydrogen peroxide from cancer cells, was developed [112], which detected the flux of H_2O_2 released from cells with high sensitivity and showed a 0.1 nM detection limit for ATP.

A hemin/G4 DNAzyme-based impedimetric biosensor was used to detect the environmental metabolite 2-hydroxyfluorene (2-HOFlu) [117], using the hemin/G4 HRP-like activity to catalyze the oxidation of 2-HOFlu by H_2O_2, with a 1.2 nM detection limit in water and a 3.6 nM detection limit in spiked lake water samples. The assay was also selective over other fluorene derivatives.

A sandwich-type electrochemical aptamer cytosensor for the detection of HepG2 cells was used [106]. On the first approach, the sensor was built up by self-assembling thiolated TLS11a aptamers on the surface of gold electrodes and a G4/hemin/aptamer and HRP-modified gold nanoparticles. The sensor detection range was from 10^2–10^7 cells·mL^{-1} and had a 30 cells·mL^{-1} low detection limit.

The system was improved by self-assembling the TLS11a aptamers with gold nanoparticles (AuNPs) on the surface of GCE [107]. Hybrid Fe_3O_4/MnO_2/Au@Pd nanoelectrocatalysts, hemin/G4 HRP-mimicking DNAzymes and the natural HRP enzyme efficiently amplified the electrochemical signal through catalyzing the oxidation of hydroquinone (HQ) by H_2O_2. This cytosensor provided a better 15 cells·mL^{-1} detection limit, good specificity and stability.

In a different approach, the hemin/G4 DNAzyme electrochemical biosensors took advantage of the hemin/G4 acting both as an NADH oxidase, assisting the oxidation of NADH to NAD$^+$ together with the generation of H_2O_2 in the presence of dissolved O_2, as well as a hemin/G4 DNAzyme to bioelectrocatalyze the reduction of the produced H_2O_2. Initially, this approach was used for the detection of thrombin [125–127]. More recently, the Pebrine disease-related *Nosema bombycis* spore wall protein was detected, using the amplification of hemin/G4 DNAzyme functionalized with Pt@Pd nanowires, the electrochemical immunosensor exhibiting a linear range from 0.001–100 ng·mL^{-1} and a 0.24 pg·mL^{-1} detection limit [128].

A DNAzyme that simultaneously served as an NADH oxidase and HRP-mimicking DNAzyme was developed to detect mercury ions (Hg^{2+}) [129], with the dynamic concentration range spanning from 1.0 ng L^{-1}–10 mg·L^{-1} Hg^{2+} and a 0.5 ng·L^{-1} (2.5 pM) detection limit, also demonstrating an excellent selectivity against other interferent metal ions.

A pseudo triple-enzyme cascade electrocatalytic electrochemical aptasensor for the determination of thrombin, using the amplification of an alcohol dehydrogenase (ADH)-Pt-Pd nanowire bionanocomposite and a hemin/G4 structure that simultaneously acted as NADH oxidase and HRP-mimicking DNAzyme, was developed [130]. The ADH immobilized on the Pt-Pd nanowires catalyzed the ethanol present in the electrolyte into acetaldehyde, accompanied by NAD^+ being converted to NADH. Then, the hemin/G4 firstly served as NADH oxidase, converting the produced NADH to NAD^+, then the hemin/G4 acting as the HRP-mimicking DNAzyme bioelectrocatalyzed the produced H_2O_2. In this way, a concentration linear range from 0.2 pM–20 nM with a low 0.067 pM detection limit for thrombin was obtained.

Another strategy for thrombin detection consisted of using porous platinum nanotubes (PtNTs) labelled with hemin/G4 and GDH [131]. Coupling with GDH and hemin/G4 as NADH oxidase and HRP-mimicking DNAzyme, the cascade signal amplification allowed the detection limit of thrombin down to the 0.15 pM level.

4. Conclusions

The detailed knowledge of G4 formation mechanism, at the surface of electrochemical transducers, is of utmost importance for the design and fabrication of G4-based electrochemical aptasensors, with applications in nanotechnology and biosensor technology. The voltammetric techniques in combination with AFM were successfully employed to study the transformation of single-strand sequences into the G4 configuration or G4-based nanostructures, in freshly-prepared solutions, for concentrations 10-times lower than usually detected by other techniques, such as UV absorbance, circular dichroism or electrospray mass spectroscopy.

The key features of the G4 conformation in nucleic acid electrochemistry and their application in G4 electrochemical biosensors that use redox labels as amplification strategies, i.e., the G4 electrochemical aptasensors and the hemin/G4 HRP-mimicking DNAzyme electrochemical biosensors, were revised.

Acknowledgments: Financial support from Fundação para a Ciência e Tecnologia (FCT), Grants SFRH/BPD/92726/2013 (Ana-Maria Chiorcea-Paquim), projects PTDC/SAU-BMA/118531/2010, PTDC/QEQ-MED/0586/2012, UID/EMS/00285/2013 (co-financed by the European Community Fund FEDER), FEDER funds through the program COMPETE—Programa Operacional Factores de Competitividade is gratefully acknowledged.

Conflicts of Interest: the authors declare no conflict of interest.

References

1. Mergny, J.-L.; De Cian, A.; Ghelab, A.; Saccà, B.; Lacroix, L. Kinetics of Tetramolecular Quadruplexes. *Nucleic Acids Res.* **2005**, *33*, 81–94.
2. Neidle, S. *Therapeutic Applications of Quadruplex Nucleic Acids*; Elsevier: Amsterdam, The Netherlands, 2012.
3. Simonsson, T. G-quadruplex DNA structures-variations on a theme. *Biol. Chem.* **2001**, *382*, 621–628.
4. Tran, P.L.T.; De Cian, A.; Gros, J.; Moriyama, R.; Mergny, J.-L. Tetramolecular Quadruplex stability and assembly. *Top. Curr. Chem.* **2013**, *330*, 243–273.
5. Zhang, S.; Wu, Y.; Zhang, W. G-quadruplex structures and their interaction diversity with ligands. *ChemMedChem* **2014**, *9*, 899–911.
6. Keniry, M.A. Quadruplex Structures in nucleic acids. *Biopolymers* **2000–2001**, *56*, 123–146.
7. Chambers, V.S.; Marsico, G.; Boutell, J.M.; Di Antonio, M.; Smith, G.P.; Balasubramanian, S. High-Throughput sequencing of DNA G-quadruplex structures in the human genome. *Nat. Biotechnol.* **2015**, *33*, 877–881.
8. Murat, P.; Balasubramanian, S. Existence and consequences of G-quadruplex structures in DNA. *Curr. Opin. Genet. Dev.* **2014**, *25*, 22–29.
9. Henderson, A.; Wu, Y.; Huang, Y.C.; Chavez, E.A.; Platt, J.; Johnson, F.B.; Brosh, R.M.; Sen, D.; Lansdorp, P.M. Detection of G-quadruplex DNA in mammalian cells. *Nucleic Acids Res.* **2014**, *42*, 860–869. [CrossRef] [PubMed]
10. Dailey, M.M.; Hait, C.; Holt, P.A.; Maguire, J.M.; Meier, J.B.; Miller, M.C.; Petraccone, L.; Trent, J.O. Structure-Based drug design: From Nucleic acid to membrane protein targets. *Exp. Mol. Pathol.* **2009**, *86*, 141–150. [CrossRef] [PubMed]
11. Todd, A.K.; Johnston, M.; Neidle, S. Highly prevalent putative quadruplex sequence motifs in human DNA. *Nucleic Acids Res.* **2005**, *33*, 2901–2907. [CrossRef] [PubMed]
12. Huppert, J.L. Hunting G-quadruplexes. *Biochimie* **2008**, *90*, 1140–1148. [CrossRef] [PubMed]

13. Huppert, J.L.; Balasubramanian, S. Prevalence of quadruplexes in the human genome. *Nucleic Acids Res.* **2005**, *33*, 2908–2916. [CrossRef] [PubMed]

14. Chiorcea-Paquim, A.-M.; Oliveira-Brett, A.M. Redox behaviour of G-quadruplexes. *Electrochim. Acta* **2014**, *126*, 162–170. [CrossRef]

15. Artandi, S.E.; DePinho, R.A. *Telomeres and Telomerase in Cancer*; Hiyama, K., Ed.; Humana Press: Totowa, NJ, USA, 2009.

16. Neidle, S. The structures of quadruplex nucleic acids and their drug complexes. *Curr. Opin. Struct. Biol.* **2009**, *19*, 239–250. [CrossRef] [PubMed]

17. Borovok, N.; Iram, N.; Zikich, D.; Ghabboun, J.; Livshits, G.I.; Porath, D.; Kotlyar, A.B. Assembling of G-strands into novel tetra-molecular parallel g4-dna nanostructures using avidin-biotin recognition. *Nucleic Acids Res.* **2008**, *36*, 5050–5060. [CrossRef] [PubMed]

18. Borovok, N.; Molotsky, T.; Ghabboun, J.; Porath, D.; Kotlyar, A. Efficient procedure of preparation and properties of long uniform G4-dna nanowires. *Anal. Biochem.* **2008**, *374*, 71–78. [CrossRef] [PubMed]

19. Gray, R.D.; Trent, J.O.; Chaires, J.B. Folding and Unfolding pathways of the human telomeric G-quadruplex. *J. Mol. Biol.* **2014**, *426*, 1629–1650. [CrossRef] [PubMed]

20. Karsisiotis, A.I.; Webba da Silva, M. Structural probes in quadruplex nucleic acid structure determination by NMR. *Molecules* **2012**, *17*, 13073–13086. [CrossRef] [PubMed]

21. Adrian, M.; Heddi, B.; Phan, A.T. NMR spectroscopy of G-quadruplexes. *Methods* **2012**, *57*, 11–24. [CrossRef] [PubMed]

22. Sun, D.; Hurley, L.H. Biochemical Techniques for the characterization of G-quadruplex structures: EMSA, DMS footprinting, and DNA polymerase stop assay. *Methods Mol. Biol.* **2010**, *608*, 65–79. [PubMed]

23. Jaumot, J.; Gargallo, R. Experimental methods for studying the interactions between G-quadruplex structures and ligands. *Curr. Pharm. Des.* **2012**, *18*, 1900–1916. [CrossRef] [PubMed]

24. Oliveira Brett, A.M.; Serrano, S.H.P.; Piedade, A.J.P. *Applications of Kinetic Modelling*; Comprehensive Chemical Kinetics; Elsevier: Amsterdam, The Netherlands, 1999; Volume 37.

25. Oliveira Brett, A.M. *Biosensors and Modern Biospecific Analytical Techniques*; Comprehensive Analytical Chemistry; Elsevier: Amsterdam, The Netherlands, 2005; Volume 44.

26. Brett, A.M.O. Electrochemistry for probing DNA damage. In *Encyclopedia of Sensors*; Grimes, C.A., Dickey, E.C., Pishko, M.V., Eds.; American Scientific Publishers: Valencia, CA, USA, 2006; p. 301.

27. Oliveira-Brett, A.M.; Paquim, A.M.C.; Diculescu, V.C.; Piedade, J.A.P. Electrochemistry of nanoscale DNA surface films on carbon. *Med. Eng. Phys.* **2006**, *28*, 963–970. [CrossRef] [PubMed]

28. Oliveira Brett, A.M.; Diculescu, V.C.; Chiorcea-Paquim, A.M.; Serrano, S.H.P. *Electrochemical Sensor Analysis*; Comprehensive Analytical Chemistry; Elsevier: Amsterdam, The Netherlands, 2007; Volume 49.

29. Oliveira Brett, A.M. Electrochemical DNA assays. In *Bioelectrochemistry: Fundamentals, Experimental Techniques and Applications*; Bartlett, P.N., Ed.; John Wiley & Sons, Ltd.: Chichester, UK, 2008; p. 411.

30. Oliveira-Brett, A.-M. Nanobioelectrochemistry. In *Electrochemistry at the Nanoscale, Nanostrutures Science and Technology*; Schmuki, P., Virtanen, S., Eds.; Nanostructure Science and Technology; Springer New York: New York, NY, USA, 2009; p. 407.

31. Diculescu, V.C.; Chiorcea-Paquim, A.-M.; Oliveira-Brett, A.M. Applications of a DNA-electrochemical biosensor. *TrAC Trends Anal. Chem.* **2016**, *79*, 23–36. [CrossRef]

32. Chiorcea–Paquim, A.-M.; Santos, P.; Diculescu, V.C.; Eritja, R.; Oliveira-Brett, A.M. Guanine Quartets. In *Guanine Quartets: Structure and Application*; Spindler, L., Fritzsche, W., Eds.; Royal Society of Chemistry: Cambridge, UK, 2012; pp. 100–109.

33. Hianik, T.; Wang, J. Electrochemical aptasensors—Recent achievements and perspectives. *Electroanalysis* **2009**, *21*, 1223–1235. [CrossRef]

34. Musumeci, D.; Montesarchio, D. Polyvalent nucleic acid aptamers and modulation of their activity: A focus on the thrombin binding aptamer. *Pharmacol. Ther.* **2012**, *136*, 202–215. [CrossRef] [PubMed]

35. Tucker, W.O.; Shum, K.T.; Tanner, J.A. G-quadruplex DNA aptamers and their ligands: Structure, function and application. *Curr. Pharm. Des.* **2012**, *18*, 2014–2026. [CrossRef] [PubMed]

36. Ni, X.; Castanares, M.; Mukherjee, A.; Lupold, S.E. Nucleic Acid Aptamers: Clinical Applications and Promising New Horizons. *Curr. Med. Chem.* **2011**, *18*, 4206–4214. [CrossRef] [PubMed]

37. Diculescu, V.C.; Chiorcea-Paquim, A.-M.; Eritja, R.; Oliveira-Brett, A.M. Thrombin-binding Aptamer quadruplex formation: Afm and voltammetric characterization. *J. Nucleic Acids* **2010**, *2010*, 841932. [CrossRef] [PubMed]
38. Diculescu, V.C.; Chiorcea-Paquim, A.-M.; Eritja, R.; Oliveira-Brett, A.M. Evaluation of the Structure–activity Relationship of Thrombin with Thrombin Binding Aptamers by Voltammetry and Atomic Force Microscopy. *J. Electroanal. Chem.* **2011**, *656*, 159–166. [CrossRef]
39. Chiorcea-Paquim, A.-M.; Santos, P.V.; Oliveira-Brett, A.M. Atomic force microscopy and voltammetric characterisation of synthetic homo-oligodeoxynucleotides. *Electrochim. Acta* **2013**, *110*, 599–607. [CrossRef]
40. Chiorcea-Paquim, A.-M.; Santos, P.V.; Eritja, R.; Oliveira-Brett, A.M. Self-Assembled G-Quadruplex Nanostructures: AFM and Voltammetric Characterization. *Phys. Chem. Chem. Phys.* **2013**, *15*, 9117–9124. [CrossRef] [PubMed]
41. Rodrigues Pontinha, A.D.; Chiorcea-Paquim, A.-M.; Eritja, R.; Oliveira-Brett, A.M. Quadruplex Nanostructures of d(TGGGGT): Influence of Sodium and Potassium Ions. *Anal. Chem.* **2014**, *86*, 5851–5857. [CrossRef] [PubMed]
42. Chiorcea-Paquim, A.-M.; Pontinha, A.D.R.; Oliveira-Brett, A.M. Time-Dependent Polyguanylic Acid Structural Modifications. *Electrochem. Commun.* **2014**, *45*, 71–74. [CrossRef]
43. Chiorcea-Paquim, A.-M.; Rodrigues Pontinha, A.D.; Oliveira-Brett, A.M. Quadruplex-Targeting Anticancer Drug BRACO-19 Voltammetric and AFM Characterization. *Electrochim. Acta* **2015**, *174*, 155–163. [CrossRef]
44. Pontinha, A.D.R.; Sparapani, S.; Neidle, S.; Oliveira-Brett, A.M. Triazole-Acridine Conjugates: Redox Mechanisms and in Situ Electrochemical Evaluation of Interaction with Double-Stranded DNA. *Bioelectrochemistry* **2013**, *89*, 50–56. [CrossRef] [PubMed]
45. Chiorcea-Paquim, A.-M.; Pontinha, A.D.R.; Eritja, R.; Lucarelli, G.; Sparapani, S.; Neidle, S.; Oliveira-Brett, A.M. Atomic Force Microscopy and Voltammetric Investigation of Quadruplex Formation between a Triazole-Acridine Conjugate and Guanine-Containing Repeat DNA Sequences. *Anal. Chem.* **2015**, *87*, 6141–6149. [CrossRef] [PubMed]
46. Oliveira Brett, A.M.; Diculescu, V.C.; Chiorcea Paquim, A.-M.; Serrano, S. Chapter 20 DNA-Electrochemical Biosensors for Investigating DNA Damage. *Compr. Anal. Chem.* **2007**, *49*, 413–437.
47. Diculescu, V.C.; Chiorcea Paquim, A.-M.; Oliveira Brett, A.M. Electrochemical DNA Sensors for Detection of DNA Damage. *Sensors* **2005**, *5*, 377–393. [CrossRef]
48. Vasilescu, A.; Marty, J.-L. Electrochemical Aptasensors for the Assessment of Food Quality and Safety. *TrAC Trends Anal. Chem.* **2016**, *79*, 60–70. [CrossRef]
49. Willner, I.; Zayats, M. Electronic Aptamer-Based Sensors. *Angew. Chem. Int. Ed.* **2007**, *46*, 6408–6418. [CrossRef] [PubMed]
50. Cheng, A.K.H.; Sen, D.; Yu, H.-Z. Design and Testing of Aptamer-Based Electrochemical Biosensors for Proteins and Small Molecules. *Bioelectrochemistry* **2009**, *77*, 1–12. [CrossRef] [PubMed]
51. Velasco-Garcia, M.; Missailidis, S. New Trends in Aptamer-Based Electrochemical Biosensors. *Gene Ther. Mol. Biol.* **2009**, *13*, 1–10.
52. Yin, X.-B. Functional Nucleic Acids for Electrochemical and Electrochemiluminescent Sensing Applications. *TrAC Trends Anal. Chem.* **2012**, *33*, 81–94. [CrossRef]
53. Hianik, T.; Ostatná, V.; Sonlajtnerova, M.; Grman, I. Influence of Ionic Strength, pH and Aptamer Configuration for Binding Affinity to Thrombin. *Bioelectrochemistry* **2007**, *70*, 127–133. [CrossRef] [PubMed]
54. Liu, J.; Cao, Z.; Lu, Y. Functional Nucleic Acid Sensors. *Chem. Rev.* **2009**, *109*, 1948–1998. [CrossRef] [PubMed]
55. Song, S.; Wang, L.; Li, J.; Fan, C.; Zhao, J. Aptamer-Based Biosensors. *TrAC Trends Anal. Chem.* **2008**, *27*, 108–117. [CrossRef]
56. Mir, M.; Vreeke, M.; Katakis, I. Different Strategies to Develop an Electrochemical Thrombin Aptasensor. *Electrochem. Commun.* **2006**, *8*, 505–511. [CrossRef]
57. Polsky, R.; Gill, R.; Kaganovsky, L.; Willner, I. Nucleic Acid-Functionalized Pt Nanoparticles: Catalytic Labels for the Amplified Electrochemical Detection of Biomolecules. *Anal. Chem.* **2006**, *78*, 2268–2271. [CrossRef] [PubMed]
58. Numnuam, A.; Chumbimuni-Torres, K.Y.; Xiang, Y.; Bash, R.; Thavarungkul, P.; Kanatharana, P.; Pretsch, E.; Wang, J.; Bakker, E. Aptamer-Based Potentiometric Measurements of Proteins Using Ion-Selective Microelectrodes. *Anal. Chem.* **2008**, *80*, 707–712. [CrossRef] [PubMed]

59.   Yang, H.; Ji, J.; Liu, Y.; Kong, J.; Liu, B. An Aptamer-Based Biosensor for Sensitive Thrombin Detection. *Electrochem. Commun.* **2009**, *11*, 38–40. [CrossRef]

60.   Peng, K.; Zhao, H.; Wu, X.; Yuan, Y.; Yuan, R. Ultrasensitive Aptasensor Based on Graphene-3,4,9,10-Perylenetetracarboxylic Dianhydride as Platform and Functionalized Hollow PtCo Nanochains as Enhancers. *Sens. Actuators B Chem.* **2012**, *169*, 88–95. [CrossRef]

61.   Centi, S.; Messina, G.; Tombelli, S.; Palchetti, I.; Mascini, M. Different Approaches for the Detection of Thrombin by an Electrochemical Aptamer-Based Assay Coupled to Magnetic Beads. *Biosens. Bioelectron.* **2008**, *23*, 1602–1609. [CrossRef] [PubMed]

62.   Centi, S.; Tombelli, S.; Minunni, M.; Mascini, M. Aptamer-Based Detection of Plasma Proteins by an Electrochemical Assay Coupled to Magnetic Beads. *Anal. Chem.* **2007**, *79*, 1466–1473. [CrossRef] [PubMed]

63.   Li, B.; Wang, Y.; Wei, H.; Dong, S. Amplified Electrochemical Aptasensor Taking AuNPs Based Sandwich Sensing Platform as a Model. *Biosens. Bioelectron.* **2008**, *23*, 965–970. [CrossRef] [PubMed]

64.   Ikebukuro, K.; Kiyohara, C.; Sode, K. Electrochemical Detection of Protein Using a Double Aptamer Sandwich. *Anal. Lett.* **2004**, *37*, 2901–2909. [CrossRef]

65.   Ikebukuro, K.; Kiyohara, C.; Sode, K. Novel Electrochemical Sensor System for Protein Using the Aptamers in Sandwich Manner. *Biosens. Bioelectron.* **2005**, *20*, 2168–2172. [CrossRef] [PubMed]

66.   Kang, Y.; Feng, K.-J.; Chen, J.-W.; Jiang, J.-H.; Shen, G.-L.; Yu, R.-Q. Electrochemical Detection of Thrombin by Sandwich Approach Using Antibody and Aptamer. *Bioelectrochemistry* **2008**, *73*, 76–81. [CrossRef] [PubMed]

67.   Rahman, M.A.; Son, J.I.; Won, M.-S.; Shim, Y.-B. Gold Nanoparticles Doped Conducting Polymer Nanorod Electrodes: Ferrocene Catalyzed Aptamer-Based Thrombin Immunosensor. *Anal. Chem.* **2009**, *81*, 6604–6611. [CrossRef] [PubMed]

68.   Lubin, A.A.; Plaxco, K.W. Folding-Based Electrochemical Biosensors: The Case for Responsive Nucleic Acid Architectures. *Acc. Chem. Res.* **2010**, *43*, 496–505. [CrossRef] [PubMed]

69.   Plaxco, K.W.; Soh, H.T. Switch-Based Biosensors: A New Approach towards Real-Time, in Vivo Molecular Detection. *Trends Biotechnol.* **2011**, *29*, 1–5. [CrossRef] [PubMed]

70.   Bai, H.-Y.; Del Campo, F.J.; Tsai, Y.-C. Sensitive Electrochemical Thrombin Aptasensor Based on Gold Disk Microelectrodearrays. *Biosens. Bioelectron.* **2013**, *42*, 17–22. [CrossRef] [PubMed]

71.   Xu, H.; Gorgy, K.; Gondran, C.; Le Goff, A.; Spinelli, N.; Lopez, C.; Defrancq, E.; Cosnier, S. Label-Free Impedimetric Thrombin Sensor Based on Poly(pyrrole-Nitrilotriacetic Acid)-Aptamer Film. *Biosens. Bioelectron.* **2013**, *41*, 90–95. [CrossRef] [PubMed]

72.   Li, X.; Shen, L.; Zhang, D.; Qi, H.; Gao, Q.; Ma, F.; Zhang, C. Electrochemical Impedance Spectroscopy for Study of Aptamer-Thrombin Interfacial Interactions. *Biosens. Bioelectron.* **2008**, *23*, 1624–1630. [CrossRef] [PubMed]

73.   Suprun, E.; Shumyantseva, V.; Bulko, T.; Rachmetova, S.; Rad'ko, S.; Bodoev, N.; Archakov, A. Au-Nanoparticles as an Electrochemical Sensing Platform for Aptamer-Thrombin Interaction. *Biosens. Bioelectron.* **2008**, *24*, 831–836. [CrossRef] [PubMed]

74.   Feng, L.; Chen, Y.; Ren, J.; Qu, X. A Graphene Functionalized Electrochemical Aptasensor for Selective Label-Free Detection of Cancer Cells. *Biomaterials* **2011**, *32*, 2930–2937. [CrossRef] [PubMed]

75.   Elahi, M.Y.; Bathaie, S.Z.; Mousavi, M.F.; Hoshyar, R.; Ghasemi, S. A New DNA-Nanobiosensor Based on G-Quadruplex Immobilized on Carbon Nanotubes Modified Glassy Carbon Electrode. *Electrochim. Acta* **2012**, *82*, 143–151. [CrossRef]

76.   Xiao, Y.; Lubin, A.A.; Heeger, A.J.; Plaxco, K.W. Label-Free Electronic Detection of Thrombin in Blood Serum by Using an Aptamer-Based Sensor. *Angew. Chem.* **2005**, *117*, 5592–5595. [CrossRef]

77.   Radi, A.-E.; Acero Sánchez, J.L.; Baldrich, E.; O'Sullivan, C.K. Reusable Impedimetric Aptasensor. *Anal. Chem.* **2005**, *77*, 6320–6323. [CrossRef] [PubMed]

78.   Radi, A.-E.; Acero Sánchez, J.L.; Baldrich, E.; O'Sullivan, C.K. Reagentless, Reusable, Ultrasensitive Electrochemical Molecular Beacon Aptasensor. *J. Am. Chem. Soc.* **2006**, *128*, 117–124. [CrossRef] [PubMed]

79.   Sánchez, J.L.A.; Baldrich, E.; Radi, A.E.-G.; Dondapati, S.; Sánchez, P.L.; Katakis, I.; O'Sullivan, C.K. Electronic "Off-On" Molecular Switch for Rapid Detection of Thrombin. *Electroanalysis* **2006**, *18*, 1957–1962. [CrossRef]

80.   Bang, G.S.; Cho, S.; Kim, B.-G. A Novel Electrochemical Detection Method for Aptamer Biosensors. *Biosens. Bioelectron.* **2005**, *21*, 863–870. [CrossRef] [PubMed]

81.   Yan, F.; Wang, F.; Chen, Z. Aptamer-Based Electrochemical Biosensor for Label-Free Voltammetric Detection of Thrombin and Adenosine. *Sens. Actuators B Chem.* **2011**, *160*, 1380–1385. [CrossRef]

82. Xiao, Y.; Piorek, B.D.; Plaxco, K.W.; Heeger, A.J. A Reagentless Signal-on Architecture for Electronic, Aptamer-Based Sensors via Target-Induced Strand Displacement. *J. Am. Chem. Soc.* **2005**, *127*, 17990–17991. [CrossRef] [PubMed]

83. Meini, N.; Farre, C.; Chaix, C.; Kherrat, R.; Dzyadevych, S.; Jaffrezic-Renault, N. A Sensitive and Selective Thrombin Impedimetric Aptasensor Based on Tailored Aptamers Obtained by Solid-Phase Synthesis. *Sens. Actuators B Chem.* **2012**, *166–167*, 715–720. [CrossRef]

84. Mir, M.; Jenkins, A.T.A.; Katakis, I. Ultrasensitive Detection Based on an Aptamer Beacon Electron Transfer Chain. *Electrochem. Commun.* **2008**, *10*, 1533–1536. [CrossRef]

85. Prabhakar, N.; Matharu, Z.; Malhotra, B.D. Polyaniline Langmuir-Blodgett Film Based Aptasensor for Ochratoxin A Detection. *Biosens. Bioelectron.* **2011**, *26*, 4006–4011. [CrossRef] [PubMed]

86. Chen, J.; Zhang, J.; Li, J.; Yang, H.-H.; Fu, F.; Chen, G. An Ultrasensitive Signal-on Electrochemical Aptasensor via Target-Induced Conjunction of Split Aptamer Fragments. *Biosens. Bioelectron.* **2010**, *25*, 996–1000. [CrossRef] [PubMed]

87. Castillo, G.; Lamberti, I.; Mosiello, L.; Hianik, T. Impedimetric DNA Aptasensor for Sensitive Detection of Ochratoxin A in Food. *Electroanalysis* **2012**, *24*, 512–520. [CrossRef]

88. Hayat, A.; Andreescu, S.; Marty, J.-L. Design of PEG-Aptamer Two Piece Macromolecules as Convenient and Integrated Sensing Platform: Application to the Label Free Detection of Small Size Molecules. *Biosens. Bioelectron.* **2013**, *45*, 168–173. [CrossRef] [PubMed]

89. Jalit, Y.; Gutierrez, F.A.; Dubacheva, G.; Goyer, C.; Coche-Guerente, L.; Defrancq, E.; Labbé, P.; Rivas, G.A.; Rodríguez, M.C. Characterization of a Modified Gold Platform for the Development of a Label-Free Anti-Thrombin Aptasensor. *Biosens. Bioelectron.* **2013**, *41*, 424–429. [CrossRef] [PubMed]

90. De Rache, A.; Kejnovská, I.; Vorlíčková, M.; Buess-Herman, C. Elongated Thrombin Binding Aptamer: A G-Quadruplex Cation-Sensitive Conformational Switch. *Chemistry* **2012**, *18*, 4392–4400. [CrossRef] [PubMed]

91. Zhang, Z.; Yang, W.; Wang, J.; Yang, C.; Yang, F.; Yang, X. A Sensitive Impedimetric Thrombin Aptasensor Based on Polyamidoamine Dendrimer. *Talanta* **2009**, *78*, 1240–1245. [CrossRef] [PubMed]

92. Cheng, G.; Shen, B.; Zhang, F.; Wu, J.; Xu, Y.; He, P.; Fang, Y. A New Electrochemically Active-Inactive Switching Aptamer Molecular Beacon to Detect Thrombin Directly in Solution. *Biosens. Bioelectron.* **2010**, *25*, 2265–2269. [CrossRef] [PubMed]

93. Zhang, S.; Zhou, G.; Xu, X.; Cao, L.; Liang, G.; Chen, H.; Liu, B.; Kong, J. Development of an Electrochemical Aptamer-Based Sensor with a Sensitive Fe3O4 Nanopaticle-Redox Tag for Reagentless Protein Detection. *Electrochem. Commun.* **2011**, *13*, 928–931. [CrossRef]

94. Li, Y.; Deng, L.; Deng, C.; Nie, Z.; Yang, M.; Si, S. Simple and Sensitive Aptasensor Based on Quantum Dot-Coated Silica Nanospheres and the Gold Screen-Printed Electrode. *Talanta* **2012**, *99*, 637–642. [CrossRef] [PubMed]

95. Radi, A.-E.; O'Sullivan, C.K. Aptamer Conformational Switch as Sensitive Electrochemical Biosensor for Potassium Ion Recognition. *Chem. Commun. (Camb.)* **2006**, *32*, 3432–3434. [CrossRef] [PubMed]

96. Wu, Z.-S.; Chen, C.-R.; Shen, G.-L.; Yu, R.-Q. Reversible Electronic Nanoswitch Based on DNA G-Quadruplex Conformation: A Platform for Single-Step, Reagentless Potassium Detection. *Biomaterials* **2008**, *29*, 2689–2696. [CrossRef] [PubMed]

97. Zhang, J.; Wan, Y.; Wang, L.; Song, S.; Li, D.; Fan, C. Switchable Charge Transport Path via a Potassium Ions Promoted Conformational Change of G-Quadruplex Probe Monolayer. *Electrochem. Commun.* **2008**, *10*, 1258–1260. [CrossRef]

98. Chen, Z.; Chen, L.; Ma, H.; Zhou, T.; Li, X. Aptamer Biosensor for Label-Free Impedance Spectroscopy Detection of Potassium Ion Based on DNA G-Quadruplex Conformation. *Biosens. Bioelectron.* **2013**, *48*, 108–112. [CrossRef] [PubMed]

99. Zhang, J.; Chen, J.; Chen, R.; Chen, G.; Fu, F. An Electrochemical Biosensor for Ultratrace Terbium Based on Tb³⁺ Promoted Conformational Change of Human Telomeric G-Quadruplex. *Biosens. Bioelectron.* **2009**, *25*, 378–382. [CrossRef] [PubMed]

100. Zhang, J.; Chen, J.; Zhang, X.; Zeng, Z.; Chen, M.; Wang, S. An Electrochemical Biosensor Based on Hairpin-DNA Aptamer Probe and Restriction Endonuclease for Ochratoxin A Detection. *Electrochem. Commun.* **2012**, *25*, 5–7. [CrossRef]

101. Han, G.; Feng, X.; Chen, Z. Hemin/G-Quadruplex DNAzyme for Designing of Electrochemical Sensors. *Int. J. Electrochem. Sci* **2015**.

102. Pelossof, G.; Tel-Vered, R.; Elbaz, J.; Willner, I. Amplified Biosensing Using the Horseradish Peroxidase-Mimicking DNAzyme as an Electrocatalyst. *Anal. Chem.* **2010**, *82*, 4396–4402. [CrossRef] [PubMed]

103. Kosman, J.; Juskowiak, B. Peroxidase-Mimicking DNAzymes for Biosensing Applications: A Review. *Anal. Chim. Acta* **2011**, *707*, 7–17. [CrossRef] [PubMed]

104. Kaneko, N.; Horii, K.; Kato, S.; Akitomi, J.; Waga, I. High-Throughput Quantitative Screening of Peroxidase-Mimicking DNAzymes on a Microarray by Using Electrochemical Detection. *Anal. Chem.* **2013**, *85*, 5430–5435. [CrossRef] [PubMed]

105. Zhang, M.; Xu, S.; Minteer, S.D.; Baum, D.A. Investigation of a Deoxyribozyme as a Biofuel Cell Catalyst. *J. Am. Chem. Soc.* **2011**, *133*, 15890–15893. [CrossRef] [PubMed]

106. Sun, D.; Lu, J.; Chen, Z.; Yu, Y.; Mo, M. A Repeatable Assembling and Disassembling Electrochemical Aptamer Cytosensor for Ultrasensitive and Highly Selective Detection of Human Liver Cancer Cells. *Anal. Chim. Acta* **2015**, *885*, 166–173. [CrossRef] [PubMed]

107. Sun, D.; Lu, J.; Zhong, Y.; Yu, Y.; Wang, Y.; Zhang, B.; Chen, Z. Sensitive Electrochemical Aptamer Cytosensor for Highly Specific Detection of Cancer Cells Based on the Hybrid Nanoelectrocatalysts and Enzyme for Signal Amplification. *Biosens. Bioelectron.* **2016**, *75*, 301–307. [CrossRef] [PubMed]

108. Li, D.; Shlyahovsky, B.; Elbaz, J.; Willner, I. Amplified Analysis of Low-Molecular-Weight Substrates or Proteins by the Self-Assembly of DNAzyme-Aptamer Conjugates. *J. Am. Chem. Soc.* **2007**, *129*, 5804–5805. [CrossRef] [PubMed]

109. Zhang, K.; Zhu, X.; Wang, J.; Xu, L.; Li, G. Strategy to Fabricate an Electrochemical Aptasensor: Application to the Assay of Adenosine Deaminase Activity. *Anal. Chem.* **2010**, *82*, 3207–3211. [CrossRef] [PubMed]

110. Yang, N.; Cao, Y.; Han, P.; Zhu, X.; Sun, L.; Li, G. Tools for Investigation of the RNA Endonuclease Activity of Mammalian Argonaute2 Protein. *Anal. Chem.* **2012**, *84*, 2492–2497. [CrossRef] [PubMed]

111. Liu, L.; Liang, Z.; Li, Y. Label Free, Highly Sensitive and Selective Recognition of Small Molecule Using Gold Surface Confined Aptamers. *Solid State Sci.* **2012**, *14*, 1060–1063. [CrossRef]

112. Wang, Z.-H.; Lu, C.-Y.; Liu, J.; Xu, J.-J.; Chen, H.-Y. An Improved G-Quadruplex DNAzyme for Dual-Functional Electrochemical Biosensing of Adenosines and Hydrogen Peroxide from Cancer Cells. *Chem. Commun. (Camb.)* **2014**, *50*, 1178–1180. [CrossRef] [PubMed]

113. Yang, Q.; Nie, Y.; Zhu, X.; Liu, X.; Li, G. Study on the Electrocatalytic Activity of Human Telomere G-Quadruplex–hemin Complex and Its Interaction with Small Molecular Ligands. *Electrochim. Acta* **2009**, *55*, 276–280. [CrossRef]

114. Zhu, X.; Zhang, W.; Xiao, H.; Huang, J.; Li, G. Electrochemical Study of a hemin–DNA Complex and Its Activity as a Ligand Binder. *Electrochim. Acta* **2008**, *53*, 4407–4413. [CrossRef]

115. Yang, C.; Lates, V.; Prieto-Simón, B.; Marty, J.-L.; Yang, X. Aptamer-DNAzyme Hairpins for Biosensing of Ochratoxin A. *Biosens. Bioelectron.* **2012**, *32*, 208–212. [CrossRef] [PubMed]

116. Zhu, Y.; Xu, L.; Ma, W.; Chen, W.; Yan, W.; Kuang, H.; Wang, L.; Xu, C. G-Quadruplex DNAzyme-Based Microcystin-LR (Toxin) Determination by a Novel Immunosensor. *Biosens. Bioelectron.* **2011**, *26*, 4393–4398. [CrossRef] [PubMed]

117. Liang, G.; Liu, X. G-Quadruplex Based Impedimetric 2-Hydroxyfluorene Biosensor Using Hemin as a Peroxidase Enzyme Mimic. *Microchim. Acta* **2015**, *182*, 2233–2240. [CrossRef]

118. Liang, G.; Liu, X.; Li, X. Highly Sensitive Detection of α-Naphthol Based on G-DNA Modified Gold Electrode by Electrochemical Impedance Spectroscopy. *Biosens. Bioelectron.* **2013**, *45*, 46–51. [CrossRef] [PubMed]

119. Zhang, Z.; Yin, J.; Wu, Z.; Yu, R. Electrocatalytic Assay of mercury(II) Ions Using a Bifunctional Oligonucleotide Signal Probe. *Anal. Chim. Acta* **2013**, *762*, 47–53. [CrossRef] [PubMed]

120. Pelossof, G.; Tel-Vered, R.; Willner, I. Amplified Surface Plasmon Resonance and Electrochemical Detection of Pb^{2+} Ions Using the Pb^{2+}-Dependent DNAzyme and hemin/G-Quadruplex as a Label. *Anal. Chem.* **2012**, *84*, 3703–3709. [CrossRef] [PubMed]

121. Yuan, Y.; Gou, X.; Yuan, R.; Chai, Y.; Zhuo, Y.; Mao, L.; Gan, X. Electrochemical Aptasensor Based on the Dual-Amplification of G-Quadruplex Horseradish Peroxidase-Mimicking DNAzyme and Blocking Reagent-Horseradish Peroxidase. *Biosens. Bioelectron.* **2011**, *26*, 4236–4240. [CrossRef] [PubMed]

122. Jiang, B.; Wang, M.; Li, C.; Xie, J. Label-Free and Amplified Aptasensor for Thrombin Detection Based on Background Reduction and Direct Electron Transfer of Hemin. *Biosens. Bioelectron.* **2013**, *43*, 289–292. [CrossRef] [PubMed]

123. Liu, S.; Wang, C.; Zhang, C.; Wang, Y.; Tang, B. Label-Free and Ultrasensitive Electrochemical Detection of Nucleic Acids Based on Autocatalytic and Exonuclease III-Assisted Target Recycling Strategy. *Anal. Chem.* **2013**, *85*, 2282–2288. [CrossRef] [PubMed]

124. Bai, L.; Yuan, R.; Chai, Y.; Yuan, Y.; Zhuo, Y.; Mao, L. Bi-Enzyme Functionlized Hollow PtCo Nanochains as Labels for an Electrochemical Aptasensor. *Biosens. Bioelectron.* **2011**, *26*, 4331–4336. [CrossRef] [PubMed]

125. Xie, S.; Chai, Y.; Yuan, R.; Bai, L.; Yuan, Y.; Wang, Y. A Dual-Amplification Aptasensor for Highly Sensitive Detection of Thrombin Based on the Functionalized Graphene-Pd Nanoparticles Composites and the hemin/G-Quadruplex. *Anal. Chim. Acta* **2012**, *755*, 46–53. [CrossRef] [PubMed]

126. Yuan, Y.; Yuan, R.; Chai, Y.; Zhuo, Y.; Ye, X.; Gan, X.; Bai, L. Hemin/G-Quadruplex Simultaneously Acts as NADH Oxidase and HRP-Mimicking DNAzyme for Simple, Sensitive Pseudobienzyme Electrochemical Detection of Thrombin. *Chem. Commun. (Camb.)* **2012**, *48*, 4621–4623. [CrossRef] [PubMed]

127. Yuan, Y.; Liu, G.; Yuan, R.; Chai, Y.; Gan, X.; Bai, L. Dendrimer Functionalized Reduced Graphene Oxide as Nanocarrier for Sensitive Pseudobienzyme Electrochemical Aptasensor. *Biosens. Bioelectron.* **2013**, *42*, 474–480. [CrossRef] [PubMed]

128. Wang, Q.; Song, Y.; Chai, Y.; Pan, G.; Li, T.; Yuan, Y.; Yuan, R. Electrochemical Immunosensor for Detecting the Spore Wall Protein of Nosema Bombycis Based on the Amplification of hemin/G-Quadruplex DNAzyme Concatamers Functionalized Pt@Pd Nanowires. *Biosens. Bioelectron.* **2014**, *60*, 118–123. [CrossRef] [PubMed]

129. Yuan, Y.; Gao, M.; Liu, G.; Chai, Y.; Wei, S.; Yuan, R. Sensitive Pseudobienzyme Electrocatalytic DNA Biosensor for mercury(II) Ion by Using the Autonomously Assembled hemin/G-Quadruplex DNAzyme Nanowires for Signal Amplification. *Anal. Chim. Acta* **2014**, *811*, 23–28. [CrossRef] [PubMed]

130. Zheng, Y.; Chai, Y.; Yuan, Y.; Yuan, R. A Pseudo Triple-Enzyme Electrochemical Aptasensor Based on the Amplification of Pt-Pd Nanowires and hemin/G-Quadruplex. *Anal. Chim. Acta* **2014**, *834*, 45–50. [CrossRef] [PubMed]

131. Sun, A.; Qi, Q.; Wang, X.; Bie, P. Porous Platinum Nanotubes Labeled with hemin/G-Quadruplex Based Electrochemical Aptasensor for Sensitive Thrombin Analysis via the Cascade Signal Amplification. *Biosens. Bioelectron.* **2014**, *57*, 16–21. [CrossRef] [PubMed]

chemosensors

MDPI

Review

Aptasensors Based on Stripping Voltammetry

Wenjing Qi [1],*, Di Wu [1], Guobao Xu [2],*, Jacques Nsabimana [3] and Anaclet Nsabimana [2],[4]

[1] Chongqing Key Laboratory of Green Synthesis and Applications, College of Chemistry,
     Chongqing Normal University, Chongqing 401331, China; qwj19840217@163.com
[2] State Key Laboratory of Electroanalytical Chemistry, Changchun Institute of Applied Chemistry,
     Chinese Academy of Sciences, Changchun 130022, China; NSABIMANA@ciac.ac.cn
[3] Faculty of Medicine and Health Sciences, University of Gitwe, 01 Nyanza, Rwanda; jcqs.ns@gmail.com
[4] University of the Chinese Academy of Sciences, Chinese Academy of Sciences, No. 19A Yuquanlu,
     Beijing 100049, China
* Correspondence: wenjingqi616@cqnu.edu.cn (W.Q.); guobaoxu@ciac.ac.cn (G.X.);
   Tel.: +86-23-65362777 (W.Q.); +86-431-85262747 (G.X.)

Academic Editors: Paolo Ugo and Ligia Moretto
Received: 26 April 2016; Accepted: 6 July 2016; Published: 15 July 2016

Abstract: Aptasensors based on stripping voltammetry exhibit several advantages, such as high sensitivity and multi-target detection from stripping voltammetric technology, and high selectivity from the specific binding of apamers with targets. This review comprehensively discusses the recent accomplishments in signal amplification strategies based on nanomaterials, such as metal nanoparticles, semiconductor nanoparticles, and nanocomposite materials, which are detected by stripping voltammetry after suitable dissolution. Focus will be put in discussing multiple amplification strategies that are widely applied in aptasensors for small biomolecules, proteins, disease markers, and cancer cells.

Keywords: aptasensors; stripping voltammetry; aptamer; signal amplification

---

1. Introduction

1.1. Electrochemical Aptasensors

Biosensors are the devices used to detect the presence of a target by using a biological recognition element in direct spatial contact with a transducer; when transduction is electrochemical we talk of electrochemical biosensors [1–3]. The key point for biosensor performance is molecular-specific recognition. Nowadays, recognition elements include receptors, enzymes, antibodies, nucleic acids, molecular imprints, etc. Aptamers are nucleic acids (DNA or RNA strands) that selectively bind to low-molecular-weight organic or inorganic molecules, macromolecules, such as proteins, and even tumor markers and cancer cells [4,5]. Aptamers are selected from a combinatorial library of synthetic nucleic acids by SELEX (systematic evolution of ligands by exponential enrichment) technology [6–8]. Generally SELEX consists in repeated binding, selection, and amplification of aptamers from the initial, synthetic combinatorial library of nucleic acids until one (or more) aptamer(s) with the desired characteristics has been isolated. SELEX procedure starts with generating nearly 10^{14}–10^{15} random nucleic acid sequences. These random nucleic acids are chemically synthesized and amplified with polymerase chain reaction (PCR). Then the nucleic acid library is incubated with the target. The targets are often immobilized onto a solid-state matrix, such as a gel or a column, so that DNA or RNA strands having affinity to the target molecule can be captured. Next, the target-bound nucleic acids are separated from the unbound strands in the pool, and then the bound DNA or RNA strands are eluted from the target and amplified via PCR to seed a new pool of nucleic acids enriched with sequences that have higher affinity to the target. The next round of the selection process is usually performed under more stringent conditions (such as lower target concentration and shorter time for binding). After nearly 10–20 rounds of the selection processes, the nucleic acids with the highest affinity to the target molecule can be obtained [9–11]. The specific binding and high affinity constants of aptamers towards their substrates are comparable to the binding constants of antibodies to antigens. In comparison with antibodies, aptamers exhibit many advantages. For example, they are designed in vitro without the need of an animal host and are lacking immunogenicity. Once selected, aptamers can be readily produced with high reproducibility and purity by chemical synthesis. More importantly, aptasensors can be used in a wide variety of sample matrixes, including non-physiological buffers and temperature conditions that would cause denaturation of typical antibodies [12]. Owing to the above advantages, many different aptasensors, such as fluorescent [13], colorimetric [14], electrochemical [4,15], and electrochemiluminescent [16–19] approaches have been developed to date. It covers nearly all biosensing strategies.

Electrochemical aptasensors are attractive since they require low-cost portable instrumentation and allow one to perform high-throughput, multi-analyte, and on-site analysis [4,20]. Therefore, electrochemical aptasensors have been extensively applied in analysis of small biomolecules (e.g., adenosine and ATP [21,22], cocaine [23,24]), proteins [25–27], disease pathogens [28–30], and cancer cells [29,31,32]. Some metal ions, such as As(III) [33] or Pb(II) [34,35], can also be detected using suitable aptasensors obtained by the SELEX procedure. In the past decades, many strategies to enhance the response of the aptasensor on the basis of electrochemical signals have been developed. Different functional nanomaterials were employed such as metal nanoparticles [36], semiconductor nanoparticles [26,37,38], magnetic nanoparticles [39], polymeric nanoparticles [31], carbon-based nanomaterials [40], and nanocomposite materials [22,41,42] in order to meet the growing demands for ultrasensitive detection [43]. Additionally, other multiple amplification strategies, such as PCR, strand-displacement amplification (SDA), hybridization chain reaction (HCR) amplification, rolling circle amplification (RCA), and cyclic target-induced primer extension (CTIPE), are also utilized to improve the sensitivity of electrochemical aptasensors [23,27,44,45].

1.2. Electrochemical Stripping Voltammetry

Stripping voltammetry is a unique voltammetric method for quantitative detection. It includes two main steps: (i) an accumulation step, whose role is to concentrate analytes at the sensor surface; and (ii) a detection step during which the reverse reaction of the first step occurs. Thanks to the preconcentration step, stripping voltammetry shows higher sensitivity than direct methods, such as cyclic voltammery, linear sweep voltammetry, or chronoamperometry [46–49]. Stripping voltammetry can be used to detect nanoparticles/nanomaterials which are often used to enhance the sensitivity of aptasensing. In these cases, the nanomaterials bound to the target or probe or electrode are first dissolved in acid solutions (such as HNO_3 and HCl) to release ions (such as Ag^+, Pb^{2+}, Cd^{2+}, Cu^{2+}) which are, afterwards, detected by stripping voltammetry. The roles of the nanomaterial are: (i) to give electrochemical signals by stripping analysis of the dissolved metals; (ii) to modify the surface of the electrode for making much sensitive electrochemical transducers; and (iii) to immobilize capturing probes [50]. Nowadays, mercury-based electrodes are scarcely used since Hg is recognized as a toxic material. Many alternative electrodes, such as carbon electrodes, gold electrodes, silver electrodes, iridium electrodes, and bismuth electrodes, have been studied for applications in stripping voltammetry. We found an alternative dioctyl phthalate-based carbon paste electrode for stripping voltammetry in 2012 [51]. It exhibits an extremely wide cathodic potential range and even higher hydrogen evolution overpotential than bismuth electrodes. Kokkinos and his group reported a novel microfabricated tin biosensor for stripping voltammetric analysis of DNA and prostate-specific antigen (PSA) in 2013 [37]. They found that a microfabricated tin electrode works better than electroplated mercury-film or bismuth-film electrodes on glassy carbon in electrochemical stripping analysis of DNA and PSA. These significantly promote the development of stripping voltammetry.

Nowadays, anodic, cathodic, and adsorptive stripping voltammetries are recognized as election techniques for trace electroanalysis of heavy metal ions, such as Pb^{2+}, Cd^{2+}, Fe^{3+}, Co^{2+}, Mn^{2+}, Zn^{2+}, As^{3+}/As^{5+}, and Sb^{5+} [52–58]. Anodic striping voltammetry is an effective approach to achieve multiple-detection of heavy metals simultaneously. Square wave and differential pulse stripping voltammetric techniques have obviously higher sensitivity than linear stripping voltammetry [25,59–61]. Stripping voltammetry has been successfully combined with aptasensing to develop electrochemical biosensors suitable to detect small biomolecules such as adenosine triphosphate (ATP) [61], drugs, such as kanamycin [62], proteins, such as thrombin [63], and cancer and tumor markers; however, further studies are required to improve and optimize the analytical performances and to widen the applications.

Herein, we focus on the combination of aptamer and stripping voltammetry technology to give a mini-review of the most recent developments on electrochemical aptasensors based on stripping voltammetry.

2. Aptasensor Based on Stripping Voltammetry

2.1. Aptasensors for Simultaneous Detection of Small Biomolecules

The development of biosensor for sensitive and selective determination of important small biomolecules, such as ATP, cocaine, and adenosine, has been extensively studied for several decades. Quite a number of biosensors can only measure a single analyte. However, the simultaneous detection of several analytes is desirable in many cases (e.g., diagnosis and environmental analysis). Aptasensors based on stripping analysis are a sensitive method for multiplex analysis. For example, Yuan and his group designed an electrochemical strategy for "signal on" and sensitive one-spot simultaneous detection of multiple analytes, such as ATP and cocaine, using two quantum dot (QD) labels [64]. It is a traditional sandwiched strategy, but it involves the self-assembly of two types of thiol-modified primary target-binding aptamers, ATP binding aptamer (ABA_1) and cocaine binding aptamer (CBA_1) on the gold substrate (Figure 1). These two aptamers contribute to the specific binding of its corresponding target ATP and cocaine, respectively. After simultaneous addition of ATP and cocaine as well as their corresponding QD-conjugated secondary binding aptamers (PbS-ABA_2 and CdS-CBA_2), ATP and cocaine form a stable sandwiched complex with a QD-conjugated secondary aptamer and primary binding aptamer. This leads to different square wave stripping voltammetric signals via the measurement of released Pb^{2+} and Cd^{2+} after an acid dissolution process, enabling simultaneous and sensitive detection of ATP and cocaine due to the inherent amplification feature of the QD labels and the "signal on" detection scheme.

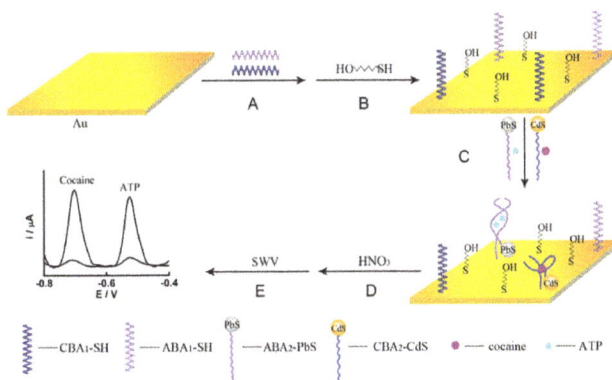

Figure 1. Scheme of one-spot simultaneous detection of two small molecules ATP and cocaine on the basis of square wave stripping voltammetry [64].

Compared with a single nanoparticle electrochemical label, nanocomposites allow better signal amplification strategies. Zhang and his group utilized QDs-AuNPs nanocomposite labels in simultaneous detection of adenosine and thrombin. The strategy involved a thiol-modified dual-aptamer DNA consisting of both an adenosine aptamer and thrombin aptamer (Figure 2) [22]. The thiol-modified dual-aptamer DNA was immobilized on the Au electrode surface via an Au–S bond. CdS QDs-AuNPs with a reporter DNA complementary to the adenosine aptamer and PbS QDs-AuNPs with a reporter DNA complementary to the thrombin aptamer were used for the detection of adenosine and thrombin by Cd^{2+} and Pd^{2+}, respectively. In the absence of adenosine and thrombin, CdS QDs-AuNPs were brought to the surface of electrode via DNA hybridization, resulting in a strong Cd stripping peak. When adenosine and thrombin were added, the aptamer could bind tightly and specifically to adenosine and form a tertiary complex. CdS QDs-AuNPs were removed from the surface of electrode because of stronger interaction of adenosine with its aptamer, and PbS QDs-AuNPs were attached to the surface of electrode due to the formation of the sandwich-type structure between thrombin and its two aptamers. As a result, Cd stripping peaks decrease with increasing adenosine concentrations and the Pb stripping peak increases with increasing thrombin concentrations, which allows simultaneous detection of adenosine and thrombin. This strategy of barcode QD tags for simultaneous detection of multiple small molecular analytes supplies an attractive route for screening of small molecules in clinical diagnosis.

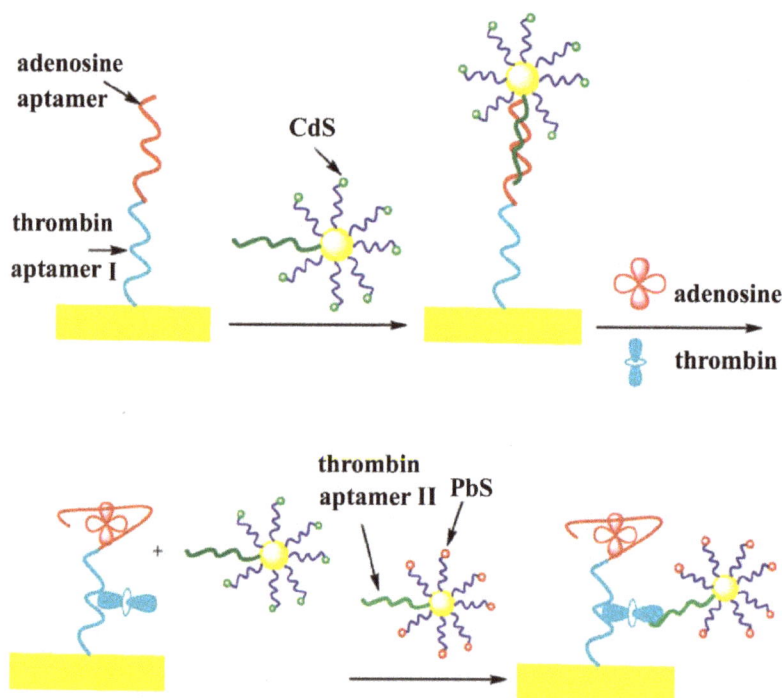

Figure 2. Scheme of the sensing interface for simultaneous detection of adenosine and thrombin via anodic stripping voltammetry [22].

These stripping voltammetric strategies mentioned above belong to the direct analysis of small biomolecules. The backfilling strategy belongs to an indirect analysis. Yuan and his group developed a multi-analyte aptasensor through a backfilling strategy for simultaneous detection of ATP and lysozyme using aptamer/nanoparticle bioconjugates as labels [21]. In this backfilling strategy, CdS

QDs and PbS QDs are firstly tagged with different DNA sequences, which are complementary with an ATP aptamer and a lysozyme aptamer, respectively (Figure 3). If no ATP and lysozyme exist, duplexes and 6-mercapto-1-hexanol are blocked on the gold substrate. After the addition of ATP and lysozyme, tertiary aptamer/targets are formed and released from the gold substrate. At this time, CdS QDs and PbS QDs tagged with different DNA are added and form backfilling hybridization with thiol-modified DNAs on the gold substrate. This achieves simultaneous detection of ATP and lysozyme by measuring the enhanced square-wave voltammetric stripping signals of the Pb^{2+} and Cd^{2+}. This aptamer/nanoparticle-based backfilling strategy can improve the sensitivity via the signal-on protocol compared with common target-induced displacement or conformational change signal-off configuration. It opens opportunities for multiplexed clinical diagnosis of different molecules with distinct sizes.

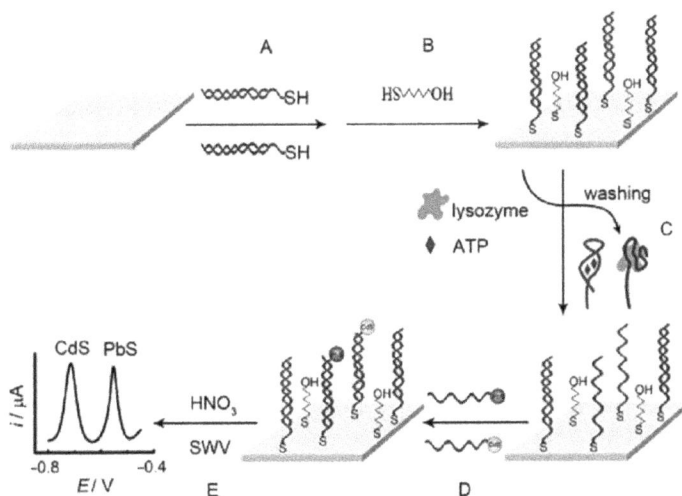

Figure 3. Scheme of the aptamer/nanoparticle-based backfilling protocol for simultaneous detection of lysozyme and ATP [21].

2.2. Aptasensors for Proteins

In bioassays, antibodies have been widely used in the analysis of clinical biofluid specimen, such as urine and blood [50,65,66]. However, antibodies are not stable and may lose activity easily. Aptamers are alternatives to antibodies for bioassays. They can specifically bind to a variety of target molecules, such as proteins [36,39,67,68], drugs [62], small biomolecules [21,22], and cells [28,31,43,69].

In protein aptasensor research area, thrombin is a typical protein model investigated by many researchers. Several works on protein aptasensors based on stripping voltammetry have been reported [26,63,70–73]. Usually only one type of nanoparticles, such as QDs, are used as labels for stripping voltammetric analysis of thrombin [73] but, compared with a single nanoparticle, nanocomposite materials can obviously achieve signal amplification. Lin and his group reported a novel electrochemical aptasensor for the determination of protein thrombin by anodic stripping voltammetric analysis of Cd^{2+} using ssDNA-labeled CdS NPs-AuNPs in 2010 [26]. In their strategy, the first step is to obtain detection probe. Linker DNA ssDNA-labeled CdS nanoparticles (NPs) is conjugated with AuNPs through one thrombin-related aptamer. Then, the sandwich type is formed after the addition of target thrombin via another aptamer modified on the surface of gold electrode. In this work, AuNPs and ssDNA-labeled CdS NPs conjugate served not only as a target recognition probe for thrombin, but also as a tool of amplification. Numerous ssDNA-labeled CdS NPs were linked with AuNPs, which enhanced the Cd^{2+} stripping signals. Therefore, it achieved the determination of thrombin in the linear range of 1.0×10^{-15} to 1.0×10^{-11} M with a low detection limit of 0.55 fM. This aptasensor was also confirmed to be able to distinguish the target thrombin from the interferents. Xu and his group also utilized nanocomposite cadmium sulfide nanoparticles (CdS NPs) functionalized with colloidal carbon particles (CPs) in an analysis of thrombin in 2011 [63]. The sandwich-type assay of thrombin was developed by in situ growing of abundant CdS NPs on the surfaces of monodisperse carbon particles (CdS/CPs) via square wave stripping voltammetric signals of the released Cd^{2+}. Both of these two works utilized nanocomposite materials as signal amplification strategy. Additionally, Shumyantseva and his group developed two similar label-free stripping voltammetric analyses of thrombin, respectively using Au^+ stripping signals 2008 [71] and using Ag^+ stripping signals in 2010 [72]. As shown in Figure 4 [72], a screen-printed electrode (SPE) modified with AgNPs served as the sensing platform and the oxidation of AgNPs ($Ag^0 \rightarrow Ag^+$) upon polarization (+100 mV) supplied the detection signals for the proposed aptasensor. Aptamers were immobilized onto the surface of SPE modified with AgNP via an S–Ag bond. In the presence of thrombin, the anodic Ag^+ stripping peak was decreased. This direct detection strategy is label-free, simple, and fast, and can be extended to other analytes or biorecognition elements.

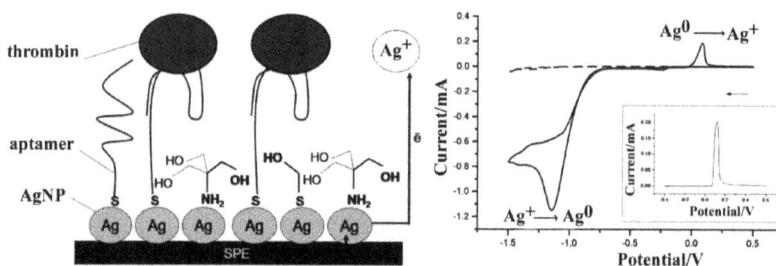

Figure 4. Scheme of label-free anodic stripping voltammetric detection of thrombin based on the oxidation of silver upon polarization by the determination of AgNPs surface status [72].

Wang and his group took advantage of the signal amplification effect of the labeled CdS NPs with the assistance of target protein-induced strand displacement and developed a sensitive electrochemical aptasensor for thrombin (Figure 5) [68]. In this electrochemical aptasensor for thrombin, a single DNA labeled with CdS nanoparticles was used as a detection probe. In the presence of thrombin, the aptamer in the dsDNA preferred to form a G quarter structure with thrombin releasing one single strand in the dsDNA sequence. In the G quarter structure, multiple guanines are organized around thrombin in a four-stranded structure. After the addition of the CdS NPs labeled probe, the concentration of thrombin is related to the amount of the captured CdS nanoparticles. After dissolving CdS particles, the released Cd^{2+} is determined by adsorptive stripping voltammeries and this amount is related to the thrombin concentration, allowing reaching a detection limit for the protein of 4.3×10^{-13} mol/L.

Xu and his group utilized AgNPs/graphene nanocomposite materials for the analysis of the protein human immunoglobulin E (IgE) [42]. IgE plays an important role in allergic reactions and other related diseases. Like other aptasensor of protein, a sandwich-type strategy was used. However, the detection probe was a streptavidin-functionalized AgNPs/graphene hybrid linked with biotinylated anti-human IgE antibody. Meanwhile, thiol-tagged IgE aptamer was used as the capture probe. Owing to specific binding between the aptamer with IgE and the antibody with IgE, a sandwich-type aptasensor for IgE was developed by a square wave anodic stripping voltammetric signal from AgNPs/graphene nanocomposite materials. The high-loading ability of graphene for AgNPs, combined with the unique electrical properties of graphene, contributed to the low detection limit. It brought a dynamic range for IgE detection from 10 to 1000 ng/mL with a low detection limit of 3.6 ng/mL. Furthermore, it is also a portable, simple, and inexpensive electrochemical biosensor for IgE since a disposable screen printed electrode was used as a sensing platform.

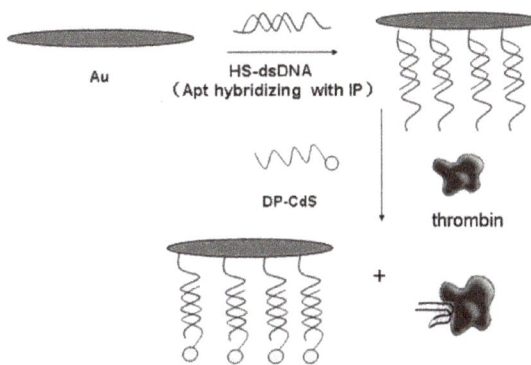

Figure 5. Scheme of the electrochemical aptasensor of thrombin based on target protein-induced strand displacement using signal amplification of CdS NPs [68].

Silver staining is commonly used as a signal amplification strategy for detection methods based on nanomaterial labels [41,66]. Xie and his group used gold label/silver staining in detection of immunoglobulin G (IgG) with the low detection limit of 0.2 fg/mL [66]. In their strategy, silver ions can be stained solely on the surface of catalytic AuNPs through chemical reduction of silver cations by hydroquinone. A beforehand "potential control" in air, and then an injection of HNO_3 for dissolution of the stained silver, enables rapid cathodic preconcentration of atomic silver onto the electrode surface for anodic stripping voltammetry measurement of silver ions.

SPE is an appropriate electrode for clinical diagnosis because it can satisfy highly sensitive and reproducible determination of target analytes, and meets the requirement for performing rapid in situ analyses. Combining the advantages of SPE and those of electrochemical array technology, SPE array technology shows great advantages in multi-analytes measurements. It has been used in various analytical methods, such as heavy metal ion detection, enzymatic biosensors, immunosensors, and DNA sensors [74,75]. Xu and his group further developed a multiplied protein aptasensor for platelet-derived growth factor (PDGF-BB) and thrombin by constructing a SPE array chip screen-printed electrode array in 2014 [25]. A sandwich-type strategy was utilized and an aptamer was tagged with DNA-functionalized AgNPs aggregate (Figure 6A). It gave an amplified differential pulse stripping voltammetry signal compared to the signal-labeled tag. Different aptamers for PDGF-BB and thrombin were chosen to construct multiplied-protein detection using a SPE array chip (Figure 6B). The novel SPE array chip achieved a wide linear range and low limit of detection in analysis of PDGF-BB and thrombin.

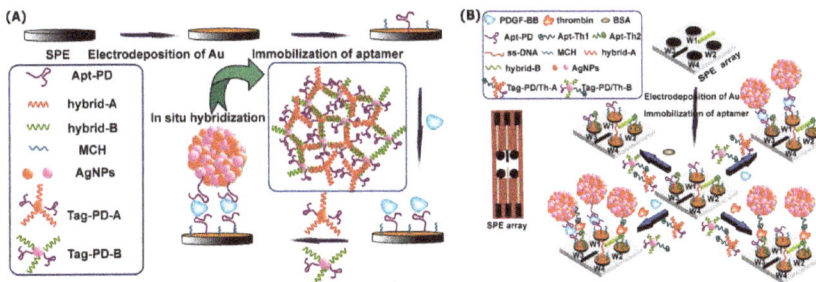

Figure 6. Scheme of (**A**) differential pulse stripping voltammetry detection and (**B**) multiplexed detection of PDGF-BB and thrombin [25].

2.3. Aptasensors for Cancer Cells and Diseases

Determining the extent of disease and planning appropriate therapies are essential for cancer treatment at an early stage. A tumor marker is a substance abnormally expressed in response to cancer. Different tumor markers can indicate different types of cancers with altered disease processes. Breast cancer is a kind of cancer mainly occurring in the inner lining of the milk ducts or lobules with different spread, aggressiveness, and genetic makeup. It has become the second most common type of cancer after lung cancer, and the fifth common cause of cancer death [69]. Li and his group developed a sensitive electrochemical aptasensor for the analysis of MCF-7 breast cancer cells by simultaneously detecting two tumor markers, human mucin-1 (MUC1) and carcinoembryonic antigen (CEA), on the surface of MCF-7 breast cancer cells in 2010 [69]. In their strategy, MCF-7 breast cancer cells were firstly recognized by its aptamer immobilized on the surface of a gold electrode. Another tumor marker, CEA, was subsequently captured by CdS nanoparticles (CdS NPs)-labeled anti-CEA. It achieved breast cancer cell analysis by anodic stripping voltammetric signals of Cd^{2+}. It efficiently improved the accuracy of the detection and avoidance of false-positive results due to low specificity. Similarly, B. Shim and his group developed an ultrasensitive and selective electrochemical diagnosis of breast cancer marker (HER2) and HER2-overexpressing breast cancer (SK-BR-3) on the basis of a hydrazine–Au NP–aptamer bioconjugate in 2013 [41]. Human epidermal growth factor receptor 2 (HER2) is a key prognostic marker and an effective therapeutic treatment target for breast cancer because it can be over-expressed in 10%–25% of breast cancers. In this strategy (Figure 7), anti-HER2 immobilized on the surface of an electrode was used to recognize breast cancer (SK-BR-3). 2,5-bis(2-thienyl)-1H-pyrrole-1-(p-benzoic acid) was self-assembled on AuNPs and the hydrazine-AuNP-aptamer bioconjugate was used as a detection probe. During the detection process, the anti-HER2-immobilized probe, HER2, or SK-BR-3 breast cancer cells, and Hyd-AuNP-Apt bioconjugate form a sandwich-type structure. In order to achieve ultrasensitive analysis of breast cancer, the silver-stained signal amplification was used. The silver ion is selectively reduced by hydrazine. The deposited silver is analyzed via square wave stripping voltammetry. This was the first report about the analysis of breast cancer cells utilizing selective silver stain through interaction with the Hyd–AuNP–Apt bioconjugate. Differently, Zhu and his group developed a competitive electrochemical sensor for anodic stripping voltammetric detection of breast cancer by using aptamer-quantum dots conjugates (Apt-QDs) [76]. Amino modified aptamers, which can specifically recognize MUC1 on the surface breast cancer cells, are firstly captured by the thiolated complementary DNA (cDNA) anchored on the gold electrode surface. Then Apt-QDs are formed through the interaction of amino groups with carboxyl groups in QDs. In the presence of breast cancer cells, MUC1 protein could compete with cDNA to conjugate with Apt-QDs conjugates. It leads to the decrease in the number of QDs retained at the electrode. After dissolving the remaining QDs, the concentration of the obtained metal species is detected via anodic stripping and the concentration of breast cancer cells is detected. The competitive electrochemical method is a good way to avoid false positive signals.

In addition to breast cancer, acute leukemia is the most common pediatric malignancy and remains the leading cause of disease-related mortality in children and adolescents. Zhu and his group developed an aptamer-based competition strategy for ultrasensitive electrochemical detection of leukemia cells (Figure 8) [77]. CCRF-CEM acute leukemia cells were chosen as model cells. The strategy was based on a dual signal amplification based on Fe_3O_4 magnetic nanoparticles (MNPs) carrying AuNPs and AuNP-catalyzed silver, which realized both as an electrosensor and cytosensor for target cells. An aptamer was designed to bind specifically to target cells. A partial cDNA was used to match the aptamer sequence. To fabricate the cytosensor, AuNPs loaded on Fe_3O_4 MNPs were prepared. Fe_3O_4 MNPs acted as both the separation tool and the strong nanocarriers for loading AuNPs and AuNP-catalyzed Ag deposition, which achieved signal amplification for the detection. After the addition of target cells, they could compete efficiently with conjugated cDNA to bind specifically with its aptamers. Therefore, the number of Fe_3O_4 MNPs carrying AuNPs and AgNPs retained at the electrode were decreased. The dissolved Ag^+, which was related with the number of target cells, were measured via square wave stripping voltammetry. The combination of the competitive, hybridization-based, and square wave anodic stripping voltammetric sensing platform supplies leukemia detection with high sensitivity, good specificity, desirable reproducibility, and acceptable stability. It exhibits great clinical value in the early diagnosis and prognosis of leukemia.

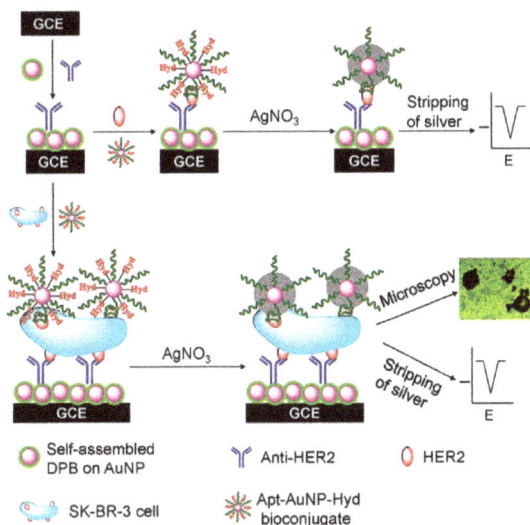

Figure 7. Scheme of square wave stripping voltammetric detection of tumor marker (human epidermal growth factor receptor 2, HER2) and HER2-overexpressing breast cancer (SK-BR-3) [41].

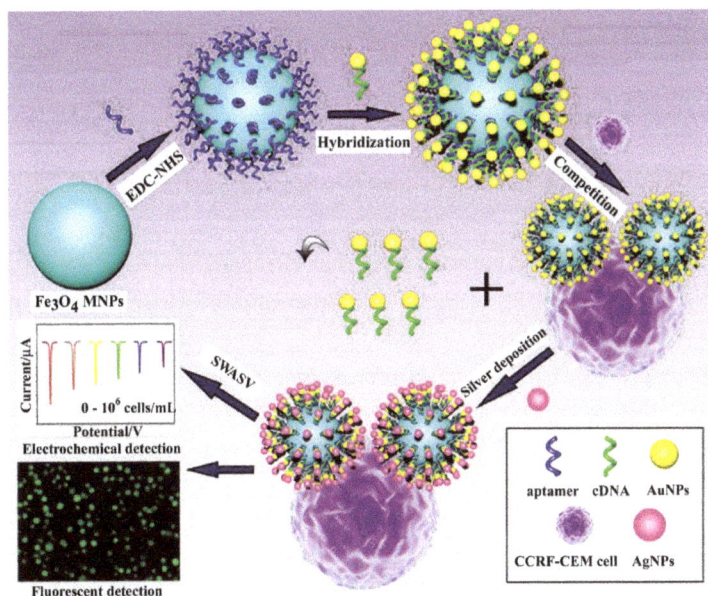

Figure 8. Scheme of competition strategy for ultrasensitive electrochemical detection of leukemia cells using dual-signal amplification based on Fe_3O_4 MNPs carrying AuNPs and AuNP-catalyzed silver [77].

Zhu and his group developed both electrochemical aptasensor and cytosensor for selective and ultrasensitive detection of cancer cells using aptamer-DNA concatamer quantum dot probes by fabricating an supersandwich structure (Figure 9) in 2013 [78]. DNA concatamers are linear polymeric structures formed by self-association of short DNA fragments through specific interactions. The proposed supersandwich structure combined with numerous CdS QDs achieves significant signal amplifications in aptasensors for cancer cells with the low detection limit of 50 cells/mL.

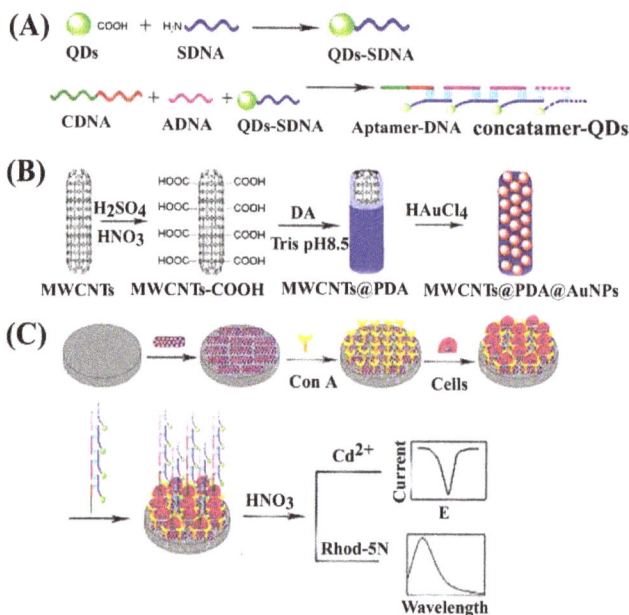

Figure 9. Procedures for (**A**) the fabrication of aptamer-DNA-concatamer-QDs probes, (**B**) MWCNTs@polydopamine(PDA)@AuNPs composites. and (**C**) a supersandwich aptasensor for cancer cells [78].

Multiple amplification strategies include PCR, SDA, HCR, RCA, etc. [43–45,79]. RCA is a simple and powerful isothermal enzymatic process. A short DNA or RNA primer is elongated to form a long, single-stranded DNA or RNA, assisted by a circular DNA template and unique DNA or RNA polymerases [80]. Since the RCA product contains thousands of tandem repeats that are complementary to the circular template, it leads to effective signal readout approaches, which are crucial to achieve various analytical purposes. Just owing to the advantages of high speed, efficiency, and specificity, RCA has been widely applied for signal amplification in analysis of DNA, protein, metal ions, and diseased cells. Ding and his group reported a cascade signal amplification strategy for ultrasensitive detection of cancer cell Ramos cells via combining RCA technique in 2012 [31]. In their strategy (Figure 10), NH_2-modified probe DNA and NH_2-modified primer DNA were immobilized on the surface of carboxyl-coated polystyrene microspheres (PSMs) to form a bio-bar code probe. The probe DNA firstly hybridized partly with the Ramos cell aptamer modified on magnetic beads (MBs) to construct a magnetic biocomplex. Carboxyl-coated MBs acted as the separation tool and the immobilization matrix. When the RCA reaction was started, a large quantity of signal DNA tagged with CdS NPs was added to the mixture. In this way, a long, single-stranded DNA which contains numerous tandem-repeat sequences was synthesized. At the same time, a large number of CdS-DNA probes for enhanced recognition were periodically assembled. But the double helixes between PSMs and MBs conjugate were opened due to much stronger binding of aptamer to Ramos cells than that of ordinary double-stranded DNA. Therefore Ramos cell were separated by magnetic separation procedure and the released Cd^{2+} form separated CdS-DNA probes was measured through anodic stripping voltammetry. This electrochemical aptasensor for cancer cells exhibited high sensitivity and specificity with the detection limits of 10 Ramos cells/mL.

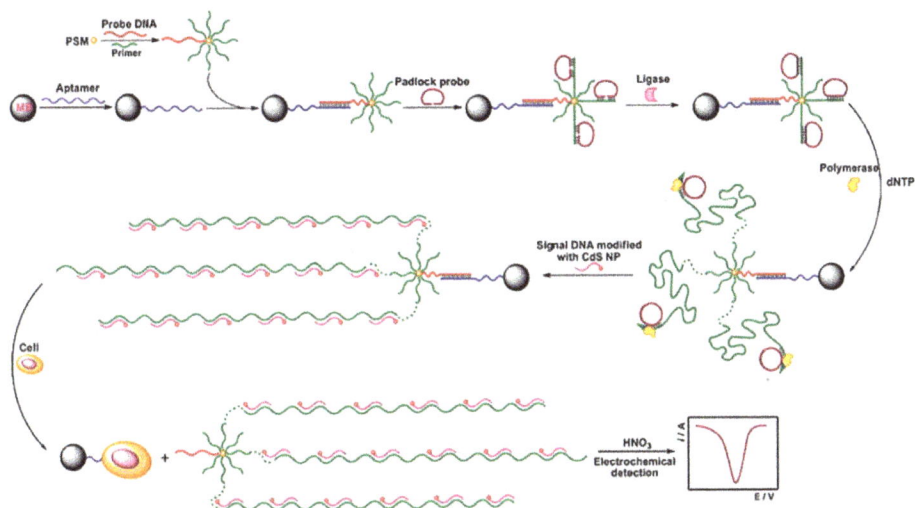

Figure 10. Scheme of aptasensor for cancer cells via anodic stripping voltammetry [31].

Jiang and his group combined the RCA reaction with poly(thymine)-templated copper nanoparticles (Cu NPs) for cascade signal amplification in ultrasensitive and highly-selective electrochemical analysis of prostate cancer biomarker PSA in 2016 [32]. DNA-templated metal nanoparticles, such as DNA-templated silver nanoclusters and DNA-templated Cu NPs, as novel-type emerging signal reporters have attracted increasing interest in biosensors owing to their advantages, including low toxicity, good biocompatibility, excellent optical properties, and facile integration with DNA-based recognition and signal amplification strategies. Though DNA-templated Cu NPs have been employed for the detection of small molecules, enzyme activity, nucleic acids, and metal ions, this was the first time of applying poly(thymine) repeats for a poly(thymine)-templated Cu NPs in electrochemical aptasensor of cancer markers. As shown in Figure 11, the RCA product contained thousands of poly(thymine) repeats which were used to synthesize DNA-templated Cu NPs. Once one PSA existed, one primer-AuNP-aptamer would be immobilized on the well and, subsequently, thousands of poly(thymine) sequences would be generated from RCA. It led to the synthesis of numerous DNA-templated Cu NPs and brought enhanced stripping voltammetric signals of released Cu^{2+}. It was a successful attempt at the ultrasensitive detection of PSA with a remarkable detection limit of 0.020 fg/mL for DNA-templated Cu NPs applied in cascade signal amplification.

In addition to cancer, some diseases also attract great attention. *Staphylococcus aureus* (*S. aureus*), one of the most important human pathogens, can cause different kinds of illnesses, from minor skin infections to life threatening diseases [28,81]. Abbaspour and his group reported a sensitive and highly-selective sandwich-type electrochemical detection of *S. aureus* using dual aptamers [28]. The primary aptamer was immobilized on the surface of MBs and the secondary aptamer, conjugated to AgNPs, acted as signal reporters to provide electrochemical stripping voltammetry characteristics. After fast magnetic separation, *S. aureus* was determined by Ag^+ released via differential pulse anodic stripping voltammetric measurement. It is a characteristic aptasensor for disease analysis, combining magnetic separation and AgNPs-based signal amplification technology. Zhang and his group developed a novel method for the determination of sulfate-reducing bacteria with the low detection limit of 1.8×10^2 cfu/mL) by combining a graphene oxide sheet-amplified assay and silver enhancement technology in 2011 [40]. They utilized graphene oxide sheets as promoters for silver reduction into AgNPs by hydroquinone. It is a good example of silver stained-based stripping voltammetry by utilizing the intrinsic catalytic property of graphene oxide sheets.

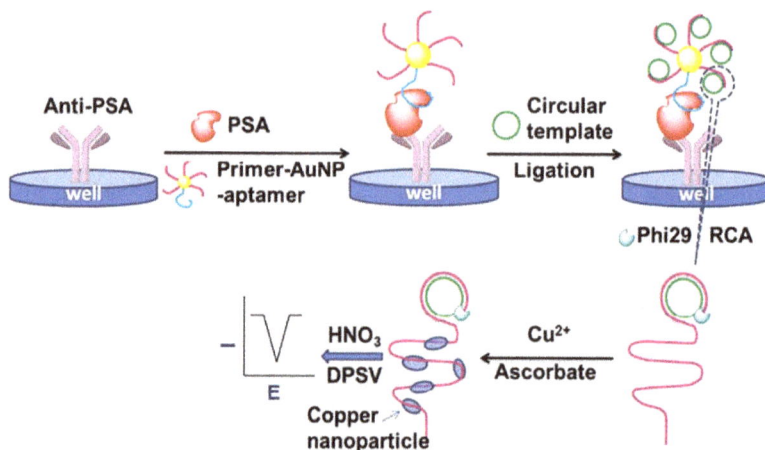

Figure 11. Scheme of rolling circle amplification for stripping voltammetric analysis of prostate cancer biomarker PSA [32].

3. Conclusions

In this article, a mini-review about the most recent developments in the field of aptasensors based on stripping voltammetry is presented. In order to improve the sensitivity of aptasensors nanomaterials, such as metal nanoparticles, semiconductor nanoparticles, and nanocomposite materials, were employed. They can give electrochemical signals by stripping analysis of the dissolved metals, modify the surface of the electrode to construct more sensitive electrochemical transducers, or immobilize capturing probes. Multiple amplifications, such as RCA, are also powerful strategies to improve the sensitivity of aptasensors. Moreover, some alternative electrodes, such as bismuth electrodes, tin electrodes, and even screen-printed electrodes, were employed by researchers. All of these will greatly promote the development of aptasensors on the basis of stripping voltammetry in single-target and multi-target analysis of small biomolecules, proteins, disease markers, and cancer cells.

Acknowledgments: It was supported by the National Natural Science Foundation of China (No. 21505011, 21475123), Chongqing Research Program of Basic Research and Frontier Technology (No. cstc2015jcyjA20019), Scientific and Technological Research Pro-gram of Chongqing Education Committee (No. KI1500306), and the Chinese Academy of Sciences (CAS)–the Academy of Sciences for the Developing World (TWAS) President's Fellowship Programme.

Conflicts of Interest: The authors declare no conflict of interest.

Abbreviations

The following abbreviations are used in this manuscript:

ABA	ATP binding aptamer
Apt-QDs	aptamer-quantum dots conjugates
ATP	adenosine triphosphate
CBA	cocaine binding aptamer

cDNA	complementary DNA
CEA	carcinoembryonic antigen
CPs	colloidal carbon particles
CTIPE	cyclic target-induced primer extension
HCR	hybridization chain reaction
HER2	breast cancer marker
IgE	human immunoglobulin E
IgG	immunoglobulin G
MBs	magnetic beads
MCF-7	The number of one type of breast cancer cells
MNPs	magnetic nanoparticles
MUC1	human mucin-1
PCR	polymerase chain reaction
PDGF-BB	platelet-derived growth factor
PSMs	polystyrene microspheres
PSA	prostate-specific antigen
QD	quantum dot
RCA	rolling circle amplification
S. aureus	Staphylococcus aureus
SDA	strand-displacement amplification
SELEX	systematic evolution of ligands by exponential enrichment
SK-BR-3	HER2-overexpressing breast cancer
SPE	screen-printed electrode

References

1. Xu, Y.; Cheng, G.; He, P.; Fang, Y. A review: Electrochemical aptasensors with various detection strategies. *Electroanalysis* **2009**, *21*, 1251–1259. [CrossRef]
2. Palchetti, I.; Mascini, M. Electrochemical nanomaterial-based nucleic acid aptasensors. *Anal. Bioanal. Chem.* **2012**, *402*, 3103–3114. [CrossRef] [PubMed]
3. Thevenot, D.R.; Tóth, K.; Durst, R.A.; Wilson, G.S. Electrochemical biosensors: Recommended definitions and classification. *Pure Appl.Chem.* **1999**, *71*, 2333–2348. [CrossRef]
4. Willner, I.; Zayats, M. Electronic aptamer-based sensors. *Angew. Chem. Int. Ed.* **2007**, *46*, 6408–6418. [CrossRef] [PubMed]
5. Famulok, M.; Mayer, G. Aptamer modules as sensors and detectors. *Acc. Chem. Res.* **2011**, *44*, 1349–1358. [CrossRef] [PubMed]
6. Sarvetnick, N.; Shizuru, J.; Liggitt, D.; Martin, L.; McIntyre, B.; Gregory, A.; Parslow, T.; Stewart, T. Loss of pancreatic islet tolerance induced by β-cell expression of interferon-γ. *Nature* **1990**, *346*, 844–847. [CrossRef] [PubMed]
7. Robertson, D.L.; Joyce, G.F. Selection in vitro of an RNA enzyme that specifically cleaves single-stranded DNA. *Nature* **1990**, *344*, 467–468. [CrossRef] [PubMed]
8. Tuerk, C.; Gold, L. Systematic evolution of ligands by exponential enrichment: RNA ligands to bacteriophage T4 DNA polymerase. *Science* **1990**, *249*, 505–510. [CrossRef] [PubMed]
9. Lee, J.H.; Yigit, M.V.; Mazumdar, D.; Lu, Y. Molecular diagnostic and drug delivery agents based on aptamer-nanomaterial conjugates. *Adv. Drug. Delivery. Rev.* **2010**, *62*, 592–605. [CrossRef] [PubMed]
10. Palchetti, I.; Mascini, M. Electrochemical nanomaterial-based nucleic acid aptasensors. *Anal. Bioanal. Chem.* **2012**, *402*, 3103–3114. [CrossRef] [PubMed]
11. Meyer, S.; Maufort, J.P.; Nie, J.; Stewart, R.; McIntosh, B.E.; Conti, L.R.; Ahmad, K.M.; Soh, H.T.; Thomson, J.A. Development of an efficient targeted cell-SELEX procedure for DNA aptamer reagents. *PLoS ONE* **2013**, *8*, e71798. [CrossRef] [PubMed]
12. Polonschii, C.; David, S.; Tombelli, S.; Mascini, M.; Gheorghiu, M. A novel low-cost and easy to develop functionalization platform. Case study: Aptamer-based detection of thrombin by surface plasmon resonance. *Talanta* **2010**, *80*, 2157–2164. [CrossRef] [PubMed]

13. Yuan, T.; Hu, L.; Liu, Z.; Qi, W.; Zhu, S.; Xu, G. A label-free and signal-on supersandwich fluorescent platform for Hg^{2+} sensing. *Anal. Chim. Acta* **2013**, *793*, 86–89. [CrossRef] [PubMed]
14. Du, J.; Jiang, L.; Shao, Q.; Liu, X.; Marks, R.S.; Ma, J.; Chen, X. Colorimetric detection of mercury ions based on plasmonic nanoparticles. *Small* **2013**, *9*, 1467–1481. [CrossRef] [PubMed]
15. Lin, L.; Liu, Y.; Zhao, X.; Li, J. Sensitive and rapid screening of T4 polynucleotide kinase activity and inhibition based on coupled exonuclease reaction and graphene oxide platform. *Anal. Chem.* **2011**, *83*, 8396–8402. [CrossRef] [PubMed]
16. Liu, Z.; Qi, W.; Xu, G. Recent advances in electrochemiluminescence. *Chem. Soc. Rev.* **2015**, *44*, 3117–3142. [CrossRef] [PubMed]
17. Hu, L.; Xu, G. Applications and trends in electrochemiluminescence. *Chem. Soc. Rev.* **2010**, *39*, 3275–3304. [CrossRef] [PubMed]
18. Liu, Z.; Zhang, W.; Qi, W.; Gao, W.; Hanif, S.; Saqib, M.; Xu, G. Label-free signal-on ATP aptasensor based on the remarkable quenching of tris(2,2′-bipyridine)ruthenium(II) electrochemiluminescence by single-walled carbon nanohorn. *Chem. Commun.* **2015**, *51*, 4256–4258. [CrossRef] [PubMed]
19. Yu, Y.; Cao, Q.; Zhou, M.; Cui, H. A novel homogeneous label-free aptasensor for 2,4,6-trinitrotoluene detection based on an assembly strategy of electrochemiluminescent graphene oxide with gold nanoparticles and aptamer. *Biosens. Bioelectron.* **2013**, *43*, 137–142. [CrossRef] [PubMed]
20. Merkoci, A. Electrochemical biosensing with nanoparticles. *FEBS J.* **2007**, *274*, 310–316. [CrossRef] [PubMed]
21. Qian, X.; Xiang, Y.; Zhang, H.; Chen, Y.; Chai, Y.; Yuan, R. Aptamer/nanoparticle-based sensitive, multiplexed electronic coding of proteins and small biomolecules through a backfilling strategy. *Chem. Eur. J.* **2010**, *16*, 14261–14265. [CrossRef] [PubMed]
22. Li, X.; Liu, J.; Zhang, S. Electrochemical analysis of two analytes based on a dual-functional aptamer DNA sequence. *Chem. Commun.* **2010**, *46*, 595–597. [CrossRef] [PubMed]
23. Shen, B.; Li, J.; Cheng, W.; Yan, Y.; Tang, R.; Li, Y.; Ju, H.; Ding, S. Electrochemical aptasensor for highly sensitive determination of cocaine using a supramolecular aptamer and rolling circle amplification. *Microchim. Acta* **2015**, *182*, 361–367. [CrossRef]
24. Zhang, D.-W.; Nie, J.; Zhang, F.-T.; Xu, L.; Zhou, Y.-L.; Zhang, X.-X. Novel homogeneous label-free electrochemical aptasensor based on functional DNA hairpin for target detection. *Anal. Chem.* **2013**, *85*, 9378–9382. [CrossRef] [PubMed]
25. Song, W.; Li, H.; Liang, H.; Qiang, W.; Xu, D. Disposable electrochemical aptasensor array by using in situ DNA hybridization inducing silver nanoparticles aggregate for signal amplification. *Anal. Chem.* **2014**, *86*, 2775–2783. [CrossRef] [PubMed]
26. Ding, C.; Ge, Y.; Lin, J.M. Aptamer based electrochemical assay for the determination of thrombin by using the amplification of the nanoparticles. *Biosens. Bioelectron.* **2010**, *25*, 1290–1294. [CrossRef] [PubMed]
27. Cheng, W.; Ding, S.; Li, Q.; Yu, T.; Yin, Y.; Ju, H.; Ren, G. A simple electrochemical aptasensor for ultrasensitive protein detection using cyclic target-induced primer extension. *Biosens. Bioelectron.* **2012**, *36*, 12–17. [CrossRef] [PubMed]
28. Abbaspour, A.; Norouz-Sarvestani, F.; Noori, A.; Soltani, N. Aptamer-conjugated silver nanoparticles for electrochemical dual-aptamer-based sandwich detection of staphylococcus aureus. *Biosens. Bioelectron.* **2015**, *68*, 149–155. [CrossRef] [PubMed]
29. Jie, G.; Zhang, J.; Wang, L. A novel quantum dot nanocluster as versatile probe for electrochemiluminescence and electrochemical assays of DNA and cancer cells. *Biosens. Bioelectron.* **2014**, *52*, 69–75. [CrossRef] [PubMed]
30. Luo, C.; Lei, Y.; Yan, L.; Yu, T.; Li, Q.; Zhang, D.; Ding, S.; Ju, H. A rapid and sensitive aptamer-based electrochemical biosensor for direct detection of Escherichia coli O111. *Electroanalysis* **2012**, *24*, 1186–1191. [CrossRef]
31. Ding, C.; Liu, H.; Wang, N.; Wang, Z. Cascade signal amplification strategy for the detection of cancer cells by rolling circle amplification and nanoparticles tagging. *Chem. Commun.* **2012**, *48*, 5019–5021. [CrossRef] [PubMed]
32. Zhu, Y.; Wang, H.; Wang, L.; Zhu, J.; Jiang, W. Cascade signal amplification based on copper nanoparticle-reported rolling circle amplification for ultrasensitive electrochemical detection of the prostate cancer biomarker. *ACS Appl. Mater. Interfaces* **2016**, *8*, 2573–2581. [CrossRef] [PubMed]

33. Cui, L.; Wu, J.; Ju, H. Label-free signal-on aptasensor for sensitive electrochemical detection of arsenite. *Biosens. Bioelectron.* **2016**, *79*, 861–865. [CrossRef] [PubMed]

34. Li, F.; Feng, Y.; Zhao, C.; Tang, B. Crystal violet as a G-quadruplex-selective probe for sensitive amperometric sensing of lead. *Chem. Commun.* **2011**, *47*, 11909–11911. [CrossRef] [PubMed]

35. Cui, L.; Wu, J.; Ju, H. Electrochemical sensing of heavy metal ions with inorganic, organic and bio-materials. *Biosens. Bioelectron.* **2015**, *63*, 276–286. [CrossRef] [PubMed]

36. Shumkov, A.A.; Suprun, E.V.; Shatinina, S.Z.; Lisitsa, A.V.; Shumyantseva, V.V.; Archakov, A.I. Gold and silver nanoparticles for electrochemical detection of cardiac troponin I based on stripping voltammetry. *BioNanoScience* **2013**, *3*, 216–222. [CrossRef]

37. Kokkinos, C.; Economou, A.; Petrou, P.S.; Kakabakos, S.E. Microfabricated tin-film electrodes for protein and DNA sensing based on stripping voltammetric detection of Cd(II) released from quantum dots labels. *Anal. Chem.* **2013**, *85*, 10686–10691. [CrossRef] [PubMed]

38. Liu, B.; Zhang, B.; Chen, G.; Yang, H.; Tang, D. Metal sulfide-functionalized DNA concatamer for ultrasensitive electronic monitoring of ATP using a programmable capillary-based aptasensor. *Biosens. Bioelectron.* **2014**, *53*, 390–398. [CrossRef] [PubMed]

39. Szymanski, M.; Noble, J.; Knight, A.; Porter, R.; Worsley, G. Aptamer-mediated detection of thrombin using silver nanoparticle signal enhancement. *Anal. Methods* **2013**, *5*, 187–191. [CrossRef]

40. Wan, Y.; Wang, Y.; Wu, J.; Zhang, D. Graphene oxide sheet-mediated silver enhancement for application to electrochemical biosensors. *Anal. Chem.* **2011**, *83*, 648–653. [CrossRef] [PubMed]

41. Zhu, Y.; Chandra, P.; Shim, Y.B. Ultrasensitive and selective electrochemical diagnosis of breast cancer based on a hydrazine-Au nanoparticle-aptamer bioconjugate. *Anal. Chem.* **2013**, *85*, 1058–1064. [CrossRef] [PubMed]

42. Song, W.; Li, H.; Liu, H.; Wu, Z.; Qiang, W.; Xu, D. Fabrication of streptavidin functionalized silver nanoparticle decorated graphene and its application in disposable electrochemical sensor for immunoglobulin E. *Electrochem. Commun.* **2013**, *31*, 16–19. [CrossRef]

43. Wu, L.; Xiong, E.; Zhang, X.; Zhang, X.; Chen, J. Nanomaterials as signal amplification elements in DNA-based electrochemical sensing. *Nano Today* **2014**, *9*, 197–211. [CrossRef]

44. Gao, F.; Zhu, Z.; Lei, J.; Geng, Y.; Ju, H. Sub-femtomolar electrochemical detection of DNA using surface circular strand-replacement polymerization and gold nanoparticle catalyzed silver deposition for signal amplification. *Biosens. Bioelectron.* **2013**, *39*, 199–203. [CrossRef] [PubMed]

45. Xia, F.; White, R.J.; Zuo, X.; Patterson, A.; Xiao, Y.; Kang, D.; Gong, X.; Plaxco, K.W.; Heeger, A.J. An electrochemical supersandwich assay for sensitive and selective DNA detection in complex matrices. *J. Am. Chem. Soc.* **2010**, *132*, 14346–14348. [CrossRef] [PubMed]

46. Rusinek, C.A.; Bange, A.; Warren, M.; Kang, W.; Nahan, K.; Papautsky, I.; Heineman, W.R. Bare and polymer-coated indium tin oxide as working electrodes for manganese cathodic stripping voltammetry. *Anal. Chem.* **2016**, *88*, 4221–4228. [CrossRef] [PubMed]

47. Herzog, G.; Beni, V. Stripping voltammetry at micro-interface arrays: A review. *Anal. Chim. Acta* **2013**, *769*, 10–21. [CrossRef] [PubMed]

48. Omanović, D.; Garnier, C.; Gibbon–Walsh, K.; Pižeta, I. Electroanalysis in environmental monitoring: Tracking trace metals—A mini review. *Electrochem. Commun.* **2015**, *61*, 78–83.

49. Mao, S.; Chang, J.; Zhou, G.; Chen, J. Nanomaterial-enabled rapid detection of water contaminants. *Small* **2015**, *11*, 5336–5359. [CrossRef] [PubMed]

50. Wang, J. Nanoparticle-based electrochemical bioassays of proteins. *Electroanalysis* **2007**, *19*, 769–776. [CrossRef]

51. Tian, Y.; Hu, L.; Han, S.; Yuan, Y.; Wang, J.; Xu, G. Electrodes with extremely high hydrogen overvoltages as substrate electrodes for stripping analysis based on bismuth-coated electrodes. *Anal. Chim. Acta.* **2012**, *738*, 41–44. [CrossRef] [PubMed]

52. Zhan, F.; Gao, F.; Wang, X.; Xie, L.; Gao, F.; Wang, Q. Determination of lead(II) by adsorptive stripping voltammetry using a glassy carbon electrode modified with β-cyclodextrin and chemically reduced graphene oxide composite. *Microchimica. Acta* **2016**, *183*, 1169–1176. [CrossRef]

53. Rosolina, S.M.; Chambers, J.Q.; Lee, C.W.; Xue, Z.-L. Direct determination of cadmium and lead in pharmaceutical ingredients using anodic stripping voltammetry in aqueous and DMSO/water solutions. *Anal. Chim. Acta* **2015**, *893*, 25–33. [CrossRef] [PubMed]

54. Fang, H.; Zhang, J.; Zhou, S.; Dai, W.; Li, C.; Du, D.; Shen, X. Submonolayer deposition on glassy carbon electrode for anodic stripping voltammetry: An ultra sensitive method for antimony in tap water. *Sens. Actuat. B Chem.* **2015**, *210*, 113–119. [CrossRef]
55. Bu, L.; Gu, T.; Ma, Y.; Chen, C.; Tan, Y.; Xie, Q.; Yao, S. Enhanced cathodic preconcentration of As(0) at Au and Pt electrodes for anodic stripping voltammetry analysis of As(III) and As(V). *J. Phys. Chem. C* **2015**, *119*, 11400–11409. [CrossRef]
56. Kang, W.; Pei, X.; Bange, A.; Haynes, E.N.; Heineman, W.R.; Papautsky, I. Copper-based electrochemical sensor with palladium electrode for cathodic stripping voltammetry of manganese. *Anal. Chem.* **2014**, *86*, 12070–12077. [CrossRef] [PubMed]
57. Laglera, L.M.; Santos-Echeandía, J.; Caprara, S.; Monticelli, D. Quantification of Iron in seawater at the low picomolar range based on optimization of bromate/ammonia/dihydroxynaphtalene system by catalytic adsorptive cathodic stripping voltammetry. *Anal. Chem.* **2013**, *85*, 2486–2492. [CrossRef] [PubMed]
58. Mirceski, V.; Hocevar, S.B.; Ogorevc, B.; Gulaboski, R.; Drangov, I. Diagnostics of anodic stripping mechanisms under square-wave voltammetry conditions using bismuth film substrates. *Anal. Chem.* **2012**, *84*, 4429–4436. [CrossRef] [PubMed]
59. Roy, E.; Maity, S.K.; Patra, S.; Madhuri, R.; Sharma, P.K. A metronidazole-probe sensor based on imprinted biocompatible nanofilm for rapid and sensitive detection of anaerobic protozoan. *RSC Adv.* **2014**, *4*, 32881–32893. [CrossRef]
60. Patra, S.; Roy, E.; Madhuri, R.; Sharma, P.K. Nano-iniferter based imprinted sensor for ultra trace level detection of prostate-specific antigen in both men and women. *Biosens. Bioelectron.* **2014**, *66*, 1–10. [CrossRef] [PubMed]
61. Chen, X.; Ge, L.; Guo, B.; Yan, M.; Hao, N.; Xu, L. Homogeneously ultrasensitive electrochemical detection of adenosine triphosphate based on multiple signal amplification strategy. *Biosens. Bioelectron.* **2014**, *58*, 48–56. [CrossRef] [PubMed]
62. Yan, J.-L. Determinationation of kanamycin by square-wave cathodic adsorptive stripping voltammetry. *Russ. J. Electrochem.* **2009**, *44*, 1334–1338. [CrossRef]
63. Dong, X.-Y.; Mi, X.-N.; Zhao, W.-W.; Xu, J.-J.; Chen, H.-Y. CdS Nanoparticles functionalized colloidal carbon particles: Preparation, characterization and application for electrochemical detection of thrombin. *Biosens. Bioelectron.* **2011**, *26*, 3654–3659. [CrossRef] [PubMed]
64. Zhang, H.; Jiang, B.; Xiang, Y.; Zhang, Y.; Chai, Y.; Yuan, R. Aptamer/quantum dot-based simultaneous electrochemical detection of multiple small molecules. *Anal. Chim. Acta* **2011**, *688*, 99–103. [CrossRef] [PubMed]
65. Zhou, S.; Wang, Y.; Zhu, J.-J. Simultaneous detection of tumor cell apoptosis regulators Bcl-2 and Bax through a dual-signal-marked electrochemical immunosensor. *ACS Appl. Mater. Interfaces* **2016**, *8*, 7674–7682. [CrossRef] [PubMed]
66. Qin, X.; Liu, L.; Xu, A.; Wang, L.; Tan, Y.; Chen, C.; Xie, Q. Ultrasensitive immunoassay of proteins based on gold label/silver staining, galvanic replacement reaction enlargement, and *in situ* microliter-droplet anodic stripping voltammetry. *J. Phys. Chem. C* **2016**, *120*, 2855–2865. [CrossRef]
67. Hansen, J.A.; Wang, J.; Kawde, A.-N.; Xiang, Y.; Gothelf, K.V.; Collins, G. Quantum-dot/aptamer-based ultrasensitive multi-analyte electrochemical biosensor. *J. Am. Chem. Soc.* **2006**, *128*, 2228–2229. [CrossRef] [PubMed]
68. Fan, H.; Chang, Z.; Xing, R.; Chen, M.; Wang, Q.; He, P.; Fang, Y. An electrochemical aptasensor for detection of thrombin based on target protein-induced strand displacement. *Electroanalysis* **2008**, *20*, 2113–2117. [CrossRef]
69. Li, T.; Fan, Q.; Liu, T.; Zhu, X.; Zhao, J.; Li, G. Detection of breast cancer cells specially and accurately by an electrochemical method. *Biosens. Bioelectron.* **2010**, *25*, 2686–2689. [CrossRef] [PubMed]
70. Xiang, Y.; Xie, M.; Bash, R.; Chen, J.J.L.; Wang, J. Ultrasensitive label-free aptamer-based electronic detection. *Angew. Chem. Int. Ed.* **2007**, *46*, 9054–9056. [CrossRef] [PubMed]
71. Suprun, E.; Shumyantseva, V.; Bulko, T.; Rachmetova, S.; Rad'ko, S.; Bodoev, N.; Archakov, A. Au-nanoparticles as an electrochemical sensing platform for aptamer–thrombin interaction. *Biosens. Bioelectron.* **2008**, *24*, 825–830. [CrossRef] [PubMed]

72. Suprun, E.; Shumyantseva, V.; Rakhmetova, S.; Voronina, S.; Radko, S.; Bodoev, N.; Archakov, A. Label-free electrochemical thrombin aptasensor based on Ag nanoparticles modified electrode. *Electroanalysis* **2010**, *22*, 1386–1392. [CrossRef]

73. Yang, H.; Ji, J.; Liu, Y.; Kong, J.; Liu, B. An aptamer-based biosensor for sensitive thrombin detection. *Electrochem. Commun.* **2009**, *11*, 38–40. [CrossRef]

74. Du, D.; Wang, L.; Shao, Y.; Wang, J.; Engelhard, M.H.; Lin, Y. Functionalized graphene oxide as a nanocarrier in a multienzyme labeling amplification strategy for ultrasensitive electrochemical immunoassay of phosphorylated p53 (S392). *Anal. Chem.* **2011**, *83*, 746–752. [CrossRef] [PubMed]

75. Du, D.; Zou, Z.; Shin, Y.; Wang, J.; Wu, H.; Engelhard, M.H.; Liu, J.; Aksay, I.A.; Lin, Y. Sensitive immunosensor for cancer biomarker based on dual signal amplification strategy of graphene sheets and multienzyme functionalized carbon nanospheres. *Anal. Chem.* **2010**, *82*, 2989–2995. [CrossRef] [PubMed]

76. Li, J.; Xu, M.; Huang, H.; Zhou, J.; Abdel-Halimb, E.S.; Zhang, J.-R.; Zhu, J.-J. Aptamer-quantum dots conjugates-based ultrasensitive competitive electrochemical cytosensor for the detection of tumor cell. *Talanta* **2011**, *85*, 2113–2120. [CrossRef] [PubMed]

77. Zhang, K.; Tan, T.; Fu, J.-J.; Zheng, T.; Zhu, J.-J. A novel aptamer-based competition strategy for ultrasensitive electrochemical detection of leukemia cells. *Analyst* **2013**, *138*, 6323–6330. [CrossRef] [PubMed]

78. Liu, H.; Xu, S.; He, Z.; Deng, A.; Zhu, J.-J. Supersandwich cytosensor for selective and ultrasensitive detection of cancer cells using aptamer-DNA concatamer-quantum dots probes. *Anal. Chem.* **2013**, *85*, 3385–3392. [CrossRef] [PubMed]

79. Zhao, Y.; Chen, F.; Li, Q.; Wang, L.; Fan, C. Isothermal amplification of nucleic acids. *Chem. Rev.* **2015**, *115*, 12491–12545. [CrossRef] [PubMed]

80. Zhao, W.A.; Cui, C.H.; Bose, S.; Guo, D.G.; Shen, C.; Wong, W.P.; Halvorsen, K.; Farokhzad, O.C.; Teo, G.S.L.; Phillips, J.A.; et al. Bioinspired multivalent DNA network for capture and release of cells. *Proc. Natl. Acad. Sci. USA* **2012**, *109*, 19626–19631. [CrossRef] [PubMed]

81. Chang, Y.-C.; Yang, C.-Y.; Sun, R.-L.; Cheng, Y.-F.; Kao, W.-C.; Yang, P.-C. Rapid single cell detection of Staphylococcus aureus by aptamer-conjugated gold nanoparticles. *Sci. Rep.* **2013**, *3*, 1863–1869. [CrossRef] [PubMed]

chemosensors

MDPI

Review

Aptamer-Based Electrochemical Sensing of Lysozyme

Alina Vasilescu [1,*], Qian Wang [2,3], Musen Li [3], Rabah Boukherroub [2] and Sabine Szunerits [2,*]

[1]   International Center of Biodynamics, 1B Intrarea Portocalelor, Sector 6, Bucharest 060101, Romania
[2]   Institute of Electronics, Microelectronics and Nanotechnology (IEMN), UMR CNRS 8520,
      Lille1 University, Avenue Poincaré-BP60069, Villeneuve d'Ascq 59652, France;
      qianwang0628@gmail.com (Q.W.); rabah.boukherroub@univ-lille1.fr (R.B.)
[3]   Key Laboratory for Liquid-Solid Structural Evolution and Processing of Materials,
      Shandong University, Jinan 250061, China; msli@sdu.edu.cn
*   Correspondence: avasilescu@biodyn.ro (A.V.); sabine.szunerits@univ-lille1.fr (S.S.);
      Tel.: +40-21-310-4354 (A.V.); +33-3-6-253-1725 (S.S.)

Academic Editors: Paolo Ugo and Ligia Moretto
Received: 10 March 2016; Accepted: 8 June 2016; Published: 15 June 2016

Abstract: Protein analysis and quantification are required daily by thousands of laboratories worldwide for activities ranging from protein characterization to clinical diagnostics. Multiple factors have to be considered when selecting the best detection and quantification assay, including the amount of protein available, its concentration, the presence of interfering molecules, as well as costs and rapidity. This is also the case for lysozyme, a 14.3-kDa protein ubiquitously present in many organisms, that has been identified with a variety of functions: antibacterial activity, a biomarker of several serious medical conditions, a potential allergen in foods or a model of amyloid-type protein aggregation. Since the design of the first lysozyme aptamer in 2001, lysozyme became one of the most intensively-investigated biological target analytes for the design of novel biosensing concepts, particularly with regards to electrochemical aptasensors. In this review, we discuss the state of the art of aptamer-based electrochemical sensing of lysozyme, with emphasis on sensing in serum and real samples.

Keywords: lysozyme; aptamers; surface modification; electrochemical methods

1. Introduction

1.1. Properties of Lysozyme and Its Importance for Daily Life

Discovered by Laschtschenko in 1909 [1] and named by Fleming in 1922 [2], lysozyme is a remarkable protein. Also called muramidase or peptidoglycan N-acetylmuramoyl-hydrolase, lysozyme is a ubiquitous enzyme (EC 3.2.1.17) present in various organisms [3–5], where it plays a vital role (Figure 1). Next to some of its physico-chemical features, such as a high isoelectric point of 10–11, a positive charge at neutral pH, excellent heat stability and stability in acid media [3], its antibacterial activity plays an important role in the defense system in the human body [6], lysozyme being suggestively called the body's own antibiotic.

Lysozyme was also the first enzyme whose tridimensional structure was solved [7], showing a globular structure with dimensions of $4.5 \times 3.0 \times 0.3$ nm and a relatively small molecular weight of approximately 14.3 kDa (human lysozyme is 14.6 kDa). It is found as a single polypeptide chain consisting of 129–130 amino acid residues, in which lysine is the N-end amino acid and leucine the C-end one. All of this together has made lysozyme a good model to study enzyme catalysis, protein structure and interactions [8–11] or amyloid-fibrillation formation [12–14]. Lysozyme from hen egg has become also a model protein for the pharmaceutical industry when it comes to the development of new drug delivery systems or the design of innovative treatment strategies [15–17] (Figure 1).

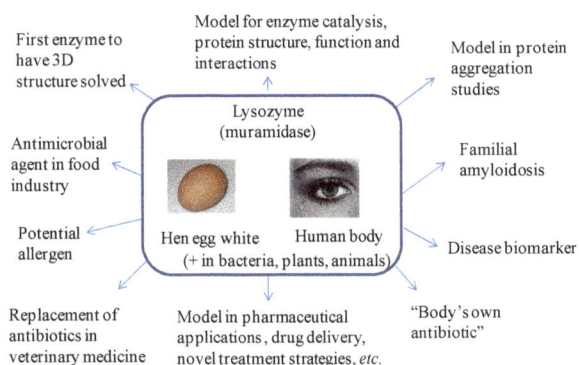

Figure 1. The relevance of lysozyme for protein science, medicine and industrial applications.

A rich and easily available source of lysozyme is the egg white of birds. In the hen egg white, lysozyme accounts for 3.5% of the total egg white proteins. Hen egg lysozyme, considered as a safe product by authorities in different countries (Austria, Australia, Belgium, Denmark, Finland, France, Germany, Italy, Japan, Spain and the United Kingdom) with antibacterial action, has been used for many years as a food preservative [4]. A major commercial use of lysozyme is in the production of some types of hard cheese, where its addition prevents the growth of butyric bacteria [18,19]. In the wine industry, lysozyme partially replaces sulfites and is added at doses of 250–500 mg/L to inhibit malolactic fermentation and to stabilize the wine afterwards [20,21]. Residual levels of 0.06–327 mg/L were found in lysozyme-treated wines [22,23], while in commercial cheeses, concentrations of 30.8–386.2 mg/kg were found to be present [24]. Lysozyme has also been used as an antibacterial agent during beer production [25], to extend the shelf-life of meat [26] and shrimp [27], as an alternative to antibiotics [28,29] in veterinary medicine or as anti-inflammatory drug in the treatment of wounds and infections [4,30].

Despite its proven utility, lysozyme can potentially trigger allergic reactions in sensitive individuals, even in trace amounts [21]. Consequently, appropriate labeling rules have been established by regulatory agencies for products containing egg white lysozyme. The importance of lysozyme sensing in real samples is furthermore associated with human lysozyme being identified as an important biomarker for several diseases [31].

In the human body, lysozyme is indeed widely distributed in tissues and body fluids with the lowest levels found in urine (1.7–123 ng/mL) [32] and cerebrospinal fluid (7.7–84 μg/L) [33] and the highest in gastric juice or mothers' milk (300 mg/L) [34] and tears (1267 ± 58 mg/L) [35]. In the serum of healthy people, concentrations in the range of 0.462–2.958 mg/L have been reported [33,35].

Increased levels of lysozyme in serum are an expression of monocyte/macrophage activity [36] and were identified in various diseases, such as AIDS [37], pulmonary tuberculosis [38], sarcoidosis [39], rheumatoid arthritis [40] and Crohn's disease [41]. Additionally, the concentration of lysozyme in cerebrospinal fluid helps distinguishing between viral and bacterial meningitis, with high lysozyme levels pointing towards bacterial meningitis [33]. Urinary lysozyme levels have become an indicator of damage to renal tubular cells with diagnostic value in monocytic or myelomonocytic leukemia [42,43]. However, only a few reports show the sensitivity and selectivity required for sensing in real biological samples, achieved using different signal amplification strategies [44–47].

1.2. Quantification Methods

The classical method for determining lysozyme is based on the ability of lysozyme to clear a suspension of *Micrococcus lysodeikticus* cells [31]. The activity of lysozyme in promoting the dissolution of the cell wall depends on its ability to catalyze the hydrolysis of the β-(1,4)-glycosidic linkage between *N*-acetyl-D-glucosamine in the polysaccharide component of the bacterial cell wall.

The decrease in turbidity correlates with the lysis of the cell wall and is recorded either by the 'lysoplate' method or by turbidimetric assays [31]. These approaches allow the quantification of <1 ng lysozyme. However, the evaluation of the enzymatic activity needs a strict control of temperature, pH and ionic strength; moreover, it is time consuming. The sensitivity of this method is strongly dependent on the assay conditions and the presence of other substances in the samples that affect the interaction of lysozyme with the medium. More recently, we developed a novel assay to sense lysozyme in serum using *Micrococcus lysodeikticus*-modified graphene oxide-coated surface plasmon resonance (SPR) interfaces [48]. In this approach, graphene oxide (GO) is integrated onto gold interfaces using layer-by-layer deposition of a polycationic polymer and GO. Adsorption of whole cells of *Micrococcus lysodeikticus* onto GO and blocking remaining GO sites with BSA (3%) gave a reproducible non-fouling surface for the sensing of lysozyme in serum samples. The detection mechanism is based on a sensitive monitoring of lysozyme-dependent desorption of *Micrococcus lysodeikticus* cells from the sensor surface. This desorption, together with a significant change in morphology of the bacterial cell, causes a characteristic decrease of the SPR signal with a limit of detection of 0.05 $\mu g \cdot mL^{-1}$ [48].

Other methods based on the quantification of the protein amount and not its enzymatic activity have been reported, including electrophoretic [49], chromatographic [50,51] and immunoenzymatic methods [52]. The ELISA technique is particularly promising for its high sensitivity, high specificity and convenience, especially for the analysis of a large number of samples [53,54] with a detection limit around 0.2 μg lysozyme/L.

Next to these approaches, in recent years, a few aptamer-based analytical methods have been developed towards lysozyme recognition and detection [47,55–63]. Aptamers are small single-stranded oligonucleotides (DNA or RNA) that are selected *in vitro* to bind with high affinity and specificity any selected target of choice, from small-sized, such as metal ions [64], to large ones, such as cells [65]. Since aptamers possess numerous advantages over antibodies, such as high stability, resistance to denaturation and degradation, as well as easy modification possibilities, they have found widespread uses for bioanalysis [56,66,67]. One of the particularities of aptamers is the fact that their affinities are not affected by labeling, offering thus generous opportunities for protein sensor designs with improved analytical performances.

The ionic charge of lysozyme renders the protein suitable for the specific binding to DNA aptamers. The first publications dealing with anti-lysozyme aptamers, selected by systematic evolution of ligands by exponential enrichment (SELEX) are from Ellington and co-workers [68–70] (Table 1). Based on these results, Kirby *et al.* [70] have developed a reusable bead-based electronic tongue sensor arrays of anti-lysozyme aptamers for the detection of proteins, where fluorescence labeling is involved. Using capillary electrophoresis-SELEX, a new lysozyme aptamer characterized by an order of magnitude increased affinity was selected (Table 1) [71]. Based on these new bioreceptors, electrochemical sensors emerged as alternative devices able to offer alternative lysozyme sensing strategies. Electrochemical biosensors represent a commercially-proven concept, capable of delivering sensitive, miniaturized and cost-effective detection of relevant analytes. Together with thrombin, lysozyme is widely used as a model analyte for developing new electrochemical aptasensor assays. The opportunities brought by electrochemical aptasensors for the bioanalysis of lysozyme are discussed in this review, putting emphasis on their use for practical applications, particularly for lysozyme analyses in serum.

Table 1. DNA sequence of different lysozyme aptamers.

DNA Sequence	K_d/nM	Reference
5′-ATCAGGGCTAAAGAGTGCAGAGTTACTTAG-3′	31	[68]
5′-GGGAATGGATCCACATCTACGAATTCATCAGGGCTAAAGAG TGCAGAGTTACTTAGTTCACTGCAGACTTGACGAAGCTT-3′	29 ± 5	[70]
5′-GCAGCTAAGCAGGCGGCTCACAAAACCATTCGCATGCGGC-3′	2.8 ± 0.3	[71]

2. Aptamer-Based Electrochemical Lysozyme Sensors

The main characteristics of the aptamers used in lysozyme biosensors are summarized in Table 1, starting from the first examples proposed by Ellington and co-workers [68,69], followed by Tran *et al.* [71]. By using electrochemical impedance spectroscopy (EIS) on screen-printed carbon electrodes [63], it was demonstrated that the lowest detection limit and wider linear range can be achieved using the aptamer proposed by Tran *et al.* [71]. More recent studies showed that using a mixture of two aptamer sequences with different affinities allows an improved control of the sensitivity and linear range of aptasensors [72]. Taking into account these developments, future aptasensors should include parallel experiments with more than one aptamer sequence present on the biosensor surface for achieving optimized sensing.

Various electrochemical aptasensors, based on different assay formats, aptamer construction strategies and detection methods have been developed for lysozyme sensing in samples, such as egg white, wine, saliva, urine, serum and cancer cells, as summarized in Table 2 (see also the relevant abbreviations' list at the end of the paper).

Table 2. Electrochemical aptasensors developed for lysozyme sensing.

Sample	Material	Method	LoD *	Linear Range	Comments	Reference
Direct Assays						
Serum	VANCNT/NA/LBA	DPV	100 fM	0.1–7 pM	2.5% decrease in signal after 2 weeks at 4 °C in buffer; RSD: 2.3%	[73]
Egg white	Au/TiO$_2$/3D-rGO/PPy/LBA	DPV	5.5 pM	0.007–3.5 nM	90% of initial signal after 1 month; RSD: 5.45%	[74]
Egg white	Au/TiO$_2$@PPAA/LBA	EIS	1.04 pM	3.5 pM–7 nM	-	[75]
Egg white	SPCE/AuNPs/LBA	SWV	21 fM	0.07–3.4 pM	RSD: 4.2%	[60]
Egg white	Au/AuNPs/LBA	EIS	0.01 pM	0.1–500 pM	84% of the original signal after 1 month in buffer at 4 °C; RSD: 2.11% (*n* = 3)	[76]
Wine	SPCE/LBA1 and LBA2	EIS	25 nM	0.025–0.8 μM	Stable several days stored dry at 4 °C; RSD: <3.8%	[63]
Chicken egg + saliva	GCE/chitosan-GR/LBA	EIS	6 fM	0.01–0.5 pM	-	[77]
Saliva + urine + plasma	Au/Cu$_2$O@rGO@PpPG	DPV	pM	0.1–200 nM	96.5% of initial activity after 15 days in buffer; RSD: 4.8%	[78]
Egg white + serum	GCE/THH Au NCs/APT	SWV	0.1 pM	0.1 pM–10 nM	7.7% decrease in signal after storage in buffer at 4 °C for 23 days	[79]
N/A	PGE/chitosan–GO/LBA [1]	EIS	28.53 nM	-	Stable 1 week at 4 °C; RSD% = 9.6%	[61]
N/A	ITO/PABA/SA/LBA	EIS	14 nM	-	-	[80]
N/A	CPE/LBA	SWV	18 nM (adenine) 36 nM (guanine)	0.06–1.4 μM (adenine) (0.11–1.4 μM (guanine)	RSD: 5.1% (guanine) and 6.8% (adenine)	[81]
N/A	MWCNT-SPE/LBA	EIS	862 nM	-	-	[82]
N/A	GR-GCE/LBA [2]	DPV	0.08 nM	0.2 nM–1040 nM	4.55% decrease in signal after storage at 4 °C for 10 days; RSD: 4.23%	[83]
N/A	Fe$_2$O$_3$-GR-GCE/LBA	EIS	11.1 pM	35 pM–350 nM	4.48% decrease in signal after storage at 4 °C for 10 days; RSD: 4.23%	[84]
N/A	GCE/O-GNs/LBA	DPV	1 pM	5.0 pM–0.7 nM	-	[85]
N/A	MB/LBA **	CPSA	7 nM	-	-	[86]

Table 2. *Cont.*

Sample	Material	Method	LoD *	Linear Range	Comments	Reference
N/A	Au/LBA	CV	-	35 nM–3.5 µM	-	[87]
N/A	ITO/(Fc-PEI/CNTs/Fc-PEI/LBA)₃	DPV	11.8 pM	13.9 pM–116 nM	7.5% decrease after 24 days at room temperature in air; 2.25% increase after 21 days in distilled water at 4 °C	[88]
Sandwich Assay						
Wine	SPCE/LBA/Lysozyme/B-AB/SA-ALP	DPV	4.3 fM	5 fM–5 nM	Stable 2 weeks at 4 °C; RSD: 5.5%	[44]
Competitive Assays						
Serum	Au/CD/DLAP1 + DLAP2	DPV	64 pM	100–1000 pM	-	[89]
Serum	Au/MeB-cDNA/LBA	SWV	16.4 pM	0.1–100 nM	Stable 3 weeks at 4 °C; RSD: <5%	[90]
Urine	Au/LBA-(DNA-Fc)	SWV	0.45 nM	7–30 nM	7.7% decrease after storage in buffer at 4 °C; for 23 days	[91]
Egg white	Au/LBA/TCA/AuNPs/cDNA	CV	0.1 pM	5 pM–1 nM	84% of the original signal after one month at 4 °C; RSD: <4.3% (n = 5)	[46]
Egg white	Au/cDNA/LBA	LSW	1 pM	1.0 pM–1.1 nM	RSD < 4.2% (n = 5)	[92]
Ramos cancer cells	DNA machine, CdS NP–DNA/LBA **	DPASV	0.52 pM	1 pM–80 nM	RSD < 6.1% (n = 3)	[93]
N/A	Au/TBA and LBA/(PbS-Lys and CdS-Thr)	SWV	-	75% signal decrease for 0.07 nM	-	[94]
N/A	Au/DNA1/BiDNA/DNA3-AuNPs	CV	0.7 nM	-	Stable for 2 weeks in distilled water at 4 °C RSD: 4.6%	[95]
N/A	Au/p-ATP-AuNPs/(LBA/Fc-cDNA)	SWV	0.1 pM	0.1 pM–1 nM	15% decrease in original signal after 1-month in buffer solution	[47]
N/A	Au/cDNA/LBA	EIS	70 pM	0.2–4.0 nM	RSD: 3.7%	[62]
N/A	GCE/Au/(Fc-cDNA/LBA TWJ)	SWV	0.2 nM	0.2–100 nM	I_{on} and I_{off} decreased by 7.9% and 18.5% after 2 weeks	[96]

* A molecular weight of 14.3 kDa for lysozyme was used to convert concentration units from g/L into M; ** not an aptasensor, but an aptamer-based assay with electrochemical detection; N/A: measurements done with lysozyme standard solutions in buffer.

2.1. Surface Immobilization of Aptamer Ligands

The immobilization of the analyte-specific ligand on the sensing interface is a key step to achieve selective detection [97]. The use of an efficient immobilization method is crucial to ensure adequate stability and optimum surface coverage, however, carefully avoiding lowering the affinity of the aptamer for its target. To ensure that K_d is not affected upon aptamer immobilization, the effective dissociation constant of the immobilized aptamer-lysozyme complex should remain constant. For instance, in [90], for the immobilized aptamer, a K_d value of 30 nM was measured, which compared well with that measured in solution, namely 31 nM [68].

The immobilization strategies used for aptamers for lysozyme are the following:

1. adsorption or π-π stacking interactions between the DNA bases of the lysozyme aptamer and graphene oxide-modified interfaces (Figure 2A) [83–85]
2. covalent linkage of the aptamer to carboxylic-acid functions present on the electrode surface (Figure 2B) [44,63,77]
3. binding of thiolated aptamers to gold electrodes or particles (Figure 2C) [46,62,76,87,94]
4. click chemistry between azide-modified gold particles and alkyne-terminated aptamers (Figure 2D) [60]
5. electrostatic interactions between the negatively-charged phosphate backbone of the aptamer and positively-charged materials (Figure 2E), such as polypyrrole, Fe_2O_3 and ferrocene-appended poly(ethyleneimine) (Fc-PEI) in a layer-by-layer approach or amine-rich films of plasma-polymerized propargylamine in a $Cu_2O@rGO@PpPG$-modified gold electrode [74,78,84,88]

6. affinity binding based on biotin-avidin [80] or host-guest interactions [89] (Figure 2F)
7. hybridization to a partially complementary DNA strand, previously immobilized on the electrode surface (Figure 2G) [96]

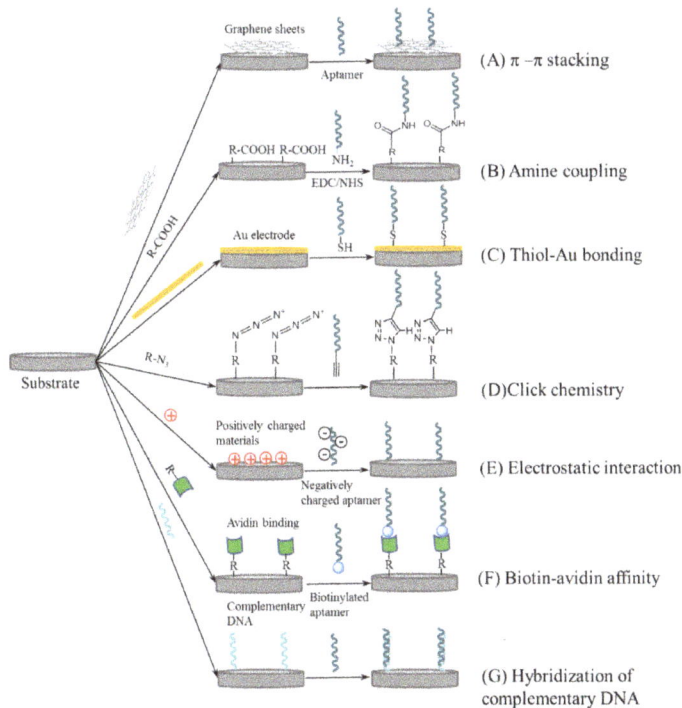

Figure 2. Schematic illustration of different strategies for the modification of electrical interfaces with aptamers.

For Case (1), polarization of carbon paste electrodes (CPE) at +0.5 V for 2 min [81] or modification via drop casting with reduced graphene oxide (rGO) [83], rGO/Fe$_2$O$_3$ or rGO/Orange II nanocomposites [84,85] results in interfaces with the ability to strongly interact with aptamers via π-π stacking.

For Strategy (2), an appropriate number of carboxylic functional groups can be introduced on the electrode surface by modification with nanocomposites of graphene oxide-chitosan [61,77], or reduction of diazonium salts of *p*-aminobenzoic acid on carbon electrodes [44,63], or by attaching vertically-aligned nitrogen-doped carbon nanotubes (VA-NCNTs) [73]. Rohrbach *et al.* modified screen-printed graphite electrodes with multi-walled carbon nanotubes (MWCNTs) containing about 5% carboxylic acid groups [82]. Another strategy relied on covering a Au electrode with hollow TiO$_2$ spheres and a film of polyacrylic acid [75], where polyacrylic acid provided the carboxyl groups for aptamer immobilization, while the hollow TiO$_2$ spheres enabled increased surface area for higher aptamer loading and improved electrical characteristics.

The majority of electrochemical aptasensors for lysozyme used thiolated aptamers on gold electrode, that is following Strategy (C) [46,62,87,91,94]. To increase the aptamer loading, a popular approach relies on modifying the electrodes with Au nanoparticles [47,76].

By using click chemistry (Strategy D), Xie *et al.* used AuNPs electrodeposited on screen-printed graphite electrodes and modified with a self-assembled monolayer of 10-azidodecane-1-thiol [60], to form a "clickable" interface with alkyne-terminated lysozyme aptamers (see Figure 3A). Compared to direct immobilization of thiolated aptamers on the AuNPs-coated electrode, this approach allowed a 3.4-fold increase in the analytical signal for a concentration of 10 pg/mL lysozyme, so increasing the sensitivity of the sensor. The higher aptamer coverage (3.3×10^{11} molecules cm^{-2} for aptamer "clicked" on SAM *versus* 2.2×10^{11} molecules cm^{-2} for simple chemisorption of thiolated aptamer) was claimed to be the origin of the improved performances [60]. The potential of the "clickable" interface as generic support for an aptasensor array was further illustrated by efficient immobilization of three different aptamers (for lysozyme, cocaine and thrombin) [60].

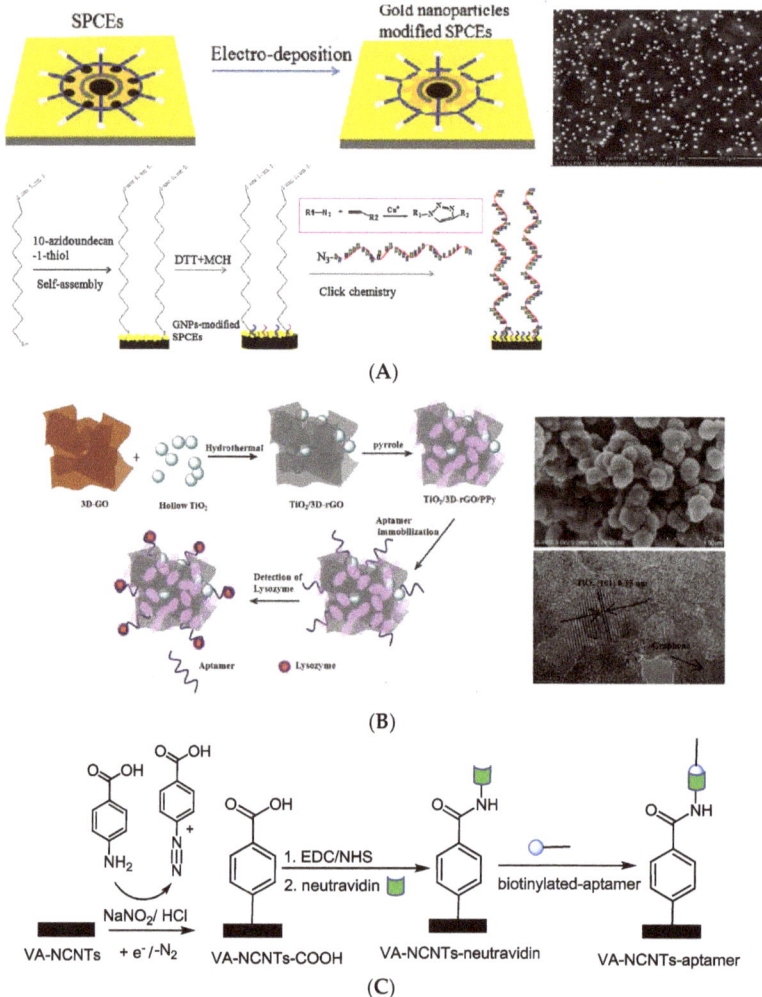

(A)

(B)

(C)

Figure 3. *Cont.*

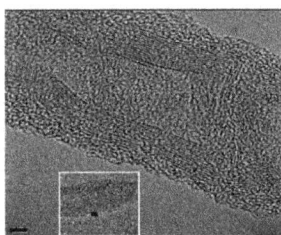

(**D**)

Figure 3. (**A**) Schematic of screen-printed electrodes modified with gold nanoparticles and subsequent surface modification (reprinted with permission from [60]); (**B**) schematic of TiO$_2$/3Dgraphene/polypyrrole aptasensor for lysozyme detection together with SEM and TEM images of the nanocomposite (reprinted with permission from [74]); (**C**) surface modification of the VA-NCNT electrode through electroreduction of 4-carboxyphenyl diazonium salt formed *in situ* followed by covalent immobilization of neutravidin and interaction with the biotinylated lysozyme aptamer; (**D**) HRTEM image of modified VA-NCNTs (scale bar 2 nm) [73].

More recently, electrodes modified with a TiO$_2$/3D graphene/polypyrrole nanocomposite [74] proved to be an interesting alternative for lysozyme sensing. The aptamer was immobilized via a combination between π-π stacking interactions with 3D graphene (Figure 2A) and adsorption onto the positively-charged polypyrrole (Figure 2E). The hollow TiO$_2$ spheres provided a large surface, porous support for 3D-graphene and polypyrrole, acting synergistically with these two materials to facilitate the immobilization of significant amounts of aptamer (Figure 3B).

Affinity interactions (Case F) were exploited for attaching a labeled lysozyme aptamer to an electrode either firmly, by taking advantage of strong biotin-avidin binding (Figure 2F) or more loosely via host-guest interactions. In the case of VA-NCNTs, a detection limit as low as 100 fM with a linear range up to 7 pM could be achieved by first grafting 4-carboxyphenyl radicals to the VA-NCNTs using a standard procedure involving the reduction of 4-carboxybenzenediazonium cations to generate aryl radicals (Figure 3C), followed by neutravidin linking and integration with biotinylated lysozyme aptamers. Interestingly, this strategy results in a functionalized layer both on the lateral sides and tip (Figure 3D) of the carbon nanotubes, suggesting that VA-NCNTs are uniformly modified throughout their length, which might be one of the reasons for their high sensitivity towards lysozyme.

A weaker binding between a dabcyl-labeled aptamer and a β-cyclodextrin-modified electrode via host-guest interactions (Figure 2G) laid the basic principle for a multiple use aptasensor, where easy regeneration of the aptasensor surface was achieved simply by re-incubation with dabcyl-labeled aptamer after each lysozyme measurement [89].

2.2. Electrochemical Assay Formats: Direct, Sandwich and Competitive Assays

In order to translate the lysozyme-aptamer binding event into a measurable electrochemical signal, various techniques have been employed, including cyclic voltammetry (CV), differential pulse voltammetry (DPV), square wave voltammetry (SWV) or electrochemical impedance spectroscopy (EIS) using direct, sandwiched or competitive assays (Figure 4). Indeed, the formation of a lysozyme-aptamer complex leads to changes in the electrical properties at the sensor/solution interface, which translate into either an improvement or a hindered access of a reporter probe to the electrode surface. This is reflected quantitatively in a change in the charge transfer resistance at the interface, which can be determined by EIS, or a change in the magnitude of the oxidation/reduction current of the reporter probe, measurable by voltammetric methods (Figure 4). Voltammetric methods, such as SWV or DPV, are generally more sensitive [47]. They have also the advantage of delivering a unique "signature" of the electroactive species used as reporters based on their oxidation/reduction potential. Provided that their oxidation/reduction potentials are separated enough so that their electrochemical signals can be resolved, several electroactive labels have been used to detect multiple proteins with the same sensor. For instance, aptasensors for dual detection of lysozyme and interferon gamma [90] or thrombin [89,94] have been developed based on this principle.

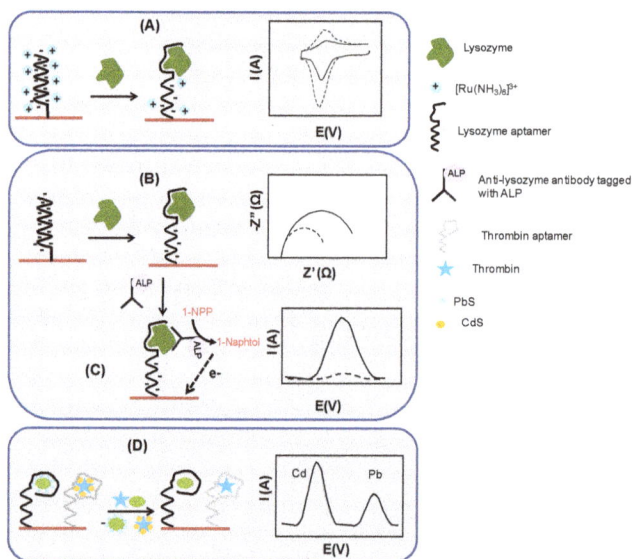

Figure 4. Electrochemical aptasensors' detection schemes of lysozyme: (**A**) direct assay by recording the conformational changes of surface-linked aptamers upon lysozyme binding, which results in a decrease in electrostatically-bound $Ru(NH_3)_6^{3+}$ detectable by CV (e.g., as illustrated in [87]); (**B**) formation of the aptamer-lysozyme complex creates a barrier for the electron transfer of $(Fe(CN)_6)^{4-/3-}$ in solution proportional to lysozyme concentration and detectable by EIS [63]; (**C**) sandwich assay for lysozyme analysis using amplification with a lysozyme antibody labeled with alkaline phosphatase [44]; (**D**) competitive assay where free lysozyme in solution displaces quantum dot-tagged lysozyme, previously bound to surface-immobilized aptamer (e.g., as illustrated in [94]).

Positively-charged ruthenium hexamine used in [87] (Figure 4A) and negatively-charged ferricyanide are widely investigated for reporting on aptamer recognition events. Ruthenium hexamine binds electrostatically to the negatively-charged phosphate backbone of DNA. The magnitude of the reduction peak of $[Ru(NH_3)_6]^{3+}$ can be used to determine the amount of surface linked aptamer [87] and additionally can be a direct measure for lysozyme: the difference in current intensity due to $[Ru(NH_3)_6]^{3+}$ reduction before and after sensor incubation with lysozyme correlates with the concentration of lysozyme in the sample. Figure 5A shows one example of the direct detection of lysozyme on aptamer-modified interfaces using EIS [63]. The change in the electrical properties at the sensor/solution interface was exploited for a comparison of the analytical performances of two aptamers and quantitative detection with detection limits of 25 nM for the aptamer selected by Tran *et al.* [71] and 100 nM for that reported by Cox and Ellington [68]. To assemble the aptasensors, screen-printed graphite electrodes were modified with diazonium salts produced *in situ* in order to introduce carboxylic groups on the electrode surface. Next, amine-ended aptamers were covalently immobilized on the surface. A comparison of the aptasensors revealed that both can be applied to lysozyme analysis in wine. However, somewhat improved sensitivity and a wider linear range were observed using the aptamer selected by Tran *et al.* [71] (linear range: 0.025–0.8 µM) with respect to the aptamer selected by Cox and Ellington [68] (linear range: 0.1–0.8 µM).

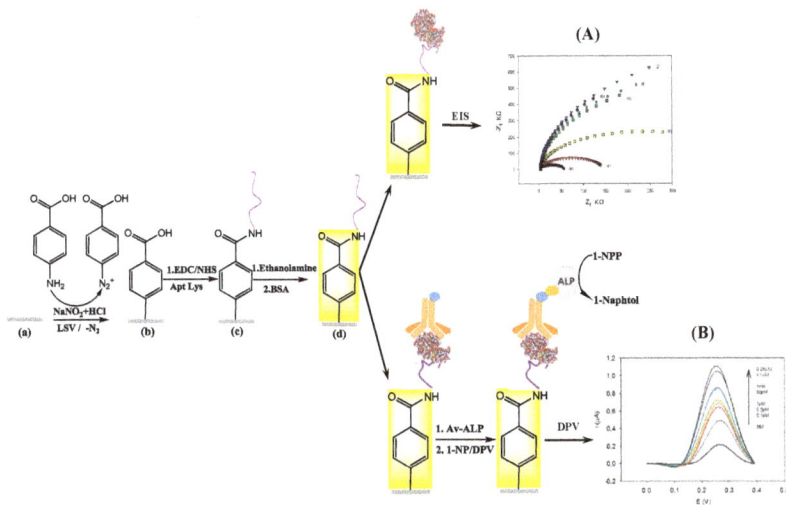

Figure 5. (**A**) Surface modification scheme and Nyquist plots for bare electrode (a), modified with diazonium salt (b), after EDC/NHS (c), functionalization with aptamer (d), after ethanolamine/BSA blocking (e) and after interaction with lysozyme (f) using PBS solutions containing $[Fe(CN)_6]^{4-/3-}$ (2 mM) [63]; (**B**) surface modification and DPV signal after incubation with anti-lysozyme antibody and avidin-ALP coupling via biotin-avidin affinity for lysozyme sensing (reprint with permission from [44]).

Although simple, practical and label-free, these aptasensors lack the required sensitivity for more demanding applications, such as the detection of lysozyme in biological samples or of trace amounts of lysozyme in wine. Moving forward from the above direct assay, a sandwich between the surface linked aptamer, the captured lysozyme and a biotinylated anti-lysozyme antibody has been recently proposed to enhance the sensitivity of the sensor (Figure 4C) [44]. Labeling the assembly with avidin-modified alkaline phosphatase ("ALP") and the addition of the enzyme specific substrate 1-naphtyl phosphate ("1-NPP") (Figure 5B) allow the indirect quantification of lysozyme, by recording the current response upon the formation of the electrochemically-active product, 1-naphtol. This assay exhibited a detection limit of 4.3 fM [44], a substantial improvement from the limit of detection of 0.1 µM reported previously for the direct assay by EIS [63]. This improvement was due mainly to the power of signal amplification based on enzymatic labels, but also to the higher sensitivity of DPV compared to EIS. The gain in sensitivity comes with a cost in complexity, price per assay and inconvenience associated with limited enzyme stability; these should be all considered before choosing the "right" aptasensor design for a particular application.

Next to the direct and sandwich assays, competitive detection schemes (Figure 4D) allowed sensitive lysozyme sensing. The sensing principles applied rely on:

1. displacement of partially complementary and labeled DNA (ferrocene, AuNPs, *etc.*) from surface linked lysozyme binding aptamer (LBA) by lysozyme (signal-off sensor) [46,47,91,98]
2. displacement of dabcyl and metallic NPs-labeled lysozyme binding aptamers forming a host-guest complex with cyclodextrin in the presence of lysozyme and subsequent release of NPs in solution [89]
3. displacement of LBA from its methylene blue-tagged DNA complex in the presence of lysozyme, resulting in a conformational change of methylene blue-tagged DNA into a hairpin structure; this brings methylene blue closer to the electrode surface, leading to an increase of its signal (signal-on sensor) [90]

4. displacement of LBA from its complex with DNA upon lysozyme addition (signal-off sensor) [46,62,95]
5. electrochemical stripping of lysozyme/quantum-dots complex [94]
6. desorption of lysozyme aptamer from rGO/Orange II-modified GCE, reversing the blocking effect and reestablishing efficient electron transfer from graphene-adsorbed aromatic dye Orange II [85]

Competitive detection schemes are typically used with regenerable sensors, where in order to revert to the characteristics of the original interface, the aptasensor needs simply to be re-incubated in fresh aptamer solution at the end of each measurement [62,89]. An ingenious detection strategy for multiple protein detection relied on immobilization of dabcyl-labeled-aptamer-modified metal nanoparticles (DLAPs) on a β-cyclodextrin-modified electrode by host-guest affinity [89]. Thrombin and lysozyme aptamers were labeled at the 3′ end with dabcyl, which is a typical "guest" for the β-cyclodextrin "host". At the 5′ end, the thrombin and lysozyme aptamers were labeled with CdS and PbS nanoparticles, respectively. The aptamers were captured at the electrode surface by host-guest interaction through their dabcyl label. Upon binding the target proteins, the aptamers were released from the β-cyclodextrin-coated surface. The detection of bound proteins was done through dissolution and electrochemical analysis of nanoparticle labels. Successful regeneration of initial aptasensing surface was demonstrated for eight regeneration cycles with no loss in activity, simply by incubating the sensors in 0.1 mM DLAP solution for 30 min. Moreover, the applicability of the aptasensor in the biomedical field was demonstrated by the analysis of human serum samples.

Note that competitive assay formats are particularly useful for devising generic interfaces that could be adapted to various analytes, allowing one to exploit generic amplification/detection schemes. All multi-analyte detection schemes developed so far that include lysozyme detection are based on competitive testing [89,90,94,95]. One of the most interesting approaches makes use simultaneously of both a "signal-on" and a "signal-off' mechanism for dual detection of lysozyme and interferon gamma (IFN-γ) [90]. The aptasensor is intended as an analytical tool in the diagnosis of acute leukemia and is built by immobilizing on a Au electrode two reporter DNA probes, one labeled with ferrocene (Fc) and the other one with methylene blue (MB). The Fc-probe and the MB-probe are designed to specifically hybridize with the IFN-γ and lysozyme aptamer, respectively. In the presence of target proteins IFN-γ and lysozyme, the aptamer is released from the respective DNA duplex, and the reporter probes suffer conformational changes, leading to decreased oxidation current for Fc and increased oxidation current for MB, according to a "signal-off" and a "signal-on" mechanism, respectively.

Sensor arrays with hybridized DNA probes allow fulfilling the requirements of high-throughput and multi-analyte testing in the biomedical field. Consequently, starting from the aptasensors for dual analyte detection of lysozyme and adenosine, thrombin and interferon gamma [89,90,94,95], the development of novel sensor arrays in conjunction with competitive detection schemes can be anticipated for the near future for the simultaneous electrochemical detection of various analytes, including lysozyme.

2.3. Evaluation of Selectivity, Reproducibility and Storage Stability

The selectivity of the electrochemical aptasensors was proven by challenging them with excess quantities of other proteins (e.g., bovine serum albumin, thrombin, immunoglobulin G, hemoglobin, myoglobin, casein, cytochrome C) tested either in individual solutions or in mixtures with lysozyme [44,46,61,74,76,77,79,82,84,89–91]. Some authors went further and proved also the selectivity of the lysozyme aptamer sequence by performing measurements with an identical sensor in which the aptamer was replaced by a scrambled DNA sequence [81,82,87]. For example, the current decrease in the adenine oxidation signal, taken as the analytical signal in the aptasensor developed by Rodriguez and Rivas [81], was only 2.7% with a scrambled sequence as compared to 87.6% for the lysozyme aptamer [81].

Most sensor platforms reported in the literature showed a good repeatability. In a typical example, the relative standard deviation of the determined charge transfer resistance was 4% for six replicate measurements of 5 µg/mL lysozyme, using the same impedimetric aptasensor [80].

Aptasensors' reproducibility was also good as proven by an RSD of around 5% for several sensors fabricated in parallel [80]. These results are particularly important for complex sensors involving nanomaterials/nanocomposites and manual steps of preparation. For example, an RSD of 4.23% (n = 6) was reported for sensors based on GCE modified with graphene/Fe$_2$O$_3$ and aptamer [84], while an RSD of 5.45% (n = 10) was obtained for Au electrodes modified with a nanocomposite of hollow titanium dioxide nanoballs, three-dimensional reduced graphene oxide and polypyrrole (TiO$_2$/3D-rGO/PPy) [74].

Moreover, most aptasensors reported satisfactory storage stability with less than a 16% decrease in the analytical signal at 4 °C over 2 weeks–1 month [44,47,74,76,91]. While in some cases, sensor development targeted single-use [63], there were a few aptasensors for which multiple regeneration cycles were performed without affecting the magnitude of the analytical signal (less than 10% decrease for 3–5 regenerations) [46,62,89,90]. The reproducibility, storage stability and regeneration figures advanced in the literature for lysozyme aptasensors support their further development for commercial applications.

2.4. Comparison of Electrochemical Lysozyme Sensors to Other Detection Schemes

To put the performance of electrochemical aptasensors for lysozyme into perspective, Table 3 summarizes the sensing characteristics of other lysozyme biosensing platforms. When compared to all of the different electrochemical assays (Table 2), it is clear that electrochemical platforms are complementary to optical methods, such as SPR and even MALDI-TOF analysis. Surface-enhanced Raman spectroscopy (SERS) has currently the lowest reported detection limit, being 1 aM. The best performing electrochemical sensor showed a detection limit of 4.3 fM [44] and is based on a sandwich format (Figure 3C). Interestingly, DPV on GCE modified with AuNPs and an iminodiacetic acid ligand proved to be also highly sensitive [98].

Table 3. Sensitivity of methods besides aptamer-based electrochemical ones for the detection of lysozyme.

Detection Method	Ligand	Limit of Detection	Reference
SPR	Aptamer	0.5 nM	[55]
DPV	IDA–Cu complex	60 fM	[98]
MALDI-TOF MS	Aptamer	1 nM	[99]
RLS	Aptamer	1 pM	[100]
ELISA	Antibody	0.1 nM	[32]
SERS	Aptamer	1 aM	[101]
Turbidimetry	*Micrococcus lysodeikticus*	0.13 nM	[102]
HPLC-FLD	-	10 nM	[103]

3. Applications of Current Electrochemical Aptasensors for Lysozyme Sensing

Choosing the optimum electrochemical detection for a particular practical application requires consideration of lysozyme levels typically encountered in the targeted type of sample, in conjunction with aptasensor simplicity, stability and any requirement for sensor regeneration. For clinical applications [73], testing the performance of the different sensors on real samples is of ultimate necessity. Approximately half of the electrochemical aptasensors for lysozyme developed so far were applied for real sample analysis (Table 2). Interestingly, all of these examples were reported in the last five years, in line with the general trend of increased focus on applications of electrochemical aptasensors observed, among others with aptasensors devoted to food safety. It is important to note that the levels of lysozyme in various types of real samples vary over a wide range, which are, however, within the detection range of aptasensors (Table 4). This proves the general applicability of the lysozyme aptasensing concept for a variety of samples.

Table 4. Level of lysozyme determined using electrochemical aptasensors.

Sample Analyzed	Lysozyme/μM	Reference
serum	0.62–0.66	[90]
serum (healthy patient)	0.22	[73]
serum (IBD patient)	0.78	[73]
saliva (3 patient samples)	5.17	[77]
urine	0.01–0.03	[91]
egg white	218	[60]
egg white	239	[77]
egg white	263	[46]

In most cases, egg white was the preferred matrix to prove the aptasensor's real-life applicability. Only two recent studies include the detection of lysozyme in spiked serum [89,90], while other reports concern saliva [77], urine [91], Ramos cancer cells [93] or wine [44,63]. This is linked to the complexity of samples: besides possible interfering compounds evaluated at an early stage during aptasensor development in buffer solutions, other challenges encountered in real matrices concern non-specific adsorption of sample components to the sensor surface; moreover, the high ionic strength of some samples can affect negatively the binding between lysozyme and aptamers. Various solutions have been devised for circumventing such problems. Efficient blocking of nonspecific adsorption is typically performed by incubation of Au-based aptasensors with short hydroxylated thiols, such as 6-mercapto-1-hexanol [60,87,90], while, for carbon-based interfaces, blocking with 0.1%–1% solutions of BSA [44,63,88] or IgG [81] is preferred. Moreover, a washing step with buffer (sometimes containing 1% Tween-20, [87,95]) performed after incubation of the aptasensor with the sample solution further alleviates problems due to non-specific binding [44,46,60]. To ensure reliable and quantitative determination of lysozyme in real samples, the simplest approach is to dilute samples with suitable electrolyte buffer in order to match the linear range of the aptasensor. However, this comes at the expense of possible errors associated with huge dilution factors. For example, many authors claimed high sensitivity for lysozyme detection, using egg white as the test sample. However, one could note that egg white contains the highest amount of lysozyme among all samples summarized in Table 3, namely around 240 μM (3.4 mg/mL). Considering an aptasensor for which the upper linearity limit is set to 0.5 pM [77], the estimated sample dilution factor required for lysozyme analysis is 4×10^8. Zhang et al. [75] compared the results obtained for several dilutions of egg white across the linear range of the aptasensor. Non-specific binding was observed for samples diluted to 50 ng/mL or higher concentrations, while diluting the sample to 1 or 10 ng/mL (dilution factor of 3×10^5 or higher) led to agreement between aptasensor responses for egg white and standard lysozyme solutions, respectively. As per the available literature data, dilution with buffer does represent an appropriate sample preparation step for serum, saliva or urine prior to analysis with the electrochemical aptasensors [78]. Instead, in wines, lysozyme binds to phenolic compounds [104], and a pre-treatment of wine with salt and surfactant was required to dissociate the complex. In another example, Ramos cancer cells analyzed by an aptamer-based electrochemical assay required trypsinization, centrifugation, washing, lysis and filtration through a 0.22-μm membrane before lysozyme analysis [93]. What is still lacking to make these sensors accepted by clinicians and analytical laboratories is that besides some notable exceptions [78,91], the studies did not include any comparison to standardized procedures, such as ELISA or HPLC, to confirm the results and validate the assay, and the accuracy was estimated solely on spiked samples and by comparison with levels reported in the literature for various types of samples.

4. Conclusions and Perspectives

Since the first aptamer for lysozyme had been proposed in 2001, a large amount of different electrochemical-based aptasensors have been developed over the years. Routinely, such sensors achieve a picomolar detection limit, with some even reaching the low femtomolar concentration range.

There is an urgent need for moving beyond research by developing new concepts for achieving even better sensitivity and selectively, in order to bring some of the current sensors into real biomedical applications. Clearly, next steps in this direction might include:

(i) Sensor designs and use of materials compatible with large-scale manufacturing technologies for producing commercial aptasensors. The good analytical characteristics and reproducibility of lysozyme aptasensors produced by manual, multiple step procedures is promising. Several types of electrodes modified with proteins, mediators and nanomaterials, produced by screen-printing and ink-jet printing, are already available commercially and could be used as a generic basis for lysozyme and other aptasensors;

(ii) Experimental confirmation of the appropriate storage stability of the aptasensors for commercial purposes. Aptamers are inherently more stable compared to antibodies, for example; however, with the lysozyme aptasensors developed so far, storage stability beyond one month remains to be investigated;

(iii) Generic approaches appropriate for high throughput, multi-analyte testing. Lysozyme analysis might prove highly beneficial in the context of the multiplexed sensing of various disease biomarkers. Going in this direction, electrochemical aptasensors have been developed for dual detection of lysozyme and interferon gamma, aiming to diagnosis acute leukemia [90]. An illustration of the potential of generic platforms was provided by an aptasensor array based on eight screen-printed electrodes modified with AuNPs, coated with azide-ended thiols, onto which three different aptamers (for lysozyme, cocaine and thrombin) were immobilized by click-chemistry [60]. Reconciling the need for a short analysis time with the simultaneous demand for a high sensitivity of detection could come from new signal amplification strategies. Among others, recent approaches based on nanomaterials, such as graphene [105] or nanoceria [106], show promising potentialities;

(iv) Validation of novel aptasensors in comparison with methods currently used in clinical and analytical laboratories, such as ELISA and HPLC. So far, only three studies reported comparative results obtained with the aptasensor and by classical methods [73,78,91]. In the particular case of lysozyme, a comparison with other methods should be made with caution, since some methods measure the amount of enzymatically-active lysozyme, while others determine the total amount of protein. Moreover, differences between results provided by methods based on very different principles, e.g., chromatographic separation and affinity, are not uncommon [107].

Although many of the studies regarding lysozyme as a disease biomarker date back to the 1970s–1980s, currently, lysozyme analysis is not routinely performed in biochemical laboratories. Moreover, the most widely-used tests rely on ELISA analysis. Recent developments in biosensing research illustrate the applications of aptasensors for the dual detection of lysozyme and other important analytes, such as adenosine [95], thrombin [89,94] and interferon gamma [90]. The usefulness of aptasensors for serum analysis of patients with inflammatory bowel disease was recently demonstrated [73]. Consequently, new studies focused on the parallel analysis and correlations between disease biomarkers and lysozyme levels in biological samples are expected to boost research efforts tapping into the applicative potential of electrochemical aptasensing of lysozyme in the biomedical field.

Acknowledgments: Financial support from the Centre National de la Recherche Scientifique (CNRS), the Université Lille 1, the Nord Pas de Calais region and the Institut Universitaire de France (IUF) is gratefully acknowledged. Q.W. thanks Chinese government for the China Scholarship Council (CSC) award for Ph.D. funding.

Author Contributions: Alina Vasilescu wrote the sensing part and parts linked to lysozyme aptamers. Qian Wang wrote Section 2 of the paper. Musen Li wrote Section 3 of the paper. Rabah Boukherroub wrote the conclusion and perspectives part and corrected the paper. Sabine Szunerits coordinated the work and wrote the introduction.

Conflicts of Interest: The authors declare no conflict of interest.

Abbreviations

The following abbreviations are used in this manuscript:

AuNPs	gold nanoparticles
B-AB	biotinylated antibody
BiDNA	bifunctional aptamer for adenosine and lysozyme, linker DNA
CD	cyclodextrin
CPE	carbon paste electrode
CPSA	chronopotentiometric stripping analysis
$Cu_2O@rGO@PpPG$	nanocomposite of reduced graphene oxide, cuprous oxide and plasma-polymerized propargylamine
CV	cyclic voltammetry
DLAP	dabcyl-labeled aptamer modified metal nanoparticles
DPASV	differential pulse adsorptive stripping voltammetry
DPV	differential pulse voltammetry
DTT	dithiothreitol
EDC	1-ethyl-3-(3-dimethylaminopropyl)carbodiimide
EIS	electrochemical impedance spectroscopy
Fc	ferrocene
FLD	fluorescence detector
GCE	glassy carbon electrode
GO	graphene oxide
GR	graphene
IDA–Cu/AuNps/GCE	iminodiacetic acid–copper ion complex immobilized on a glassy carbon electrode modified with gold nanoparticles
ITO	indium tin oxide
IFN-γ	interferon gamma
LBA	lysozyme binding aptamer
MCH	mercaptohexanol
MeB-cDNA	methylene blue-tagged complementary DNA
MWCNTs-CS	multiwalled carbon nanotubes-chitosan nanocomposites
NHS	N-hydroxysuccinimide
O-GNs	Orange II functionalized graphene nanosheets
p–ATP	p-aminothiophenol
PABA	poly-aminobenzoic acid
PEI	polyethyleneimine
PGE	pencil graphite electrode
RLS	resonance light scattering
SA-ALP	streptavidin-conjugate of alkaline phosphatase
SERS	surface-enhanced Raman scattering
SPCE	screen-printed carbon electrode
SWV	square wave voltammetry
TBA	thrombin binding aptamer
TCA/AuNP/ssDNA	thiocyanuric acid (TCA)/gold nanoparticles (AuNPs) modified with ssDNA
(THH) Au NCs	tetrahexahedral gold nanocrystals
$TiO_2@PPAA$	composite made of polyacrylic acid and hollow TiO_2 spheres
TiO_2/3D-rGO/PPy	hollow titanium dioxide nanoball, three-dimensional reduced graphene oxide and polypyrrole
TPA	tripropylamine
TWJ	three-way junction
VANCNT	vertically-aligned nitrogen-doped carbon nanotubes

References

1. Laschtschenko, P. Uber die keimtötende und entwicklungshemmende Wirkung von Hühnereiweiß. *Z. Hyg. Infekt. Krankh.* **1909**, *64*, 419–427. [CrossRef]
2. Fleming, A. On a Remarkable Bacteriolytic Element Found in Tissues and Secretions. *Proc. R. Soc. B Biol. Sci.* **1922**, *93*, 306–317. [CrossRef]
3. Jollès, P. Recent Developments in the Study of Lysozymes. *Angew. Chem. Int. Ed. Engl.* **1964**, *3*, 28–36. [CrossRef] [PubMed]

4. Cunningham, F.E.; Proctor, V.A.; Goetsch, S.J. Egg-white lysozyme as a food preservative: An overview. *World's Poult. Sci. J.* **1991**, *47*, 141–163. [CrossRef]

5. Callewaert, L.; Michiels, C.W. Lysozymes in the animal kingdom. *J. Biosci.* **2010**, *35*, 127–160. [CrossRef] [PubMed]

6. Jolles, P.; Jolles, J. What's new in lysozyme research? Always a model system, today as yesterday. *Mol. Cell Biochem.* **1984**, *63*, 165–189. [PubMed]

7. Blake, C.C.F.; Koenig, D.F.; Mair, G.A.; North, A.C.T.; Phillips, D.C.; Sarma, V.R. Structure of Hen Egg-White Lysozyme: A Three-dimensional Fourier Synthesis at 2 Å Resolution. *Nature* **1965**, *206*, 757–761. [CrossRef] [PubMed]

8. Blake, C.C.; Mair, G.A.; North, A.C.; Phillips, D.C.; Sarma, V.R. On the conformation of the hen egg-white lysozyme molecule. *Proc. R. Soc. Lond. B Biol. Sci.* **1967**, *167*, 365–377. [CrossRef] [PubMed]

9. Miranker, A.; Radford, S.E.; Karplus, M.; Dobson, C.M. Demonstration by NMR of folding domains in lysozyme. *Nature* **1991**, *349*, 633–636. [CrossRef] [PubMed]

10. Dobson, C.M.; Evans, P.A.; Radford, S.E. Understanding how proteins fold: The lysozyme story so far. *Trends Biochem. Sci.* **1994**, *19*, 31–37. [CrossRef]

11. Merlini, G.; Bellotti, V. Lysozyme: a paradigmatic molecule for the investigation of protein structure, function and misfolding. *Clin. Chim. Acta.* **2005**, *357*, 168–172. [CrossRef] [PubMed]

12. Goda, S.; Takano, K.; Yamagata, Y.; Nagata, R.; Akutsu, H.; Maki, S.; Namba, K.; Yutani, K. Amyloid protofilament formation of hen egg lysozyme in highly concentrated ethanol solution. *Protein Sci.* **2000**, *9*, 369–375. [CrossRef] [PubMed]

13. Booth, D.R.; Sunde, M.; Bellotti, V.; Robinson, C.V.; Hutchinson, W.L.; Fraser, P.E.; Hawkins, P.N.; Dobson, C.M.; Radford, S.E.; Blake, C.C.; *et al.* Instability, unfolding and aggregation of human lysozyme variants underlying amyloid fibrillogenesis. *Nature* **1997**, *385*, 787–793. [CrossRef] [PubMed]

14. Swaminathan, R.; Ravi, V.K.; Kumar, S.; Kumar, M.V.; Chandra, N. Lysozyme: A model protein for amyloid research. *Adv. Protein Chem. Struct. Biol.* **2011**, *84*, 63–111. [PubMed]

15. Dumoulin, M.; Last, A.M.; Desmyter, A.; Decanniere, K.; Canet, D.; Larsson, G.; Spencer, A.; Archer, D.B.; Sasse, J.; Muyldermans, S.; *et al.* A camelid antibody fragment inhibits the formation of amyloid fibrils by human lysozyme. *Nature* **2003**, *424*, 783–788. [CrossRef] [PubMed]

16. Haas, M.; Moolenaar, F.; Meijer, D.K.; de Zeeuw, D. Specific drug delivery to the kidney. *Cardiovasc. Drugs Ther.* **2002**, *16*, 489–496. [CrossRef] [PubMed]

17. Dolman, M.E.; Harmsen, S.; Storm, G.; Hennink, W.E.; Kok, R.J. Drug targeting to the kidney: Advances in the active targeting of therapeutics to proximal tubular cells. *Adv. Drug Deliv. Rev.* **2010**, *62*, 1344–1357. [CrossRef] [PubMed]

18. Wasserfall, F.; Teuber, M. Action of egg white lysozyme on Clostridium tyrobutyricum. *Appl. Environ. Microbiol.* **1979**, *38*, 197–199. [PubMed]

19. Masschalck, B.; Michiels, C.W. Antimicrobial properties of lysozyme in relation to foodborne vegetative bacteria. *Crit. Rev. Microbiol.* **2003**, *29*, 191–214. [CrossRef] [PubMed]

20. Chen, K.; Han, S.-Y.; Zhang, B.; Li, M.; Sheng, W.-J. Development of lysozyme-combined antibacterial system to reduce sulfur dioxide and to stabilize Italian Riesling ice wine during aging process. *Food Sci. Nutr.* **2015**, *3*, 453–465. [CrossRef] [PubMed]

21. Peñas, E.; di Lorenzo, C.; Uberti, F.; Restani, P. Allergenic Proteins in Enology: A Review on Technological Applications and Safety Aspects. *Molecules* **2015**, *20*, 13144–13164. [CrossRef] [PubMed]

22. Weber, P.; Steinhart, H.; Paschke, A. Investigation of the allergenic potential of wines fined with various proteinogenic fining agents by ELISA. *J. Agric. Food Chem.* **2007**, *55*, 3127–3133. [CrossRef] [PubMed]

23. Weber, P.; Kratzin, H.; Brockow, K.; Ring, J.; Steinhart, H.; Paschke, A. Lysozyme in wine: A risk evaluation for consumers allergic to hen's egg. *Mol. Nutr. Food Res.* **2009**, *53*, 1469–1477. [CrossRef] [PubMed]

24. Schneider, N.; Werkmeister, K.; Becker, C.M.; Pischetsrieder, M. Prevalence and stability of lysozyme in cheese. *Food Chem.* **2011**, *128*, 145–151. [CrossRef] [PubMed]

25. Silvetti, T.; Brasca, M.; Lodi, R.; Vanoni, L.; Chiolerio, F.; Groot, M.; Bravi, A. Effects of Lysozyme on the Microbiological Stability and Organoleptic Properties of Unpasteurized Beer. *J. Inst. Brew.* **2010**, *116*, 33–40. [CrossRef]

26. Zimoch-Korzycka, A.; Jarmoluk, A. The use of chitosan, lysozyme, and the nano-silver as antimicrobial ingredients of edible protective hydrosols applied into the surface of meat. *J. Food Sci. Technol.* **2015**, *52*, 5996–6002. [CrossRef] [PubMed]

27. Chander, R.; Lewis, N.F. Effect of lysozyme and sodium EDTA on shrimp microflora. *Eur. J. Appl. Microbiol. Biotechnol.* **1980**, *10*, 253–258. [CrossRef]

28. Humphrey, B.D.; Huang, N.; Klasing, K.C. Rice expressing lactoferrin and lysozyme has antibiotic-like properties when fed to chicks. *J. Nutr.* **2002**, *132*, 1214–1218. [PubMed]

29. Oliver, W.T.; Wells, J.E. Lysozyme as an alternative to antibiotics improves growth performance and small intestinal morphology in nursery pigs. *J. Anim. Sci.* **2013**, *91*, 3129–3136. [CrossRef] [PubMed]

30. Proctor, V.A.; Cunningham, F.E.; Fung, D.Y.C. The chemistry of lysozyme and its use as a food preservative and a pharmaceutical. *CRC Crit. Rev. Food Sci. Nutr.* **1988**, *26*, 359–395. [CrossRef] [PubMed]

31. Osserman, E.F.; Lawlor, D.P. Serum and urinary lysozyme (muramidase) in monocytic and monomyelocytic leukemia. *J. Exp. Med.* **1966**, *124*, 921–952. [CrossRef] [PubMed]

32. Immundiagnostik AG. Manual-Lysozyme ELISA for the *in vitro* Determination of Lysozyme in Serum, Urine and Liquor. Avaiable online: www.immundiagnostik.com/fileadmin/pdf/LYSOZYM_Serum%20Urin%20Liq_K6902.pdf (accessed on 22 February 2016).

33. Portsmann, B.; Jung, K.; Schmechta, H.; Evers, U.; Pergande, M.; Porstmann, T.; Kramm, H.J.; Krause, H. Measurement of lysozyme in human body fluids: Comparison of various enzyme immunoassay techniques and their diagnostic application. *Clin. Biochem.* **1989**, *22*, 349–355. [CrossRef]

34. Lonnerdal, B. Biochemistry and physiological function of human milk proteins. *Am. J. Clin. Nutr.* **1985**, *42*, 1299–1317. [PubMed]

35. Hankiewicz, J.; Swierczek, E. Lysozyme in human body fluids. *Clin. Chim. Acta* **1974**, *57*, 205–209. [CrossRef]

36. Venge, P.; Foucard, T.; Henriksen, J.; Hakansson, L.; Kreuger, A. Serum-levels of lactoferrin, lysozyme and myeloperoxidase in normal, infection-prone and leukemic children. *Clin. Chim. Acta* **1984**, *136*, 121–130. [CrossRef]

37. Grieco, M.H.; Reddy, M.M.; Kothari, H.B.; Lange, M.; Buimovici-Klein, E.; William, D. Elevated beta 2-microglobulin and lysozyme levels in patients with acquired immune deficiency syndrome. *Clin. Immunol. Immunopathol.* **1984**, *32*, 174–184. [CrossRef]

38. Perillie, P.E.; Khan, K.; Finch, S.C. Serum lysozyme in pulmonary tuberculosis. *Am. J. Med. Sci.* **1973**, *265*, 297–302. [CrossRef] [PubMed]

39. Pascual, R.S.; Gee, J.B.; Finch, S.C. Usefulness of serum lysozyme measurement in diagnosis and evaluation of sarcoidosis. *N. Engl. J. Med.* **1973**, *289*, 1074–1076. [CrossRef] [PubMed]

40. Syrjanen, S.M.; Syrjanen, K.J. Lysozyme in the labial salivary glands of patients with rheumatoid arthritis. *Z. Rheumatol.* **1983**, *42*, 332–336. [PubMed]

41. Falchuk, K.R.; Perrotto, J.L.; Isselbacher, K.J. Serum lysozyme in Crohn's disease. A useful index of disease activity. *Gastroenterology* **1975**, *69*, 893–896. [PubMed]

42. Levinson, S.S.; Elin, R.J.; Yam, L. Light chain proteinuria and lysozymuria in a patient with acute monocytic leukemia. *Clin. Chem.* **2002**, *48*, 1131–1132. [PubMed]

43. Guder, W.G.; Hofmann, W. Clinical role of urinary low molecular weight proteins: their diagnostic and prognostic implications. *Scand. J. Clin. Lab Investig. Suppl.* **2008**, *241*, 95–98. [CrossRef] [PubMed]

44. Ocaña, C.; Hayat, A.; Mishra, R.; Vasilescu, A.; del Valle, M.; Marty, J.L. A novel electrochemical aptamer-antibody sandwich assay for lysozyme detection. *Analyst* **2015**, *140*, 4148–4153. [CrossRef] [PubMed]

45. Hun, X.; Chen, H.; Wang, W. Design of Ultrasensitive Chemiluminescence Detection of Lysozyme in Cancer Cells Based on Nicking Endonuclease Signal Amplification Technology. *Biosens. Bioelectron.* **2010**, *26*, 248–254. [CrossRef] [PubMed]

46. Chen, Z.; Li, L.; Tian, Y.; Mu, X.; Guo, L. Signal amplification architecture for electrochemical aptasensor based on network-like thiocyanuric acid/gold nanoparticle/ssDNA. *Biosens. Bioelectron.* **2012**, *38*, 37–42. [CrossRef] [PubMed]

47. Li, L.-D.; Chen, Z.-B.; Zhao, H.-Z.; Guo, L.; Mu, X. An Aptamer-Based Biosensor for the Detection of Lysozyme With Gold Nanoparticles Amplification. *Sens. Actuators B Chem.* **2010**, *149*, 110–115. [CrossRef]

48. Vasilescu, A.; Gaspar, S.; Gheorghiu, M.; David, S.; Dinca, V.; Peteu, S.; Wang, Q.; Li, M.; Boukherroub, R.; Szunerits, S. Surface Plasmon Resonance based sensing of lysozyme in serum on Micrococcus lysodeikticus-modified graphene oxide surfaces. *Biosens. Bioelectron.* **2016**. in press. [CrossRef] [PubMed]

49. Weth, F.; Schroeder, T.; Buxtorf, U.P. Determination of lysozyme content in eggs and egg products using SDS-gel electrophoresis. *Z. Lebensm. Unters. Forsch.* **1988**, *187*, 541–545. [CrossRef] [PubMed]

50. Galyean, R.D.; Cotterill, O.J. Ion-Exchange Chromatographic Determination of Lysozyme in Egg White. *J. Food Sci.* **1981**, *46*, 1827–1834. [CrossRef]

51. Daeschel, M.A.; Musafija-Jeknic, T.; Wu, Y.; Bizzarri, D.; Villa, A. High-Performance Liquid Chromatography Analysis of Lysozyme in Wine. *Am. J. Enol. Vitic.* **2002**, *53*, 154–157.

52. Francina, A.; Cloppet, H.; Guinet, R.; Rossi, M.; Guyotat, D.; Gentilhomme, O.; Richard, M. A rapid and sensitive non-competitive avidin-biotin immuno-enzymatic assay for lysozyme. *J. Immunol. Methods* **1986**, *87*, 267–272. [CrossRef]

53. Vidal, M.L.; Gautron, J.; Nys, Y. Development of an ELISA for quantifying lysozyme in hen egg white. *J. Agric. Food Chem.* **2005**, *53*, 2379–2385. [CrossRef] [PubMed]

54. Lacorn, M.G.C.; Haas-Lauterbach, S.; Immer, U. Sensitive lysozyme testing in red and white wine using the RIDASCREEN FAST Lysozyme ELISA. *Bulletin de l'OIV* **2010**, *83*, 507–511.

55. Subramanian, P.; Lesniewski, A.; Kaminska, I.; Vlandas, A.; Vasilescu, A.; Niedziolka-Jonsson, J.; Pichonat, E.; Happy, H.; Boukherroub, R.; Szunerits, S. Lysozyme detection on aptamer functionalized graphene-coated SPR interfaces. *Biosens. Bioelectron.* **2013**, *50*, 239–243. [CrossRef] [PubMed]

56. Mascini, M.; Palchetti, I.; Tombelli, S. Nucleic acid and peptide aptamers: fundamentals and bioanalytical aspects. *Angew. Chem. Int. Ed. Engl.* **2012**, *51*, 1316–1332. [CrossRef] [PubMed]

57. Sener, G.; Uzun, L.; Say, R.; Denizli, A. Use of molecular imprinted nanoparticles as biorecognition element on surface plasmon resonance sensor. *Sens. Actuators B Chem.* **2011**, *160*, 791–799. [CrossRef]

58. Wang, L.; Zhu, C.; Han, L.; Jin, L.; Zhou, M.; Dong, S. Label-free, regenerative and sensitive surface plasmon resonance and electrochemical aptasensors based on graphene. *Chem. Commun.* **2011**, *47*, 7794–7796. [CrossRef] [PubMed]

59. Mihai, I.; Vezeanu, A.; Polonschii, C.; Albu, C.; Radu, G.-L.; Vasilescu, A. Label-free detection of lysozyme in wines using an aptamer based biosensor and SPR detection. *Sens. Actuators B Chem.* **2015**, *206*, 198–204. [CrossRef]

60. Xie, D.; Li, C.; Shangguan, L.; Qi, H.; Xue, D.; Gao, Q.; Zhang, C. Click chemistry-assisted self-assembly of DNA aptamer on gold nanoparticles-modified screen-printed carbon electrodes for label-free electrochemical aptasensor. *Sens. Actuators B Chem.* **2014**, *192*, 558–564. [CrossRef]

61. Erdem, A.; Eksin, E.; Muti, M. Chitosan-graphene oxide based aptasensor for the impedimetric detection of lysozyme. *Colloids Surf. B Biointerfaces* **2014**, *115*, 205–211. [CrossRef] [PubMed]

62. Peng, Y.; Zhang, D.; Li, Y.; Qi, H.; Gao, Q.; Zhang, C. Label-free and sensitive faradic impedance aptasensor for the determination of lysozyme based on target-induced aptamer displacement. *Biosens. Bioelectron.* **2009**, *25*, 94–99. [CrossRef] [PubMed]

63. Ocaña, C.; Hayat, A.; Mishra, R.K.; Vasilescu, A.; Del Valle, M.; Marty, J.L. Label free aptasensor for Lysozyme detection: A comparison of the analytical performance of two aptamers. *Bioelectrochemistry* **2015**, *105*, 72–77. [CrossRef] [PubMed]

64. Rajendran, M.; Ellington, A.D. Selection of fluorescent aptamer beacons that light up in the presence of zinc. *Anal. Bioanal. Chem.* **2008**, *390*, 1067–1075. [CrossRef] [PubMed]

65. Shangguan, D.; Li, Y.; Tang, Z.; Cao, Z.C.; Chen, H.W.; Mallikaratchy, P.; Sefah, K.; Yang, C.J.; Tan, W. Aptamers evolved from live cells as effective molecular probes for cancer study. *Proc. Natl. Acad. Sci. USA* **2006**, *103*, 11838–11843. [CrossRef] [PubMed]

66. Bunka, D.H.; Stockley, P.G. Aptamers come of age at last. *Nat. Rev. Microbiol.* **2006**, *4*, 588–596. [CrossRef] [PubMed]

67. Iliuk, A.B.; Hu, L.; Tao, W.A. Aptamer in bioanalytical applications. *Anal. Chem.* **2011**, *83*, 4440–4452. [CrossRef] [PubMed]

68. Cox, J.C.; Ellington, A.D. Automated selection of anti-protein aptamers. *Bioorg. Med. Chem.* **2001**, *9*, 2525–2531. [CrossRef]

69. Cox, J.C.; Hayhurst, A.; Hesselberth, J.; Bayer, T.S.; Georgiou, G.; Ellington, A.D. Automated selection of aptamers against protein targets translated *in vitro*: From gene to aptamer. *Nucleic. Acids Res.* **2002**, *30*, e108. [CrossRef] [PubMed]

70. Kirby, R.; Cho, E.J.; Gehrke, B.; Bayer, T.; Park, Y.S.; Neikirk, D.P.; McDevitt, J.T.; Ellington, A.D. Aptamer-based sensor arrays for the detection and quantitation of proteins. *Anal. Chem.* **2004**, *76*, 4066–4075. [CrossRef] [PubMed]

71. Tran, D.T.; Janssen, K.P.; Pollet, J.; Lammertyn, E.; Anne, J.; Van Schepdael, A.; Lammertyn, J. Selection and characterization of DNA aptamers for egg white lysozyme. *Molecules* **2010**, *15*, 1127–1140. [CrossRef] [PubMed]

72. Schoukroun-Barnes, L.R.; Glaser, E.P.; White, R.J. Heterogeneous Electrochemical Aptamer-Based Sensor Surfaces for Controlled Sensor Response. *Langmuir* **2015**, *31*, 6563–6569. [CrossRef] [PubMed]

73. Wang, Q.; Subramanian, P.; Schlechter, A.; Teblum, E.; Yemini, R.; Nessim, G.D.; Vasilescu, A.; Li, M.; Boukherroub, R.; Szunerits, S. Vertically aligned nitrogen-doped carbon nanotube carpet electrodes: Highly sensitive interfaces for the analysis of serum from patients with inflammatory bowel disease. *ACS Appl. Mater. Interfaces* **2016**, *8*, 9600–9609. [CrossRef] [PubMed]

74. Wang, M.; Zhai, S.; Ye, Z.; He, L.; Peng, D.; Feng, X.; Yang, Y.; Fang, S.; Zhang, H.; Zhang, Z. An electrochemical aptasensor based on a TiO_2/three-dimensional reduced graphene oxide/PPy nanocomposite for the sensitive detection of lysozyme. *Dalton Trans.* **2015**, *44*, 6473–6479. [CrossRef] [PubMed]

75. Zhang, Z.; Zhang, S.; He, L.; Peng, D.; Yan, F.; Wang, M.; Zhao, J.; Zhang, H.; Fang, S. Feasible electrochemical biosensor based on plasma polymerization-assisted composite of polyacrylic acid and hollow TiO_2 spheres for sensitively detecting lysozyme. *Biosens. Bioelectron.* **2015**, *74*, 384–390. [CrossRef] [PubMed]

76. Chen, Z.; Li, L.; Zhao, H.; Guo, L.; Mu, X. Electrochemical impedance spectroscopy detection of lysozyme based on electrodeposited gold nanoparticles. *Talanta* **2011**, *83*, 1501–1506. [CrossRef] [PubMed]

77. Xiao, Y.; Wang, Y.; Wu, M.; Ma, X.; Yang, X. Graphene-based lysozyme binding aptamer nanocomposite for label-free and sensitive lysozyme sensing. *J. Electroanal. Chem.* **2013**, *702*, 49–55. [CrossRef]

78. Fang, S.; Dong, X.; Ji, H.; Liu, S.; Yan, F.; Peng, D.; He, L.; Wang, M.; Zhang, Z. Electrochemical aptasensor for lysozyme based on a gold electrode modified with a nanocomposite consisting of reduced graphene oxide, cuprous xide, and plasma-polymerized propargylamine. *Microchim. Acta* **2016**, *183*, 633–642. [CrossRef]

79. Chen, Z.; Guo, J.; Li, J.; Guo, L. Tetrahexahedral Au nanocrystals/aptamer based ultrasensitive electrochemical biosensor. *RSC Adv.* **2013**, *3*, 14385–14389. [CrossRef]

80. Rodriguez, M.C.; Kawde, A.-N.; Wang, J. Aptamer biosensor for label-free impedance spectroscopy detection of proteins based on recognition-induced switching of the surface charge. *Chem. Commun.* **2005**, *34*, 4267–4269. [CrossRef] [PubMed]

81. Rodriguez, M.C.; Rivas, G.A. Label-free electrochemical aptasensor for the detection of lysozyme. *Talanta* **2009**, *78*, 212–216. [CrossRef] [PubMed]

82. Rohrbach, F.; Karadeniz, H.; Erdem, A.; Famulok, M.; Mayer, G. Label-free impedimetric aptasensor for lysozyme detection based on carbon nanotube-modified screen-printed electrodes. *Anal. Biochem.* **2012**, *421*, 454–459. [CrossRef] [PubMed]

83. Du, M.; Yang, T.; Zhao, C.; Jiao, K. Electrochemical logic aptasensor based on graphene. *Sens. Actuators B Chem.* **2012**, *169*, 255–260. [CrossRef]

84. Du, M.; Yang, T.; Guo, X.; Zhong, L.; Jiao, K. Electrochemical synthesis of Fe_2O_3 on graphene matrix for indicator-free impedimetric aptasensing. *Talanta* **2013**, *105*, 229–234. [CrossRef] [PubMed]

85. Guo, Y.; Han, Y.; Guo, Y.; Dong, C. Graphene-Orange II composite nanosheets with electroactive functions as label-free aptasensing platform for "signal-on" detection of protein. *Biosens. Bioelectron.* **2013**, *45*, 95–101. [CrossRef] [PubMed]

86. Kawde, A.-N.; Rodriguez, M.C.; Lee, T.M.H.; Wang, J. Label-free bioelectronic detection of aptamer–protein interactions. *Electrochem. Commun.* **2005**, *7*, 537–540. [CrossRef]

87. Cheng, A.K.; Ge, B.; Yu, H.Z. Aptamer-based biosensors for label-free voltammetric detection of lysozyme. *Anal. Chem.* **2007**, *79*, 5158–5164. [CrossRef] [PubMed]

88. Du, Y.; Chen, C.; Li, B.; Zhou, M.; Wang, E.; Dong, S. Layer-by-layer electrochemical biosensor with aptamer-appended active polyelectrolyte multilayer for sensitive protein determination. *Biosens. Bioelectron.* **2010**, *25*, 1902–1907. [CrossRef] [PubMed]

89. Cheng, L.; Zhang, J.; Lin, Y.; Wang, Q.; Zhang, X.; Ding, Y.; Cui, H.; Fan, H. An electrochemical molecular recognition-based aptasensor for multiple protein detection. *Anal. Biochem.* **2015**, *491*, 31–36. [CrossRef] [PubMed]

90. Xia, J.; Song, D.; Wang, Z.; Zhang, F.; Yang, M.; Gui, R.; Xia, L.; Bi, S.; Xia, Y.; Li, Y.; *et al.* Single electrode biosensor for simultaneous determination of interferon gamma and lysozyme. *Biosens. Bioelectron.* **2015**, *68*, 55–61. [CrossRef] [PubMed]

91. Chen, Z.; Guo, J. A reagentless signal-off architecture for electrochemical aptasensor for the detection of lysozyme. *Electrochim. Acta* **2013**, *111*, 916–920. [CrossRef]

92. Liu, D.Y.; Zhao, Y.; He, X.W.; Yin, X.B. Electrochemical aptasensor using the tripropylamine oxidation to probe intramolecular displacement between target and complementary nucleotide for protein array. *Biosens. Bioelectron.* **2011**, *26*, 2905–2910. [CrossRef] [PubMed]

93. Zhang, H.; Fang, C.; Zhang, S. Ultrasensitive electrochemical analysis of two analytes by using an autonomous DNA machine that works in a two-cycle mode. *Chemistry* **2011**, *17*, 7531–7537. [CrossRef] [PubMed]

94. Hansen, J.A.; Wang, J.; Kawde, A.N.; Xiang, Y.; Gothelf, K.V.; Collins, G. Quantum-dot/aptamer-based ultrasensitive multi-analyte electrochemical biosensor. *J. Am. Chem. Soc.* **2006**, *128*, 2228–2229. [CrossRef] [PubMed]

95. Deng, C.; Chen, J.; Nie, L.; Nie, Z.; Yao, S. Sensitive bifunctional aptamer-based electrochemical biosensor for small molecules and protein. *Anal. Chem.* **2009**, *81*, 9972–9978. [CrossRef] [PubMed]

96. Xia, Y.; Gan, S.; Xu, Q.; Qiu, X.; Gao, P.; Huang, S. A three-way junction aptasensor for lysozyme detection. *Biosens. Bioelectron.* **2013**, *39*, 250–254. [CrossRef] [PubMed]

97. Vasilescu, A.; Marty, J.-L. Electrochemical aptasensors for the assessment of food quality and safety. *Trends Anal. Chem.* **2016**, *79*, 60–70. [CrossRef]

98. Arabzadeh, A.; Salimi, A. Novel voltammetric and impedimetric sensor for femtomolar determination of lysozyme based on metal-chelate affinity immobilized onto gold nanoparticles. *Biosens. Bioelectron.* **2015**, *74*, 270–276. [CrossRef] [PubMed]

99. Zhang, X.; Zhu, S.; Xiong, Y.; Deng, C.; Zhang, X. Development of a MALDI-TOF MS strategy for the high-throughput analysis of biomarkers: On-target aptamer immobilization and laser-accelerated proteolysis. *Angew. Chem. Int. Ed. Engl.* **2013**, *52*, 6055–6058. [CrossRef] [PubMed]

100. Chen, F.; Cai, C.; Chen, X.; Chen, C. "Click on the bidirectional switch": The aptasensor for simultaneous detection of lysozyme and ATP with high sensitivity and high selectivity. *Sci. Rep.* **2016**, *6*, 18814. [CrossRef] [PubMed]

101. He, P.; Zhang, Y.; Liu, L.; Qiao, W.; Zhang, S. Ultrasensitive SERS Detection of Lysozyme by a Target-Triggering Multiple Cycle Amplification Strategy Based on a Gold Substrate. *Chem. Eur. J.* **2013**, *19*, 7452–7460. [CrossRef] [PubMed]

102. Liao, Y.H.; Brown, M.B.; Martin, G.P. Turbidimetric and HPLC assays for the determination of formulated lysozyme activity. *J. Pharm. Pharmacol.* **2001**, *53*, 549–554. [CrossRef] [PubMed]

103. Determination of Lysozyme in Wine Using High-Performance Liquid Chromatography. Available online: http://www.oiv.int/public/medias/2553/oiv-ma-as315--25-en.pdf (accessed on 22 March 2016).

104. Guzzo, F.; Cappello, M.S.; Azzolini, M.; Tosi, E.; Zapparoli, G. The inhibitory effects of wine phenolics on lysozyme activity against lactic acid bacteria. *Int. J. Food Microbiol.* **2011**, *148*, 184–190. [CrossRef] [PubMed]

105. Loo, A.H.; Bonanni, A.; Pumera, M. Mycotoxin aptasensing amplification by using inherently electroactive graphene-oxide nanoplatelet labels *ChemElectroChem* **2015**, *2*, 743–747. [CrossRef]

106. Bulbul, G.; Hayat, A.; Andreescu, S. A generic amplification strategy for electrochemical aptasensors using a non-enzymatic nanoceria tag. *Nanoscale* **2015**, *7*, 13230–13238. [CrossRef] [PubMed]

107. Barbara Kerkaert, B.; Mestdagh, F.; De Meulenaer, B. Detection of hen's egg white lysozyme in food: Comparison between a sensitive HPLC and a commercial ELISA method. *Food Chem.* **2010**, *120*, 580–584. [CrossRef]

chemosensors

Review

Recent Trends in Field-Effect Transistors-Based Immunosensors

Ana Carolina Mazarin de Moraes and Lauro Tatsuo Kubota *

Department of Analytical Chemistry, Institute of Chemistry, University of Campinas, P.O. Box 6154, Campinas, São Paulo 13083-970, Brazil; anacmmo@gmail.com
* Correspondence: kubota@iqm.unicamp.br; Tel.: +55-19-3521-2134

Academic Editors: Paolo Ugo and Ligia Moretto
Received: 29 July 2016; Accepted: 12 October 2016; Published: 21 October 2016

Abstract: Immunosensors are analytical platforms that detect specific antigen-antibody interactions and play an important role in a wide range of applications in biomedical clinical diagnosis, food safety, and monitoring contaminants in the environment. Field-effect transistors (FET) immunosensors have been developed as promising alternatives to conventional immunoassays, which require complicated processes and long-time data acquisition. The electrical signal of FET-based immunosensors is generated as a result of the antigen-antibody conjugation. FET biosensors present real-time and rapid response, require small sample volume, and exhibit higher sensitivity and selectivity. This review brings an overview on the recent literature of FET-based immunosensors, highlighting a diversity of nanomaterials modified with specific receptors as immunosensing platforms for the ultrasensitive detection of various biomolecules.

Keywords: biosensors; immunosensors; field-effect transistors; immunoFET; antigen-antibody; nanomaterials

1. Introduction to Immunosensors

Biosensors are defined as analytical devices combining functional materials or biological elements for the selective detection of an analyte. The recognition system consists of biological receptors, for instance: enzymes, cellular receptors, antibodies, nucleic acids, microorganisms, or artificial biomimetic materials, all equipped with biochemical mechanisms for recognition. Moreover, these biodevices are capable of providing quantitative or semi-quantitative analytical information by forming specific biological recognition complexes such as enzyme-substrate, antigen-antibody, etc., which are able to convert the biochemical changes into a measurable signal by means of a physicochemical transducer system [1,2]. Immunosensors are biosensors that detect either the sensitive biological elements antibodies or antigens through the event of formation of specific antigen-antibody complexes. The detection of antibodies is preferred and these biological elements are often properly immobilized on the surface of the transducer [3].

An antibody (Figure 1, top-left inset) is a large protein in a "Y" shape that consists of highly ordered sequences of hundreds of amino acids. Specifically, the antibody molecule structure is composed of two heavy polypeptide chains (molecular weights of 50 kDa) and two light chains (molecular weights of 25 kDa) linked by disulfide bonds. The chains have both constant and variable regions and the antigen-binding takes place at light and heavy variable domains, forming the hyper-variable regions of the antibody, known as the complementarity determining regions (CDR). The constant region is generally preserved from one antibody to another class, while the CDR show a great diversity of sequences [4,5].

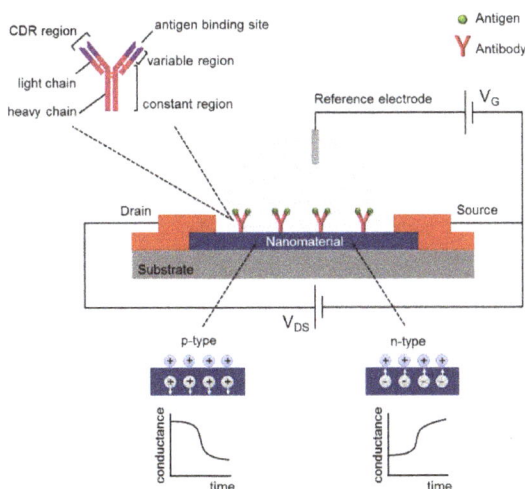

Figure 1. Scheme outlining an immunoFET operating in solution. Top-left inset represents an antibody structure, which is basically composed of two light and two heavy chains linked by disulfide bonds (red lines). Nanomaterials are embedded on the gate region as the semiconductor channel of a FET-based biosensor. Specific antibodies for antigen recognition are immobilized on the nanomaterial surface. The binding of positively charged target molecules on a p-type channel causes depletion of charge carriers (holes) and decrease in conductance, while in an n-type channel, the positively charged molecules result in accumulation of charge carriers (electrons) and increase in conductance.

Antibodies are glycoproteins produced by the immune system to identify and neutralize foreign substances to the body, known as antigens. Antigens are molecules or particles capable of initiating an immune response, therefore triggering the production of antibodies. The antibodies, in turn, identify and bind to antigens with extremely high specificity, and a minor change in the chemical structure of an antigen can dramatically reduce its affinity for an antibody. Thus, the antibody must be specific, have a stable and strong interaction with the antigen, and be able to detect the target analyte at the required concentration range, ensuring the sensitivity and reliability of an immunosensor [5–8].

Polyclonal antibodies are generated by multiple immune cells and can bind to the antigen target through a large number of epitopes (specific antigen regions where antibodies bind). Polyclonal antibodies are cheap and facile to produce, however they lack specificity. Monoclonal antibodies are produced from immune cells identical to the parent cells and recognize specific epitopes of an antigen, thus conferring higher selectivity than polyclonal. There is a plethora of commercially available antibodies and their production is generally mediated by immunizing an animal with a foreign infection to produce an immune response [5,9].

Immunosensors may be useful to monitor and quantify the functioning of the immune system, detecting the presence of specific antibodies or antigens in body fluids; therefore, they can be considered as promising tools in clinical diagnosis applications. The analytes may also include microorganisms such as viruses and bacteria, which find an important role in food safety, and pollutants such as pesticides and herbicides, that are valuable in environmental monitoring.

Immunosensors can be divided in two groups: labeled and label-free. The labeled immunosensors are designed in a way that the immunoreaction is measured through the detection of the labeling of a molecule with labels such as luminescent, fluorescent, radioactive, and epitope tags [3]. For example, the most utilized enzyme immunoassay in clinical analysis and residual analysis in environmental and food samples is the label-based enzyme-linked immunosorbent assay (ELISA) [10]. The ELISA test basically consists of translating the recognition event between a specific enzyme-labeled antibody and an antigen into a colorimetric, chemiluminescent, or electrochemical signal [11,12]. Radioimmunoassay (RIA) is also used in clinical diagnosis and involves the use of radiolabeled antigens or antibodies [13]. Despite the reliability and high sensitivity of conventional immunoassays, they suffer from drawbacks such as the demand for relatively large volumes of sample, long incubation periods, complicated laboratory apparatus, needing to be carried out by highly trained laboratory staff, and requiring multiple labeling strategies that would not interfere with the antigen-antibody interactions [5]. On the other hand, the label-free immunosensors allow the direct monitoring of immunoreactions by measuring physicochemical changes induced by the antigen-antibody complex formation, being more attractive as alternatives amongst traditional immunoassays.

The transduction platforms convert the chemical or physical changes induced by selective antigen-antibody interactions into a quantifiable signal. Signal transducing can be performed by different transducing mechanisms. The most employed transducers in immunosensors are related to the measurement of electrons, photons, and masses, thus including electrochemical, optical, and piezoelectric systems, respectively. Biodevices based on electrochemical transducing platforms are the most exploited so far because of their simple principle of measurement. The antigen-antibody complex formation generates an electrical signal that can be measured by different means. In general terms, electrochemical transducers comprise: amperometric/voltammetric, conductimetric, potentiometric, impedance, and semiconductor field-effect principles; which measure changes in: current, modulation of conductance, potential or charge accumulation, electrical impedance, and either the current or the potential across a semiconductor in response of a binding process at the gate surface, respectively [9].

The optical transducers operate by means of an optical signal (color or fluorescence) or changes in optical properties such as absorption, reflectance, emission, etc. promoted by the biorecognition event. Lastly, piezoelectric transducers detect a mass increase induced by the antigen-antibody complex, which can be detectable by piezoelectric devices such as a quartz crystal balance or a microcantilever [2,14].

Although there are many kinds of transducing mechanisms, this review is concerned with immunosensors whose transducing system is based on the field-effect transistors (FET). Such transducers are considered charge-sensitive devices and present several associated advantages such as excellent sensitivity and selectivity, label-free detection, real-time response, cost-effective fabrication, and ease of miniaturization and integration in electronic chips, making them excellent interfaces towards point-of-care (PoC) systems, as well as reusable and portable immunosensing devices. The focus of this review is the description of recent developments and current trends in FET-based immunosensors using nanomaterials as sensing platforms for different application fields. For a more thorough comprehension, we refer to several reviews on immunosensor principles [3,5,14], antibody engineering [4,6], transduction systems in immunosensing [5,14–16], basic mechanisms of FET sensors [17], and FET-based biosensors [18,19].

2. FET-Based Immunosensors

FET-based sensors have been exploited for decades since the first reports on ion-sensitive field effect transistors (ISFET) by Bergveld et al. [20–23]. FET-based biosensors operate by means of an electrical field modulating charge carriers across a semiconductor material. Such sensors are able to directly convert specific biological interactions into electrical signals. In the standard configuration of a FET, the electric current flows along a semiconductor channel connected to the source and drain electrodes. A third electrode, the gate contact, which is capacitively coupled to the device through a thin dielectric layer (typically SiO_2), modulates the conductance between these two electrodes [20,24].

A FET-based sensor detects potential changes on its gate surface. The FET device characteristics can be assessed by the transfer curve, which is the plot of the current across drain-source electrodes (I_{DS}) as a function of the gate voltage (V_G) at a constant drain-source voltage (V_{DS}) [25–27]. A FET-based biosensor usually operates in solution. In this case, the conductance between source and drain electrodes is modulated by a gate voltage applied in the electrolyte solution by means of a reference electrode placed on the top of the channel. In this configuration, the surface charge in an electrolyte attracts counter-ions forming the electric double layer (Debye layer) [26,28].

The semiconductor determines the type of charge carriers that can accumulate or deplete in the channel, thus the current flow can either be the result of movement of holes ("p-type") or electrons ("n-type"). For example, the application of positive voltage on the gate of a p-type channel FET results in depletion of charge carriers and a decrease in conductivity, whereas the application of negative voltage leads to accumulation of charge carriers and an increase in conductivity [28,29]. There is a great diversity of FET biosensors but we focus in this review on the FET-based immunosensors, whose concept was introduced by Shenck in 1978 [30]. In an immunoFET, which is shown schematically in Figure 1, the gate region is biofunctionalized by immobilizing antibodies or antigens on its surface, allowing direct analyte-binding upon the device surface. Since antibodies and antigens are mostly charged molecules, the biorecognition event imparts an electrical field, which modulates the charge carrier flow between source and gate, generating an electrically detectable signal [31,32]. Thus, the changes in conductance induced by the antigen-antibody binding can be measured and correlated to the analyte concentration [33].

ImmunoFET features simplicity of use and production, portability, high sensitivity (detecting picomolar and femtomolar levels of an analyte), utilizes a small volume of sample (in the microliters range), has low operating costs, and enables real-time analyte detection and quantification [34]. The exquisite sensibility of immunoFET is due to the fact that these sensors act as a combination of a sensor and an amplifier, in which the biorecognition channel is in direct contact with the analyte and the occurrence of only a single biological event is capable of causing a pronounced current change in the sensing channel [35,36].

Under ideal conditions, an immunoFET is able to detect target biomolecules in biochemically or clinically meaningful range of concentrations, which can vary in the order of sub-fM to μM, achieving very low detection limits. Ideal conditions include full antibody coverage, capacitive interface, highly charged antigens, and low ionic strength buffers [18]. However, in real conditions, samples contain rich levels of salts and species of non-interest. Therefore, these immunosensors present practical issues in transducing the antigen-antibody immunoreaction into a measurable signal and the direct sensing has proven to be challenging [37].

It was argued that an immunoFET is able to detect only charge or potential changes occurring within the Debye layer [21]. According to the Debye theory, an increase in ion concentration reduces the Debye length due to charge screening by counter-ions [38]. Thus, the Debye screening length, which is a physical distance where the charged analyte is electrically screened by the ions in the medium, strongly affects the immunosensor sensitivity in high ionic strength buffers. The Debye length (λ_D) in an electrolyte is given by the following equation [39]:

$$\lambda_D = \sqrt{\frac{\varepsilon_0\,\varepsilon_r\,k_B\,T}{2\,N_A q^2\,I}}\,, \tag{1}$$

where ε_0 corresponds to the vacuum permittivity; ε_r is the relative permittivity of the medium; k_B is the Boltzmann constant; T is the absolute temperature; N_A is the Avogadro number; q is the charge on an electron; and I is the ionic strength of the solution.

Therefore, λ_D is dependent on the ionic strength of the electrolytic buffer solution. To summarize, it decreases as the ionic strength increases, and a short Debye length implies a lower sensitivity because the FET sensor cannot detect the analyte-binding beyond the Debye length, as illustrated in Figure 2a. For instance, in a physiological solution (biological fluids such as blood, serum, urine, etc.) or undiluted phosphate-buffered saline solution (PBS buffer, ion concentration of 0.15 M), the dimensions of some antibodies (10–12 nm) are large enough to surpass the thickness of the Debye length (~0.8 nm) at the electrolyte-immunosensor interface. In this case, the analyte charges will be at a farther distance from the surface than the Debye length and will be shielded from the sensing channel by buffer counter-ions [17,40]. However, the poor sensitivity caused by the enlargement of the Debye layer can be overcome by pretreating the analyte solution, which may include conditioning steps of centrifugation, filtration, and dilution to low ionic strength (ion concentration lower than 10^{-2}–10^{-3} M) prior to analysis [41], and then a measurable effect by an immunoFET can be observed.

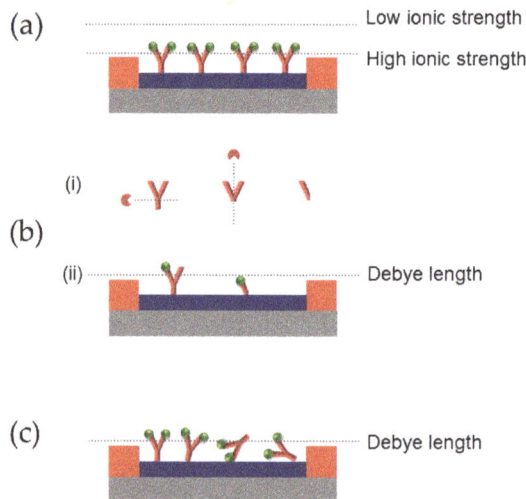

Figure 2. Antibodies (Y shape) immobilized on the gate region. Antigens (spheres) are bound to antibodies and the dashed lines indicate the Debye length position. (**a**) The Debye length is dependent on the ionic strength of the electrolytic buffer and it increases with the reduction of solution ionic strength. A measurable analyte-binding may occur only within the Debye length; (**b**,**c**) show alternative methods for antigen detection in high ionic strength solutions, without sample desalting; (**b**) (i) the antibody was cleaved by enzymatic digestion; and (ii) the antigen was brought within the Debye length [40]; (**c**) a more realistic representation of antibody alignment on the surface of an immunoFET showing random orientation of antibodies relative to the sensing surface, implying that some analyte-binding may occur within the Debye length [42].

The classical assessment described above evidences the dependence on proximity of the analyte to the device surface for an immunoFET proper function [19–23,43]. However, conceptual flaws concerning this model have been revised. In this context, Elnathan et al. demonstrated the direct detection of cardiac troponin I (cTnI) myocardial infarction biomarker without the need for sample desalting by bringing the antigen within the Debye length [40]. Their approach was based on fragmentation of antibody units, allowing the biorecognition event to occur in closer proximity to the silicon nanowire (SiNW) sensing surface, and consequently within the Debye screening length (Figure 2b). The authors cleaved antibodies by enzymatic digestion in order to remove parts that were not required for the antigen-binding and the fragments were able to directly detect cTnI with a sensitivity limit down to pM, without the need of biosample desalting. Casal et al. discussed that antibodies were highly flexible structures that can adsorb to surfaces via a nearly random distribution of their structural domains, and can also be variably oriented in relation to the surface [42]. Therefore, the analytes can bind to antibodies in a distribution of orientations and distances from the sensing surface, implying that some analyte charges would be held within the Debye length, and consequently, the analyte should be detected by the immunoFET even in a high ionic strength condition (Figure 2c). The authors supported their assumption demonstrating multiple immunoFETs able to detect human and murine chemokines in physiologically relevant high ionic strength buffers [42].

3. Trends in FET-Based Immunosensors Using Nanomaterials as Sensing Platforms

Nanostructured materials have attracted a great deal of attention in their application for developing novel FET-based biosensing devices with enhanced sensibility and selectivity [5,34,44]. Owing to nanoscale dimensions (1–100 nm), nanomaterials exhibit tunable and unique physicochemical properties which are not observed in their bulk counterparts, including high specific surface area, high aspect ratio, increased chemical and thermal stability, and remarkable electrical properties. In addition, various biocompatible nanomaterials have their dimensions comparable to a range of biomolecules, facilitating their use for immunocompounds immobilization.

Nanomaterials are capable of modifying the exposed gate region of FET-based immunosensors, thus improving the immobilization of bioreceptors and the signal transducing [44]. Properties like the high aspect ratio allow the atoms to be located at or close to the gate surface, and as the biological recognition is related to the physical signal across the whole device, the device sensitivity is improved as a result of the increased surface-to-volume ratio. In addition, some nanomaterials can easily be chemically modified with specific sensing elements. Therefore, the use of nanomaterials as sensing platforms offers excellent interface perspectives for biological recognition, facilitating the attachment of biomolecules and allowing the scaling of microelectronics down to the nano level, which consequently enables the design of immunosensors with enhanced performance.

Recently, an increasing number of immunosensors using nanomaterials such as carbon nanotubes, graphene, nanowires, metal and metal oxide, and nanohybrids as semiconducting channel have been reported [44–46]. In this review, we highlight recent advances on the incorporation of nanomaterials on the gate surface of FET-based immunosensors applied to biomedical clinical diagnosis, environmental monitoring, and food safety fields. We consider the most important semiconducting nanomaterials for immunosensing purpose over the past five years. Selected reports outlining the exploitation of these nanomaterials in composing the sensing platform of FET immunosensors are discussed in the following sections.

Tables 1 and 2 assess the successful application of various types of nanomaterials in the fabrication of FET-based immunosensors. Recent reports were carefully examined on the basis of target analyte, electroactive nanomaterial, range of concentrations, and detection limit. Table 1 shows several FET immunosensors relevant to clinical biomedical diagnosis. It is evident that a great diversity of disease biomarkers can be detected through immunoFET devices, opening new paths for early-stage diagnosis of serious diseases such as cancer. For instance, prostate specific antigen (PSA), a prostate cancer biomarker, was ultrasensitively detected by using immunoFET devices based on different nanomaterials such as silicon nanowires (SiNW) [47,48], graphene [49], molybdenum disulfide (MoS$_2$) [50,51], etc. The lowest detection limit of PSA was achieved by a FET device based on SiNW (1 fg·mL^{-1}) [47]. Detection of tumor biomarkers for breast cancer [52,53], hepatic carcinoma [54], oral squamous cell carcinoma [55], pancreatic cancer [56], and bladder cancer [57] have also been reported. Moreover, other disease biomarkers related to cardiovascular diseases [58–60], thyroid hormone [61], venous thromboembolism [62], Alzheimer [63], and diabetes [64,65] have been detected with very low detection limits.

Table 2 evidences the ultrasensitive detection of pathogenic microorganisms associated with human health and food safety, such as *Salmonella* [66,67] and *Escherichia coli* [68–71] bacteria, rotavirus [72] and bacteriophage [69] viruses, and parasitic protozoan [73]. For example, graphene-based FET devices were found to detect only 10 colony-forming units per milliliter (CFU·mL^{-1}) of *E. coli* bacteria [68,70] whereas a cerium oxide (CeO$_x$)-based FET device was able to achieve a limit of detection (LOD) as low as 2–3 cells of *Salmonella typhimurium* per mL [67]. Table 2 also depicts the application of immunoFET devices for detecting environmental contaminants such as pesticides [74], herbicides [75], toxins [76], and phytopathogens [77].

Table 1. Recent reports on immunoFET sensors based on various types of nanomaterials applied to clinical biomedical diagnosis.

Target Analyte	Electroactive Nanomaterial	Range	Detection Limit	Reference
–H5N2 avian influenza virus	SiNW	10^{-17}–10^{-12} M	10^{-17} M	[78]
Hepatitis B marker (HBsAg) and cancer marker α-fetoprotein (AFP)	SiNW	-	10^{-14} M for HBsAg and 10^{-15} M for AFP	[79]
Human thyroid stimulating hormone	SiNW	0.02–30 mIU·L^{-1}	0.02 mIU·L^{-1}	[61]
Biomarkers of oral squamous cell carcinoma Interleukin-8 (IL-8) and tumor necrosis factor α (TNF-α)	SiNW	1 fg·mL^{-1}–1 ng·mL^{-1}	10 fg·mL^{-1} in PBS and 100 fg·mL^{-1} in saliva	[55]
Cancer biomarkers cytokeratin 19 fragment (CYFRA21-1) and prostate specific antigen (PSA)	SiNW	1 fg·mL^{-1}–1 ng·mL^{-1}	1 fg·mL^{-1} in buffer solution and 10 fg·mL^{-1} in undiluted human serums	[47]
Prostate cancer biomarker PSA	SiNW	5 fg·mL^{-1}–500 pg·mL^{-1}	5 fg·mL^{-1} in buffer and desalted serum	[48]
AFP and carcinoembryonic antigen (CEA) primary hepatic carcinoma biomarkers	SiNW	500 fg·mL^{-1}–50 ng·mL^{-1} for AFP and 50 fg·mL^{-1}– 10 ng·mL^{-1} for CEA	500 fg·mL^{-1} for AFP and 50 fg·mL^{-1} for CEA	[54]
Cardiac troponin I (cTnI) biomarker for acute myocardial infarction	SiNW	0.092 ng·mL^{-1}–46 ng·mL^{-1}	0.092 ng·mL^{-1}	[60]
Cardiac disease biomarker cTnI	SiNW	5 pg·mL^{-1}–5 ng·mL^{-1}	5 pg·mL^{-1}	[59]
C-reactive protein (CRP) inflammatory biomarker related with cardiovascular diseases	SWCNT	10^{-4} to 10^2 μg·mL^{-1}	10^{-4} μg·mL^{-1}	[58]
Prostate cancer biomarker osteopontin (OPN)	SWCNT	1 pg·mL^{-1}–1 μg·mL^{-1}	0.3 pg·mL^{-1}	[80]
Pro-inflammatory cytokine and anti-inflammatory myokine interleukin-6 (IL-6)	SWCNT	1 pg·mL^{-1}–100 pg·mL^{-1}	1.37 pg·mL^{-1}	[81]
Stress biomarker cortisol in saliva	SWCNT	1 pg·mL^{-1}–1000 ng·mL^{-1}	1 pg·mL^{-1}	[82]
Lyme disease antigen	SWCNT	1 ng·mL^{-1}–3000 ng·mL^{-1}	1 ng·mL^{-1}	[83]
Prostate cancer biomarker OPN	SWCNT	1 pg·mL^{-1}–1 μg·mL^{-1}	1 pg·mL^{-1} or 30 fM	[84]
Alzheimer biomarker, amyloid-β	SWCNT	10^{-12}–10^{-9} g·mL^{-1}	1 pg·mL^{-1} in human serum	[63]

Table 1. *Cont.*

Target Analyte	Electroactive Nanomaterial	Range	Detection Limit	Reference
Chondroitin sulfate proteoglycan 4, multiple cancer types biomarker	Graphene	0.01 fM–10 pM	0.01 fM	[85]
Pancreatic cancer biomarker, carbohydrate antigen 19-9 (CA 19-9)	Graphene	0.01 unit·mL^{-1}–1000 unit·mL^{-1}	0.01 unit·mL^{-1}	[56]
Prostate specific antigen/α1-antichymotrypsin (PSA-ACT) complex	Graphene	100 fg·mL^{-1}–1 µg·mL^{-1}	100 fg·mL^{-1}	[49]
Breast cancer biomarkers human epidermal growth factor receptor 2 (HER2) and epidermal growth factor receptor (EGFR)	SiO$_2$/graphene	100 pM–1 µM	1 pM for HER2 and 100 pM for EGFR	[52]
Human immunodeficiency virus (HIV)	Graphene/CPPyNP	1 pM–10 nM	1 pM	[86]
Bladder cancer biomarker, urinary APOA2 protein	Graphene/SiNW	19.5 fg·mL^{-1}–1.95 mg·mL^{-1}	6.7 pg·mL^{-1}	[57]
Prostate cancer biomarker PSA–ACT complex	Graphene/ZnO nanorods/TiO$_2$	100 fg·mL^{-1}–100 ng·mL^{-1}	1 fM	[87]
D-Dimer, biomarker of venous thromboembolism	Graphene/TiO$_2$	10 pg·mL^{-1}–100 ng·mL^{-1}	10 pg·mL^{-1} in buffer and 100 pg·mL^{-1} in serum sample	[62]
Prostate cancer biomarker PSA	MoS$_2$	3.75 nM, 37.5 pM, and 375 fM	375 fM	[51]
Prostate cancer biomarker PSA	MoS$_2$	1 pg·mL^{-1}–10 ng·mL^{-1}	1 pg·mL^{-1}	[50]
Breast cancer biomarker EGFR	ZnO nanofilm	10 fM–10 nM	10 fM	[53]
Prostate cancer biomarker PSA	Si nanobelt	50 fg·mL^{-1}–500 pg·mL^{-1}	5 pg·mL^{-1}	[88]
Index for diabetes, Hemoglobin-A1c	Au nanoparticles	1.67 ng·mL^{-1}–170.5 ng·mL^{-1}	in the order of ng·mL^{-1}	[65]
Diabetes related hormone, insulin	Si nanogratings	1 fM–1 nM	10 fM in buffer and diluted human serum	[64]
CRP inflammatory biomarker	P3HT polymer	4 pM–2 µM	2 pM	[89]

Table 2. Recent reports on immunoFET sensors based on various types of nanomaterials listed along with their application in food safety and environmental monitoring fields.

Target Analyte	Electroactive Nanomaterial	Range	Detection Limit	Reference
Food safety				
Salmonella bacteria	SWCNT	10^3–10^8 CFU·mL^{-1}	10^3 CFU·mL^{-1}	[66]
Escherichia coli foodborne pathogen	SWCNT	10^2–10^5 CFU·mL^{-1}	10^2 CFU·mL^{-1}	[71]
Escherichia coli O157:H7 and bacteriophage viruses	SWCNT	10^3–10^7 CFU·mL^{-1} for *E. coli* and 10^2–10^7 PFU·mL^{-1} for bacteriophage	10^5 CFU·mL^{-1} for *E. coli* and 10^3 PFU·mL^{-1} for bacteriophage	[69]
Cryptosporidium parvum intestinal parasitic protozoan	Graphene	10^2–10^4 Cp. oocysts per 4 mL buffer	25 Cp. oocysts per mL buffer	[73]
Escherichia coli bacteria	Graphene	10–10^5 CFU·mL^{-1}	10 CFU·mL^{-1}	[70]
Escherichia coli O157:H7 bacteria	Graphene	10–10^4 CFU·mL^{-1}	10 CFU·mL^{-1}	[68]
Food toxin aflatoxin B1	Graphene	10^{-4} ppt^{-1} ppt	0.1 fg·mL^{-1}	[90]
Rotavirus	Graphene	0–10^5 PFU·mL^{-1}	10^2 PFU·mL^{-1}	[72]
Salmonella typhimurium bacteria	CeO$_x$	2–5×10^5 cells·mL^{-1}	2–3 cells·mL^{-1}	[67]
Environmental monitoring				
Atrazine pesticide	SWCNT	0.001–10 ng·mL^{-1}	0.001 ng·mL^{-1}	[74]
2,4-Dichlorophenoxyacetic acid herbicides	SWCNT	5 fM–500 μM	500 fM in soil sample and 50 pM in buffer	[75]
2,4,6-Trinitrotoluene (TNT) contamination	SWCNT	0.5 ppb–5000 ppb	0.5 ppb	[76]
Microcystin-LR (cyanotoxin in surface waters)	SWCNT	1–1000 ng·L^{-1}	0.6 ng·L^{-1}	[91]
Citrus tristeza virus and *Xylella fastidiosa* phytopathogens	InP	60–340 ng·mL^{-1} for *Citrus tristeza virus* and 34–250 ng·mL^{-1} for *Xylella fastidiosa*	2 nM for both phytopathogens	[77]

3.1. Silicon Nanowires

Silicon nanowires (SiNW) are semiconducting one-dimensional (1D) nanostructures with cross-sectional diameters of nanometers and length of micrometers, which can exhibit either p-type or n-type conductivity [92–94]. SiNW are attractive materials in FET-based biosensors design due to their high surface-to-volume ratio, which enables the sensors ultrasensitive detection capability [95,96]. SiNW FET-based biosensors were firstly reported in 2011 by Cui et al. [97]. The authors demonstrated that p-type boron-doped SiNW modified with biotin detected streptavidin protein down to picomolar concentration range.

FET biosensors based on SiNW have been employed as promising clinical diagnostic platforms. For example, Huang et al. developed a novel SiNW-FET biosensor for the detection of PSA biomarker in human serum [48]. The SiNW were fabricated by the polysilicon sidewall spacer technique, which is cheaper than the electron beam lithography. Figure 3a depicts top-view scanning electron microscopy (SEM) images of such SiNW-FET devices. The SiNW surface was modified with 3-aminopropyltriethoxysilane (APTES) to convert silanol groups into amines for glutaraldehyde functionalization. Then, the aldehyde groups present on glutaraldehyde molecules were connected to the amine groups to form the linker between APTES and anti-PSA antibodies. The transfer curve of such n-type SiNW-FET sensor showed that the drain-source current (I_{DS}) was controlled by the gate voltage (V_G) (Figure 3b). The device exhibited excellent electrical performance, and the I_{DS} versus drain-source voltage (V_{DS}) depended on the applied V_G (Figure 3c). Human serum was pretreated by filtration, desalting, and buffer exchange prior to PSA detection in order to keep proper pH and ionic strength of serum proteome. The electrical responses from the measurement of various PSA concentrations in desalted human serum are shown in Figure 3d and the immunosensor was able to detect PSA levels in concentrations as low as 5 fg·mL^{-1}.

Alpha-fetoprotein (AFP) and carcinoembryonic antigen (CEA) markers high expression levels have been associated with primary hepatic carcinoma (PHC). In this context, Zhu et al. developed SiNW-FET with polydimethylsiloxane (PDMS) microfluidic channels for the simultaneous detection of both AFP and CEA biomarkers in desalted human serum [54]. SiNW surface was treated with APTES and glutaraldehyde before the AFP and CEA antibodies immobilization by covalent bonding. Bovine serum albumin (BSA) was used as a blocking agent for non-specific binding. The negatively charged AFP and CEA enhanced the conductivity of the p-type SiNW. The dual-channel setup demonstrated the potential use of SiNW-FET for multiple tumor markers detection in concentrations down to fg·mL^{-1} and ng·mL^{-1}.

Cardiac troponin I, a very sensitive biomarker for acute myocardial infarction, was detected by a label-free SiNW-FET biosensor [60]. Anti-cTnI antibodies were immobilized on the SiNW surfaces for measuring the cTnI in a range of concentrations from 0.092 ng·mL^{-1} to 46 ng·mL^{-1}. Higher sensitivity of cTnI detection was reported by Kim et al. [59]. The authors demonstrated a SiNW honeycomb-like structure for nanowire configuration. The geometry of honeycomb nanowires provided increased surface area rather than straight nanowire configuration, improving the probability of binding events between antigen and antibody, and enabling superior intrinsic electrical performance of biosensors. The devices showed n-type behavior and presented LOD of 5 pg·mL^{-1}, which was about 8-fold smaller than the value previously reported by Kong et al. [60].

As aforementioned, the detection mechanism of FET-based biosensors is hampered in samples of high ionic strength and efforts for bringing the antigen-antibody binding within the Debye layer have been reported [40]. Puppo et al. got rid of the Debye screening problem by performing the electrical measurements on dried samples after the antigen-binding coupling [98]. The authors developed a SiNW-FET sensor for the detection of vascular endothelial growth factor (VEGF), a pathological angiogenesis factor. Increasing concentrations of VEGF caused increase in conductance, leading to detection in the concentration range of fM.

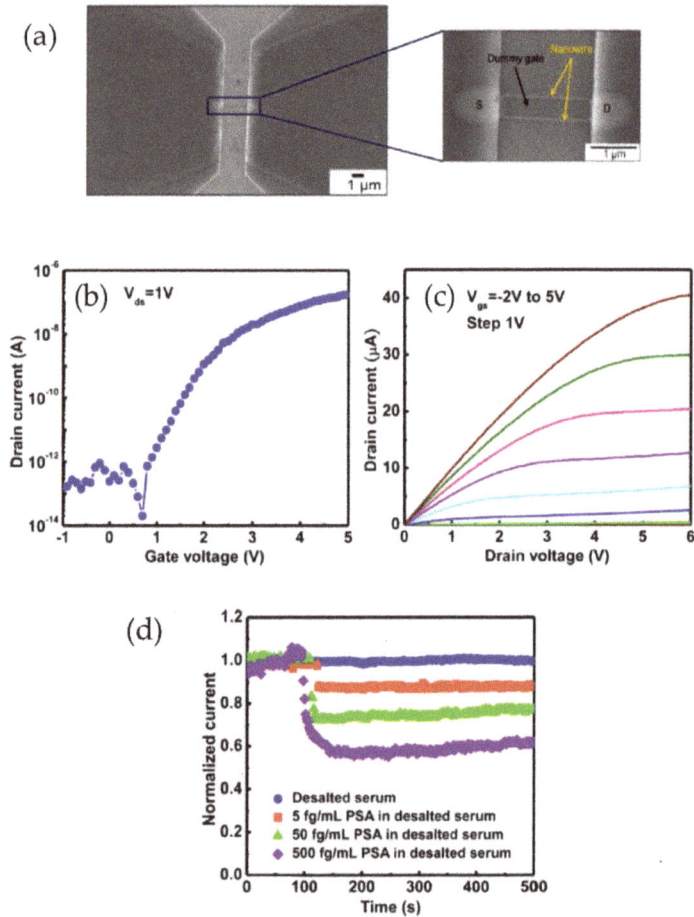

Figure 3. n-type SiNW-based FET immunosensor for detection of PSA biomarker. (**a**) SEM images of a SiNW-FET sensor; (**b**) Electrical characteristics ($I_{DS} \times V_G$ curves) of the SiNW-FET device; (**c**) Output characteristics ($I_{DS} \times V_{DS}$ curves) showing a dependence on the controlled V_G; (**d**) Normalized current × time for SiNW devices to detect various PSA concentrations in desalted human serum. No electrical response was observed for serum without PSA. Reprinted with permission from Huang et al. [48]. Copyright 2016 American Chemical Society.

3.2. Carbon Nanotubes

Carbon nanotubes are cylindrical carbon structures with diameters of nanometers and high length/diameter ratio [99]. Multi-walled carbon nanotubes (MWCNT) consist of multiple rolled graphite sheets [100] while single-walled carbon nanotubes (SWCNT) are single-atom rolled graphitic layers [101,102]. Carbon nanotubes are exciting 1D materials due to their structure-dependent electronic and mechanical properties. Their physical properties are strongly dependent on the way that graphitic sheets are wrapped to form the tubes (chirality), causing them to exhibit metallic or semiconducting characteristics [99]. In particular, SWCNTs have been explored as promising building blocks in the construction of FET-based immunosensors with improved sensibility [58,69,76,80–82,103]. SWCNTs display high electron transfer because all carbon atoms are present in their surface. In addition,

SWCNTs surface can be easily modified, giving various possibilities for non-covalent and covalent immobilization of antigens or antibodies.

For instance, Justino et al. reported an immunoFET based on SWCNT functionalized with C-reactive protein (CRP) antibodies for detecting CRP, an inflammatory biomarker related with cardiovascular diseases [58]. The anti-CRP antibodies were non-covalently immobilized directly on the surface of SWCNT and the devices were capable of detecting CRP in the broad range of concentrations from 10^{-4} to 10^2 µg·mL^{-1}. In addition, the LOD of such devices (10^{-4} µg·mL^{-1}) was 2–3 orders lower than conventional immunoassays. Sharma et al. reported a SWCNT-FET sensor for detecting the prostate cancer biomarker osteopontin (OPN) [80]. SWCNTs were deposited on transparent glass substrates by dielectrophoresis, allowing the direct alignment of the nanotubes at room temperature. The SWCNT surface was treated with 1-ethyl-3-(3-dimethylaminopropyl) carbodiimide (EDC)/N-hydroxysuccinimide (NHS). Then, the covalent immobilization of monoclonal antibodies specific for OPN occurred through the binding with NH$_2$ groups of NHS succinimide ester on the SWCNTs. These devices were incubated with Tween 20 to avoid non-specific binding. The electrical measurements showed a linear behavior after each step of functionalization of the SWCNT-FET device, indicating a good ohmic contact between SWCNTs and source/drain electrodes. Immunosensors exhibited a highly linear resistance change over a range of concentrations (1 pg·mL^{-1} to 1 µg·mL^{-1}) of the prostate cancer biomarker OPN in human serum and PBS buffer, being 3-fold more sensitive than the conventional ELISA immunoassay, with a LOD of 0.3 pg·mL^{-1}.

Horizontally aligned SWCNTs grown on quartz substrates were applied as sensing platform for measuring interleukin-6 (IL-6), a protein that acts as pro-inflammatory cytokine and anti-inflammatory myokine [81]. The highly specific binding of the IL-6 analyte to the antibodies in the gate region caused a change in the drain current, which was measured as an electrical signal. The devices exhibited low detection limit (1.37 pg·mL^{-1}), good selectivity (no responses for BSA and cysteine), and excellent stability (no electronic degradation after storage for up to three months).

Carbon nanotubes-based FET immunosensors are also able to detect microorganisms such as bacteria and viruses. García-Aljaro et al. reported an immunoFET based on SWCNT for the detection of human pathogens *Escherichia coli* O157:H7 bacteria and bacteriophage T7 viruses [69]. The sensing platform was composed of parallel aligned SWCNTs functionalized with 1-pyrene butanoic acid succinimidyl ester (PBASE), in which monoclonal antibodies were covalently attached. The immunosensor exhibited a linear response for both bacteria and viruses increasing concentrations, achieving LOD of 10^5 colony forming units per milliliters (CFU·mL^{-1}) and 10^3 plaque forming units per milliliter (PFU·mL^{-1}), respectively. The biosensor showed a better performance for bacteriophage. This result can be attributed to the decreased size and different morphology of such viruses in comparison to bacteria, which enabled a better and faster diffusion of bacteriophage in the solution towards the antibodies.

Small molecules, either charged or uncharged, cannot be detected by the conventional configuration of an immunoFET, where the biorecognition event between the analyte and the immobilized molecules takes place at the gate sensing surface. In this case, the analyte binding does not produce a measurable conductance change because of the lack of charge or depletion in the semiconductor channel [103]. Therefore, the displacement represents a successful strategy for detecting such target analytes. For example, Tliti et al. described a label-free FET immunosensor based on SWCNT for the ultrasensitive detection of stress biomarker cortisol in saliva [82]. SWCNTs were covalently functionalized with a cortisol analog (cortisol-3-CMO-NHS ester) followed by covalent conjugation with monoclonal anti-cortisol antibodies (large size and charged proteins). This chemiresistive biosensor was able to detect small size and uncharged cortisol molecules because the antibody bound to the SWCNT functionalized with cortisol analog was stripped/displaced, provoking a large change of the device resistance/conductance and a LOD of 1 pg·mL^{-1} [82]. Tan et al. also fabricated a label-free immunoFET based on SWCNT by the displacement assay of the immobilized antibodies [91]. The immunosensors were able to detect the small molecular mass microcystin-LR (MCRL), a toxin released by cyanobacteria in surface waters, with high sensitivity and specificity along with a detection limit of 0.6 ng·L^{-1}. Park and co-workers applied a SWCNT immunoFET for the detection of a small molecule of 2,4,6-trinitrotoluene (TNT), a compound used as ammunition/explosive [76]. It is a harmful chemical to soil and groundwater and can cause severe environmental contaminations. The authors also employed the displacement mode/format to develop the SWCNT-based immunosensor. The biosensor detected TNT with good selectivity and LOD of 0.5 ppb.

3.3. Graphene

Graphene consists of a honeycomb lattice of carbon atoms arranged in a two-dimensional (2D) array in which atoms are covalently linked through sp^2 hybridization. Actually, graphene is the building block of different carbon allotropes: single layer sheets can be enrolled in carbon nanotubes, wrapped in fullerenes, and stacked in three-dimensional graphite [104,105]. Graphene has attracted much attention because of its remarkable properties, including high surface area (~2630 m^2·g^{-1}), carrier mobility (~200,000 cm^2·V^{-1}·s^{-1}), electrical conductivity (~10^4 S·cm^{-1}), optical transmittance, and Young's modulus of ~1 TPa [106–108]. Its notable electronic and optical properties are due to electron confinement and absence of interlayer interactions, whereas its distinct mechanical and chemical properties are explained by geometrical effects and high surface-to-volume ratio [109]. These exceptional set of properties are particularly useful for the development of electronic sensors with high signal-to-noise ratio. Moreover, graphene is a zero bandgap semiconductor and shows ambipolar field-effect when incorporated in the configuration of a FET, since the Fermi level can be modulated by the application of a voltage between graphene and the transistor source [110–112].

This 2D nanomaterial presents some advantages beyond other nanostructures in FET-based sensors design. For example, graphene exhibits high specific surface area that can be easily functionalized for specific interactions, and providing an increased contact area for detection. Graphene is biocompatible, which helps to maintain the activity of antibodies; has excellent conductivity, which enhances the electron transfer at the electrode surface, improving the sensibility; and exhibits ambipolar characteristics [105], being able to detect both positive and negatively charged biomolecules. Additionally, as each carbon atom of the 2D structure is directly exposed to the environment, any biological recognition event occurring at the gate surface will generate an electrical perturbation in the surrounding, thus improving the sensibility of the biosensor [113].

Graphene has been employed as sensing platform of FET immunosensors for detecting various disease biomarkers. For example, Jung et al. developed a graphene FET sensor for detecting a pancreatic cancer biomarker, carbohydrate antigen 19-9 (CA 19-9). Graphene was grown by chemical vapor deposition (CVD) technique and transferred by a novel method to the sensor substrate, free from remaining polymer residues. The cleaner surface resulted in higher p-doping, higher channel mobility, and significantly enhanced sensitivity [56]. Prostate cancer biomarker, PSA/α-1-antichymotrypsin (PSA-ACT) complex, was detected by a FET sensor based on reduced graphene oxide (RGO), a chemically derived graphene [49]. RGO nanosheets were self-assembled on an aminated substrate and monoclonal PSA antibodies were immobilized on this channel. Immunoreactions caused a linear shift of gate voltage, with a LOD as low as 100 fg·mL^{-1}.

Graphene-based FET sensors are also capable of detecting foodborne and waterborne pathogens such as bacteria. Huang et al. demonstrated an immunosensor based on CVD graphene to detect *E. coli* bacteria [70]. Graphene was deposited on a quartz substrate. Anti-*E. coli* antibodies were immobilized on the graphene surface through a linker molecule (PBASE), which provided an ester group to react with amino groups of antibodies. Non-specific binding was prevented by ethanolamine and Tween 20 was used to passivate the uncoated regions of graphene (Figure 4a). Figure 4b exhibits the transfer curves of the graphene-based FET sensor after each functionalization step, demonstrating the ambipolar characteristics of graphene. Such a biosensor operating at the p-type region exhibited a significant conductance increase after exposure to various *E. coli* concentrations (from

0 to 10^5 CFU·mL^{-1}) (Figure 4c). No response was triggered after exposure to *Pseudomonas aeruginosa* bacteria, indicating high specificity (Figure 4d).

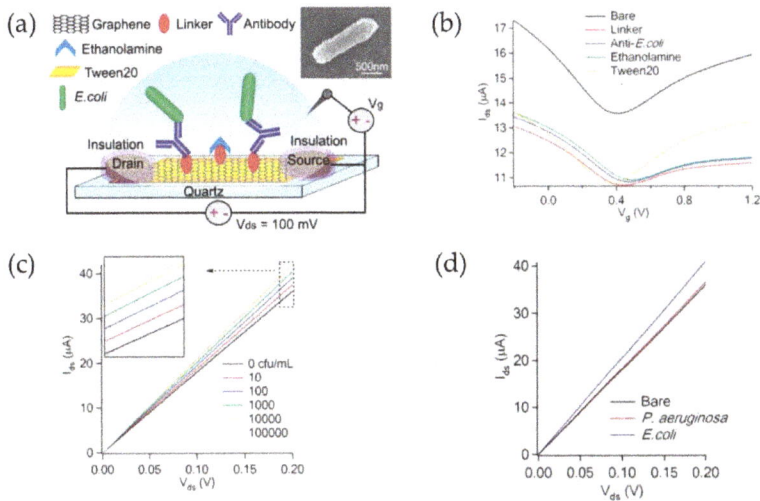

Figure 4. (**a**) Scheme showing a graphene-based FET sensor operating in an electrolyte solution for detecting *E. coli* bacteria. Bacterial cells were attached to antibodies immobilized onto the graphene surface; (**b**) Transfer curves ($I_{DS} \times V_G$) of the graphene-FET sensor. Graphene exhibited ambipolar characteristics; (**c**) $I_{DS} \times V_{DS}$ curves of graphene FET devices on various *E. coli* concentrations; (**d**) $I_{DS} \times V_{DS}$ characteristics of devices after incubation with *P. aeruginosa*, demonstrating high specificity of detection. Reproduced from Huang et al. [70] with permission from The Royal Society of Chemistry.

Chang et al. reported a graphene-based FET device to detect the most pathogenic strain of *E. coli* bacteria, known as *E. coli* O157:H7 [68]. The authors fabricated FET devices by a solution process in which graphene oxide nanosheets were self-assembled on the substrate and subsequently reduced by thermal annealing. The conductance of the devices increased with increasing concentrations of *E. coli* cells, achieving a detection limit as low as 10 CFU·mL^{-1}. Such a device operated in the p-type region and the conductance increase was induced by the highly negatively charged bacterial cells. Pathogenic rotavirus was detected by a p-type FET biosensor based on micropatterned RGO (MRGO-FET) [72]. Specific antibodies for rotavirus were covalently immobilized all over the RGO surface. As the virus concentration increased, the drain-source current proportionally decreased. The device was able to detect rotavirus in a concentration as low as 10^2 PFU·mL^{-1}, which is lower than the concentrations detected by the conventional ELISA method.

The most toxic and carcinogenic food toxin, aflatoxin (AFB1), was sensitively detected by a graphene-based FET sensor developed by Basu et al. [90]. Electrophoretically deposited RGO films were integrated as active channel into the FET device. The high sensitivity of devices was possible because of enhanced biomolecule immobilization capability of RGO. AFB1 was detected in the sub-fM range, measured with a LOD of 0.1 fg·mL^{-1}.

Okamoto et al. reported a graphene-based FET sensor that allows the antigen-antibody reaction to occur within the electrical double layer in a buffer solution of high ionic strength [114]. Graphene was produced through the mechanical cleavage of graphite and the monolayer's surface was modified with antigen-binding fragments (Fab). Fab is the binding site component of conventional antibodies and presents a size of approximately 3 nm. Thus, it is considered that the immunoreaction may occur inside the Debye length. Heat-shock proteins were used as target proteins to interact with Fab and were detected with high specificity and sensitivity by the immunoFET device.

3.4. Molybdenum Disulfide

The 2D layered material molybdenum disulfide (MoS_2) belongs to the class of transition metal dichalcogenides and features unique optical and electronic properties that have triggered great interest in the application of FET-based sensors [115]. The layers are held together by weak van der Waals forces and a pristine MoS_2 monolayer presents only ~0.65 nm thick [116]. Unlike graphene, a gapless material, monolayer MoS_2 exhibits a direct energy bandgap of ~1.9 eV which lowers leakage current and turns it an emerging material for designing highly sensitive FET biosensors [51,108,117]. Furthermore, the 2D MoS_2 semiconductor offers high surface-to-volume ratio, facilitating surface functionalization and doping; high transparency, flexibility, and mechanical strength, making it an appealing material for flexible and transparent biodevices [116].

The first demonstration of the biofunctionalization of MoS_2 nanosheets for designing a liquid-phase FET biosensor was reported by Wang et al. [51]. The authors fabricated a FET device for real-time detection of PSA cancer biomarker. Specific antibodies for PSA protein recognition were immobilized onto the MoS_2 film surface. The n-type device conductance increased upon PSA binding to antibody receptors. The novel immunosensor showed LOD of 375 fM as well as high selectivity, exhibiting no response towards non-target proteins.

The detection of the same cancer biomarker was also reported by Lee et al. [50]. The team presented a MoS_2-FET biosensor that does not require a dielectric layer due to the hydrophobicity nature of MoS_2, which allows the direct adsorption of antibodies and an improved sensitivity of the device. Figure 5a displays the schematic representation of this FET immunosensor, showing antibodies immobilized on the MoS_2 nanosheets sensing area and PSA antigen selectively bound to the antibodies. Figure 5b exhibits the sensor response to various concentrations of PSA, which was detected with a LOD as low as 1 pg·mL^{-1}, a value much lower than the clinical cut-off.

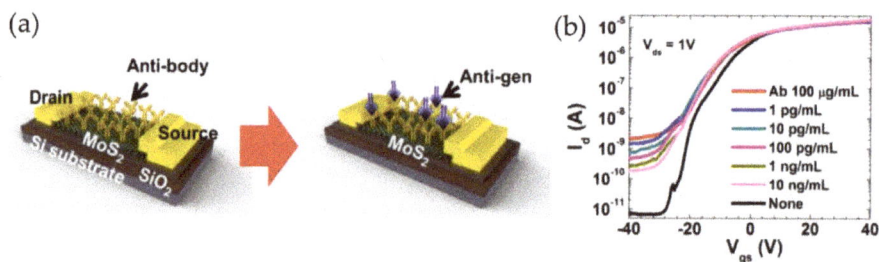

Figure 5. (**a**) Scheme showing the structure of the FET immunosensor based on MoS_2 nanosheets as sensing channel. PSA antibodies are physiosorbed onto MoS_2 and the antigen is selectively bound to the antibodies; (**b**) Transfer curves of the MoS_2-FET immunosensor under different PSA concentrations. The analyte binding caused current increase. Lee et al. are fully acknowledged for the images [50].

3.5. Titanium Dioxide

The biocompatibility combined with the environmentally friendly character of nanostructured titanium dioxide (TiO_2) make this material an excellent perspective interface for the development of biosensing devices [118]. Moreover, TiO_2 nanomaterial has a large specific surface area, shows a wide bandgap energy (between 1.8 and 4.1 eV), and possesses the ability of accepting electrons, thus the electrons resulting from the bioreceptor-analyte coupling can be gathered by TiO_2 [119]. Chu et al. demonstrated a FET immunosensor based on TiO_2 nanowires for the detection of rabbit immunoglobulin G (IgG) protein [120]. Specific antibodies for rabbit IgG were encapsulated on the 1D TiO_2 surface by the electrochemical polymerization of polypyrrole propylic acid in order to immobilize antibodies on specific regions. Biomolecules needed to be encapsulated because the material surface stability hinders the directly immobilization of antibodies or antigens. The target protein was detected in the range from 119 pg·mL^{-1} to 5.95 ng·mL^{-1}, with the application of a drain-source voltage of 5 V.

3.6. Zinc Oxide

Nanostructured zinc oxide (ZnO) is a semiconductor with a wide bandgap energy of 3.37 eV [121]. ZnO is a suitable material for biosensing applications due to its biocompatibility with low toxicity to humans and a high isoelectric point. It has been explored in the detection of enzymes, antibodies, DNA, etc. [122,123]. Recently, a highly sensitive and selective FET immunosensor based on ZnO was reported [53]. A thin ZnO nanofilm (50 nm thick) was grown onto the gate region to act as an n-type channel and monoclonal antibodies were biofunctionalized on it. The drain current increased as a function of the antigen concentration increase and EGFR, a biomarker overexpressed by breast cancer tumors, was detected with a LOD as low as 10 fM.

3.7. Hybrid Nanomaterials

The performance of FET-based biosensors can be tuned by incorporating other nanomaterials (metal, oxides and semiconductor nanoparticles) on the nanostructured channel to obtain a hybrid structure. Nanoparticles-based hybrid nanomaterials have gaining significant attention because they offer the possibility to combine the individual properties in one material, resulting in a novel material that may exhibit synergistic properties, contributing to improve the selectivity and sensitivity of biosensors [24,31].

The presence of nanoparticles on the nanomaterial surface increases the specific surface area, providing an even larger surface for recognizing an analyte, thus resulting in amplified signal transduction response and higher conductivity. The improved sensitivity of these hybrid-based sensors is also associated with increased interfacial capacitance caused by the capacitive coupling between the nanoparticles and the nanomaterial platform [124]. For instance, Mao et al. demonstrated the electrical protein binding detection by an immunosensor based on MWCNT decorated with gold nanoparticles-antibody conjugates [125]. The gold nanoparticles labeled with anti-horseradish antibodies were attached onto MWCNT surface through non-covalent binding. Biological recognition events between antibodies and horseradish peroxidase antigens caused changes in the drain current and proteins were found to be accurately detected in the order of 1 fM.

In a further work, Mao and co-workers reported the first highly sensitive and selective FET-based biosensor using RGO decorated with gold nanoparticles-antibody conjugates for protein detection (Figure 6a). Gold nanoparticles of 20 nm average size conjugated with IgG antibodies were immobilized on a thermally obtained RGO surface. The binding event of recognizing IgG target induced significant changes in electrical measurements of the device, achieving a detection limit as low as 2 ng·mL^{-1} [126]. In another effort to develop a new method to design a FET graphene-hybrid biosensor, the same research team switched the RGO sensing platform to vertically-oriented graphene sheets which were directly grown on the substrate (Figure 6b). Graphene was also functionalized with gold nanoparticles conjugated with antibodies and the devices were able to detect IgG protein again with high sensitivity (down to 2 ng·mL^{-1}) [127].

Hybrid sensors have a great potential to detect a wide variety of proteins for disease diagnosis. In this direction, Myung et al. demonstrated a novel FET immunosensor based on RGO encapsulated nanoparticle for detecting breast cancer biomarker [52]. Graphene oxide layers enwrapped silicon oxide nanoparticles functionalized with APTES by electrostatic interactions. Then, the hybrid graphene-nanoparticles were assembled at the gate region and specific monoclonal antibodies for human epidermal growth factor receptor 2 (HER2) and EGFR protein biomarkers were immobilized on its surface. The p-type device presented a decrease in conductance upon biomolecule target binding, with LOD of 1 pM for HER2 and 100 pM for EGFR. Kwon et al. fabricated a FET immunosensor for human immunodeficiency virus (HIV) detection using graphene-conducting polymer nanoparticle arrays nanohybrids [86]. The close-packed carboxylated polypyrrole nanoparticles (CPPyNP) of approximately 20 nm in size increased the surface area, resulting in a synergistic effect. The biomarker HIV-2 gp36 antigen (HIV-2 Ag) was covalently anchored to the nanoparticles surface and allowed HIV antibodies recognition in a concentration as low as 1 pM.

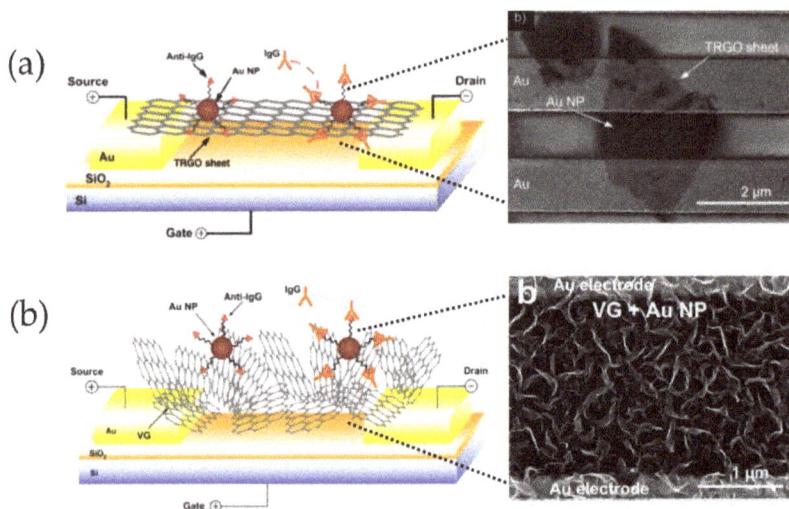

Figure 6. Representative schemes of FET immunosensors based on hybrid nanomaterials combined with corresponding SEM images of nanohybrids proposed by Mao and co-workers in different reports: (**a**) reduced graphene oxide decorated with gold nanoparticles-antibody conjugates [126]; and (**b**) vertically-oriented graphene sheets functionalized with gold nanoparticles conjugated with antibodies [127].

Zhang and co-workers developed a renewable FET immunosensor based on a graphene-TiO$_2$ hybrid for detection of D-Dimer, a venous thromboembolism biomarker [62]. The FET device was fabricated by assembling a nanocomposite of RGO functionalized with TiO$_2$ nanoparticles (TiO$_2$@RGO) onto a RGO surface in order to form a sandwich architecture on the sensing channel. The biodevice was capable of detecting the D-Dimer biomarker with excellent sensibility and specificity, achieving LOD of 10 pg·mL^{-1} in PBS buffer and 100 pg·mL^{-1} in serum, respectively. The reusability of the hybrid immunosensor was performed by irradiation of ultraviolet light to photocatalytically clean the organic molecules on the surface.

3.8. Other Nanomaterials/Other Reports of Interest

Starodub et al. fabricated a FET immunosensor based on CeO$_x$ as the gate surface to detect *Salmonella typhimurium*, a bacterium that causes food poisoning [67]. Thin CeO$_x$ layers were deposited on the gate region by electron beam evaporation. *S. typhimurium* was detected with sensitivity of 2–3 cells·mL^{-1}. The overall time of analysis was reduced from 30 min to 15–20 min by immobilizing specific antibodies labeled with horseradish peroxidase enzymes onto the gate surface, without altering the device sensitivity. Moreover, the devices were found to be reusable for up to five times without signal decrease by simply treating them with an acidic solution for the destruction of the antigen-antibody bindings.

Wu et al. proposed a FET sensor based on silicon nanobelt to detect prostate cancer biomarkers [88]. The n-type Si nanobelt immunoFET device was fabricated by functionalizing the nanobelt surface with APTES, glutaraldehyde, and antibodies specific for PSA recognition. The authors observed a decrease in drain current as a function of increasing PSA concentration, achieving a LOD of 5 pg·mL^{-1}.

The sensitivity of the device was enhanced by introducing arginine molecules between APTES and glutaraldehyde, providing a more effective space region for antigen binding, and PSA was detected in concentration levels as low as 50 fg·mL^{-1}.

A FET immunosensor based on gold nanoparticles capped with self-assembled monolayers (SAM) of alkanethiol molecules deposited onto a gold surface was fabricated for detecting hemoglobin-A1c (HbA1c), an important index for diabetes [65]. HbA1c antibodies were stably immobilized on SAM by the end functional groups of thiols. Gold nanoparticles played an important role in the sensor's sensitivity because they possess high surface-to-volume ratio, thus offering more sites to immobilize biomolecules. The immuno device was capable of detecting the antigen in concentrations in the order of ng·mL^{-1}.

Conventional FET sensors based on SiNW exhibit very low LOD, however they suffer from device instability, device-to-device variations, and discrete dopant fluctuations. An alternative to these drawbacks was demonstrated by using silicon nanogratings (SiNG) rather than SiNW [64]. The SiNG devices were fabricated using p-doped silicon-on-insulator substrates that were subjected to etching steps, thermal surface oxidation, and annealing. The biodevices presented higher electrical stability and reproducibility when exposed to buffer solution, being capable of detecting insulin, a diabetes-related hormone, with LOD down to 10 fM, both in buffer and diluted human serum.

C-reactive protein was ultrasensitively detected by a FET immunosensor based on an organic semiconductor surface composed of poly-3-hexyl thiophene (P3HT) [89]. Monoclonal anti-CRP antibodies were immobilized by physical adsorption onto the gate without any previous surface treatment. The transfer characteristics of the immunosensor showed a reduction in current with increasing concentrations of CRP and a LOD of 2 pM.

Multiplexed FET immunosensors are able to give insights on differential diagnosis. For instance, Cheng et al. reported a multianalyte immunoFET capable of detecting two lung cancer tumor markers [128]. Both cytokeratin 19 fragments (CYFRA21-1) and neuron-specific enolase (NSE) were quantitatively detected at the same time. The microdevice was fabricated through two FET integrated on the same chip and each transistor gate was biofunctionalized with different specific antibodies. CYFRA21-1 and NSE were detected with LOD of 1 and 100 ng·mL^{-1}, respectively. These results suggest a potential to easily identify different lung cancer types.

Label-free FET sensing also plays an important role on the early infection detection of agriculture plagues. *Citrus tristeza virus* and *Xylella fastidiosa* bacterium were detected by a FET biosensor using n-type indium phosphide (InP) as a biosensing platform [77]. InP substrate was aminated and PEGylated prior to the antibodies immobilization. The immunoFET detected phytopathogens with sensitivity of 2 nM, this value being comparable with highly sensitive biosensing electrochemical approaches.

4. Aptamers Instead of Antibodies

Aptamers consist of artificial oligonucleotide sequences (peptides or nucleic acids) which can recognize and bind to a wide range of targets including amino acids, proteins, enzymes, peptides, metal ions, small chemicals, viruses, and even cells with remarkable specificity and affinity [129]. Aptamers are produced through the in vitro selection and amplification of populations of random sequence oligonucleotide libraries, known as the SELEX process (selection evolution of ligands by exponential enrichment) [130].

This new class of synthetic molecules was first reported in 1990 [130,131]. Aptamers are known as antibody mimics and they have drawn much attention as promising alternatives to conventional antibodies in the design of novel FET-based immunosensors. Aptamers are capable of offering some advantages over antibodies: as they are synthetic molecules, they are chemically produced with high accuracy and reproducibility, without the use of animals or cell cultures; their production is less expensive and time-consuming than the whole process to generate specific monoclonal antibodies; and they exhibit high chemical stability in several buffer conditions, without losing bioactivity [129,132,133]. The greatest advantage of using aptamers instead of antibodies is their small size (approximately

1–2 nm), a property that makes them shorter than the Debye length. Therefore, the biorecognition event between the target analyte and the aptamer may occur within the electrical double layer, even in physiological solutions of high ionic strength, resulting in improved sensitivity and a broader range of analytes [134].

Recently, various FET-based immunosensors using aptamers immobilized on nanomaterial sensing channels were reported [135–143]. For instance, So et al. firstly demonstrated a FET biosensor based on SWCNT using aptamers as bioreceptors. Thrombin aptamers were covalently immobilized onto the surface of SWCNT previously treated with carbodiimidazole-activated Tween 20 (CDI-Tween). The SWCNT-FET sensor presented a conductance decrease upon aptamer-target binding, being able to detect thrombin, a coagulation protein, with a LOD of 10 ng·mL^{-1} [140]. Maehashi et al. fabricated a SWCNT-FET immunosensor for immunoglobulin E (IgE) detection [137]. IgE is a protein overexpressed by individuals with immune deficiency diseases. SWCNT channel was covalently modified with anti-IgE aptamers using PBASE as a linker molecule. The presence of IgE caused a sharp decrease in the drain-source current, indicating that the recognition event occurred inside the Debye layer. The dissociation constant for the reactions between aptamers and IgE was found to be 1.9×10^{-9} M. Ohno et al. reported the detection of the same protein (IgE) by a graphene-FET immunosensor [138]. Anti-IgE aptamer DNA oligonucleotides of approximately 3 nm in size were covalently immobilized on the graphene channel surface through a linker molecule (PBASE). The drain current was found to be directly dependent on the IgE concentration and a dissociation constant of 47 nM was estimated.

FET sensors using aptamers have been employed for the detection of disease biomarkers. For example, the detection of VEGF for cancer diagnosis was demonstrated by a SiNW-FET sensor modified with VEGF RNA aptamers [136]. Charged VEGF molecules on the surface of SiNW acted as electrically positive point-charges in both p-type and n-type SiNW-FET sensors. They were detected with LOD of 1.04 nM and 104 pM for n-type and p-type SiNW-FET, respectively. Interferon-gamma (IFN-γ) can be used as a biomarker to diagnose infectious diseases like tuberculosis [135]. This cytokine was sensitively detected through a graphene-FET immunosensor using IFN-γ DNA aptamers immobilized on the graphene surface. The binding of IFN-γ caused an increase in current across the graphene channel with increasing concentrations of IFN-γ, achieving LOD as low as 83 pM. Abnormal levels of CEA tumor marker were detected by a FET sensor fabricated using carboxylated polypyrrole multidimensional nanotubes (C-PPy MNTs) conjugated with CEA-binding aptamers [139]. The FET-aptasensor presented a p-type behavior since the drain-source current increased with the application of negative gate voltages. CEA was ultrasensitively detected, with a LOD of 1 fg·mL^{-1}, being a value 2–3-fold lower than those previously reported.

Aptamer-based sensors are able to directly detect small molecules and weakly charged analytes. For example, Wang et al. reported a graphene–FET device to detect the small molecule steroid hormone dehydroepiandrosterone sulfate (DHEA-S) [141]. The sensing surface was prepared by anchoring a short DNA sequence complementary to the aptamer onto the graphene surface, thus forming an aptamer–DNA anchor hybrid layer. The analyte (DHEA-S)-aptamer binding changes the aptamer conformation, releasing the aptamer from the graphene surface, and consequently inducing changes in graphene conductance. DHEA-S biomarker was detected with high specificity, achieving a clinically relevant detection limit (44.7 nM). Their findings demonstrate the potential of an aptamer-based FET sensor to detect other important low-charged small molecules in the biomedical field.

5. Conclusions and Future Perspectives

This review presents recent trends on FET-based immunosensors for the label-free detection of a broad range of analytes. The most recent reports have demonstrated a growing interest on the application of various nanomaterials such as silicon nanowires, carbon nanotubes, graphene, molybdenum disulfide, and others as sensing channels of FET-based devices. Nanostructured materials exhibit excellent physicochemical properties including high specific surface area and chemical

Chemosensors **2016**, *4*, 20

stability, which make them attractive platforms for immobilizing specific antibodies to design novel FET immunosensors with improved specificity and sensitivity. These nanomaterials-based FET immunosensors have shown very low detection limits towards biomolecules such as protein disease biomarkers, pathogenic microorganisms like bacteria and viruses, and environmental pollutants like toxins, pesticides, and herbicides.

The detection of analytes in low ionic strength buffers and desalted serum by FET immunosensors has been described. However, the direct detection of antigen-antibody reactions in physiological solutions without sample pretreatment has proven to be a challenging undertaking. Therefore, the use of aptamers instead of antibodies has been reported as an alternative to bring the biorecognition event within the electrical double layer even under high ionic strength conditions.

In summary, research regarding FET-based immuno devices appreciably increased in the last years, demonstrating countless opportunities to explore these biosensors as promising alternatives to conventional immunoassays, especially for the early-stage detection of disease biomarkers. The use of nanomaterials as sensing channels enables the design of immunosensors with enhanced performance, opening new prospects in the development of highly sensitive, miniaturized, and unlabeled immuno devices for PoC applications and simultaneous multiplexed immunoassays.

Acknowledgments: The authors acknowledge the financial assistance provided by The São Paulo Research Foundation (FAPESP, project #2013/22127-2 and grant #2016/04739-9) and the National Council for Scientific and Technological Development (CNPq).

Conflicts of Interest: The authors declare no conflict of interest.

Abbreviations

The following abbreviations are used in this manuscript:

λ_D	Debye length
AFB1	aflatoxin
AFP	α-fetoprotein
APTES	3-aminopropyltriethoxysilane
BSA	bovine serum albumin
CA 19-9	carbohydrate antigen 19-9
CDR	complementarity determining regions
CEA	carcinoembryonic antigen
CeO_x	cerium oxide
CPPyNP	carboxylated polypirrole nanoparticles
CRP	C-reactive protein
cTnI	cardiac troponin I
CVD	chemical vapor deposition
CYFRA21-1	cytokeratin 19 fragments
DHEA-S	dehydroepiandrosterone sulfate
EDC	1-ethyl-3-(3-dimethylaminopropyl) carbodiimide
EGFR	epidermal growth factor receptor
ELISA	enzyme-linked immunosorbent assay
Fab	antigen-binding fragments
FET	field-effect transistor
HER2	human epidermal growth factor receptor 2
HBsAg	hepatitis B marker
HIV	human immunodeficiency virus
InP	indium phosphide
I_{DS}	drain-source current
IFN-γ	interferon-gamma
IgE	immunoglobulin E

IgG	immunoglobulin G
IL-6	interleukin-6
IL-8	interleukin-8
ISFET	ion-sensitive field effect transistor
LOD	limit of detection
MCRL	microcystin-LR
MoS_2	molybdenum disulfide
MWCNT	multi-walled carbon nanotubes
NHS	*N*-hydroxysuccinimide
NSE	neuron-specific enolase
OPN	osteopontin
P3HT	poly-3-hexyl thiophene
PBASE	1-pyrene butanoic acid succinimidyl ester
PBS	phosphate-buffered saline
PDMS	polydimethylsiloxane
PoC	point-of-care
PSA	prostate specific antigen
PSA-ACT	Prostate specific antigen/α1-antichymotrypsin
RIA	radioimmunoassay
RGO	reduced graphene oxide
SAM	self-assembled monolayers
SELEX	selection evolution of ligands by exponential enrichment
SEM	scanning electron microscopy
SiNG	silicon nanogratings
SiNW	silicon nanowire
SWCNT	single-walled carbon nanotubes
TiO_2	titanium dioxide
TNF-α	tumor necrosis factor α
TNT	2,4,6-trinitrotoluene
V_{DS}	drain-source voltage
VEGF	vascular endothelial growth factor
V_G	gate voltage
ZnO	zinc oxide

References

1. Freire, R.S.; Pessoa, C.A.; Mello, L.D.; Kubota, L.T. Direct electron transfer: An approach for electrochemical biosensors with higher selectivity and sensitivity. *J. Braz. Chem. Soc.* **2003**, *14*, 230–243. [CrossRef]

2. Thevenot, D.R.; Toth, K.; Durst, R.A.; Wilson, G.S. Electrochemical biosensors: Recommended definitions and classification. *Biosens. Bioelectron.* **2001**, *16*, 121–131. [CrossRef] [PubMed]

3. Luppa, P.B.; Sokoll, L.J.; Chan, D.W. Immunosensors—Principles and applications to clinical chemistry. *Clin. Chim. Acta* **2001**, *314*, 1–26. [CrossRef]

4. Conroy, P.J.; Hearty, S.; Leonard, P.; O'Kennedy, R.J. Antibody production, design and use for biosensor-based applications. *Semin. Cell Dev. Biol.* **2009**, *20*, 10–26. [CrossRef] [PubMed]

5. Viguier, C.; Lynam, C.; O'Kennedy, R. Trends and perspectives in immunosensors. In *Antibodies Applications and New Developments*; 2012; pp. 184–208. Available online: http://ebooks.benthamscience.com/book/9781608052646/ (accessed on 20 October 2016).

6. Hock, B. Antibodies for immunosensors—A review. *Anal. Chim. Acta* **1997**, *347*, 177–186. [CrossRef]

7. Wang, J. Electrochemical biosensors: Towards point-of-care cancer diagnostics. *Biosens. Bioelectron.* **2006**, *21*, 1887–1892. [CrossRef] [PubMed]

8. Song, Y.; Luo, Y.N.; Zhu, C.Z.; Li, H.; Du, D.; Lin, Y.H. Recent advances in electrochemical biosensors based on graphene two-dimensional nanomaterials. *Biosens. Bioelectron.* **2016**, *76*, 195–212. [CrossRef] [PubMed]

9. Luo, X.L.; Davis, J.J. Electrical biosensors and the label free detection of protein disease biomarkers. *Chem. Soc. Rev.* **2013**, *42*, 5944–5962. [CrossRef] [PubMed]

10. Porstmann, T.; Kiessig, S.T. Enzyme-immunoassay techniques—An overview. *J. Immunol. Methods* **1992**, *150*, 5–21. [CrossRef]

11. Lee, T.M.H. Over-the-counter biosensors: Past, present, and future. *Sensors* **2008**, *8*, 5535–5559. [CrossRef]

12. Ivnitski, D.; Abdel-Hamid, I.; Atanasov, P.; Wilkins, E. Biosensors for detection of pathogenic bacteria. *Biosens. Bioelectron.* **1999**, *14*, 599–624. [CrossRef]
13. Yalow, R.S.; Berson, S.A. Immunoassay of endogenous plasma insulin in man. *J. Clin. Investig.* **1960**, *39*, 1157–1175. [CrossRef] [PubMed]
14. Holford, T.R.J.; Davis, F.; Higson, S.P.J. Recent trends in antibody based sensors. *Biosens. Bioelectron.* **2012**, *34*, 12–24. [CrossRef] [PubMed]
15. Centi, S.; Laschi, S.; Mascini, M. Strategies for electrochemical detection in immunochemistry. *Bioanalysis* **2009**, *1*, 1271–1291. [CrossRef] [PubMed]
16. Marco, M.P.; Gee, S.; Hammock, B.D. Immunochemical techniques for environmental-analysis. 1. Immunosensors. *Trends Anal. Chem.* **1995**, *14*, 341–350. [CrossRef]
17. Zachariah, E.S.; Gopalakrishnakone, P.; Neuzil, P. Immunologically sensitive field-effect transistors. In *Encyclopedia of Medical Devices and Instrumentation*; John Wiley & Sons: New York, NY, USA, 2006.
18. Poghossian, A.; Schoning, M.J. Label-free sensing of biomolecules with field-effect devices for clinical applications. *Electroanal* **2014**, *26*, 1197–1213. [CrossRef]
19. Schoning, M.J.; Poghossian, A. Bio feds (field-effect devices): State-of-the-art and new directions. *Electroanal* **2006**, *18*, 1893–1900. [CrossRef]
20. Bergveld, P. Development, operation, and application of ion-sensitive field-effect transistor as a tool for electrophysiology. *IEEE Trans. Biomed. Eng.* **1972**, *19*, 342–351. [CrossRef] [PubMed]
21. Bergveld, P. A critical-evaluation of direct electrical protein-detection methods. *Biosens. Bioelectron.* **1991**, *6*, 55–72. [CrossRef]
22. Bergveld, P. The future of biosensors. *Sens. Actuat. A Phys.* **1996**, *56*, 65–73. [CrossRef]
23. Schasfoort, R.B.M.; Bergveld, P.; Kooyman, R.P.H.; Greve, J. Possibilities and limitations of direct detection of protein charges by means of an immunological field-effect transistor. *Anal. Chim. Acta* **1990**, *238*, 323–329. [CrossRef]
24. Yin, P.T.; Shah, S.; Chhowalla, M.; Lee, K.B. Design, synthesis, and characterization of graphene-nanoparticle hybrid materials for bioapplications. *Chem. Rev.* **2015**, *115*, 2483–2531. [CrossRef] [PubMed]
25. Ohno, Y.; Maehashi, K.; Yamashiro, Y.; Matsumoto, K. Electrolyte-gated graphene field-effect transistors for detecting ph protein adsorption. *Nano Lett.* **2009**, *9*, 3318–3322. [CrossRef] [PubMed]
26. Ang, P.K.; Chen, W.; Wee, A.T.S.; Loh, K.P. Solution-gated epitaxial graphene as ph sensor. *J. Am. Chem. Soc.* **2008**, *130*, 14392–14393. [CrossRef] [PubMed]
27. Di Bartolomeo, A.; Rinzan, M.; Boyd, A.K.; Yang, Y.; Guadagno, L.; Giubileo, F.; Barbara, P. Electrical properties and memory effects of field-effect transistors from networks of single-and double-walled carbon nanotubes. *Nanotechnology* **2010**, *21*, 115204. [CrossRef] [PubMed]
28. Cramer, T.; Campana, A.; Leonardi, F.; Casalini, S.; Kyndiah, A.; Murgia, M.; Biscarini, F. Water-gated organic field effect transistors—opportunities for biochemical sensing and extracellular signal transduction. *J. Mater. Chem. B* **2013**, *1*, 3728–3741. [CrossRef]
29. Patolsky, F.; Zheng, G.F.; Lieber, C.M. Nanowire-based biosensors. *Anal. Chem.* **2006**, *78*, 4260–4269. [CrossRef] [PubMed]
30. Schenk, J.F. *Theory, Design, and Biomedical Applications of Solid State Chemical Sensors*; Cheung, P.W., Ed.; CRC Press: West Palm Beach, FL, USA, 1978; p. 296.
31. Yin, P.T.; Kim, T.H.; Choi, J.W.; Lee, K.B. Prospects for graphene-nanoparticle-based hybrid sensors. *Phys. Chem. Chem. Phys.* **2013**, *15*, 12785–12799. [CrossRef] [PubMed]
32. Pumera, M. Graphene in biosensing. *Mater. Today* **2011**, *14*, 308–315. [CrossRef]
33. Ramnani, P.; Saucedo, N.M.; Mulchandani, A. Carbon nanomaterial-based electrochemical biosensors for label-free sensing of environmental pollutants. *Chemosphere* **2016**, *143*, 85–98. [CrossRef] [PubMed]
34. Nehra, A.; Singh, K.P. Current trends in nanomaterial embedded field effect transistor-based biosensor. *Biosens. Bioelectron.* **2015**, *74*, 731–743. [CrossRef] [PubMed]
35. Lin, P.; Yan, F. Organic thin-film transistors for chemical and biological sensing. *Adv. Mater.* **2012**, *24*, 34–51. [CrossRef] [PubMed]
36. He, R.X.; Lin, P.; Liu, Z.K.; Zhu, H.W.; Zhao, X.Z.; Chan, H.L.W.; Yan, F. Solution-gated graphene field effect transistors integrated in microfluidic systems and used for flow velocity detection. *Nano Lett.* **2012**, *12*, 1404–1409. [CrossRef] [PubMed]

37. Mu, L.; Chang, Y.; Sawtelle, S.D.; Wipf, M.; Duan, X.X.; Reed, M.A. Silicon nanowire field-effect transistors-a versatile class of potentiometric nanobiosensors. *IEEE Access* **2015**, *3*, 287–302. [CrossRef]

38. Debye, P. Reaction rates in ionic solutions. *J. Electrochem. Soc.* **1942**, *82*, 265–272. [CrossRef]

39. Russel, W.B.; Saville, D.A.; Schowalter, W.R. *Colloidal Dispersions*; Cambridge University Press: Cambridge, UK, 1989.

40. Elnathan, R.; Kwiat, M.; Pevzner, A.; Engel, Y.; Burstein, L.; Khatchtourints, A.; Lichtenstein, A.; Kantaev, R.; Patolsky, F. Biorecognition layer engineering: Overcoming screening limitations of nanowire-based fet devices. *Nano Lett.* **2012**, *12*, 5245–5254. [CrossRef] [PubMed]

41. Stern, E.; Wagner, R.; Sigworth, F.J.; Breaker, R.; Fahmy, T.M.; Reed, M.A. Importance of the debye screening length on nanowire field effect transistor sensors. *Nano Lett.* **2007**, *7*, 3405–3409. [CrossRef] [PubMed]

42. Casal, P.; Wen, X.J.; Gupta, S.; Nicholson, T.; Wang, Y.J.; Theiss, A.; Bhushan, B.; Brillson, L.; Lu, W.; Lee, S.C. Immunofet feasibility in physiological salt environments. *Philos. Trans. R. Soc. A* **2012**, *370*, 2474–2488. [CrossRef] [PubMed]

43. Schoning, M.J.; Poghossian, A. Recent advances in biologically sensitive field-effect transistors (biofets). *Analyst* **2002**, *127*, 1137–1151. [CrossRef] [PubMed]

44. Ansari, A.A.; Alhoshan, M.; Alsalhi, M.S.; Aldwayyan, A.S. Prospects of nanotechnology in clinical immunodiagnostics. *Sensors* **2010**, *10*, 6535–6581. [CrossRef] [PubMed]

45. Zhu, C.Z.; Yang, G.H.; Li, H.; Du, D.; Lin, Y.H. Electrochemical sensors and biosensors based on nanomaterials and nanostructures. *Anal. Chem.* **2015**, *87*, 230–249. [CrossRef] [PubMed]

46. Hu, W.H.; Li, C.M. Nanomaterial-based advanced immunoassays. *Wires Nanomed. Nanobiotechnol.* **2011**, *3*, 119–133. [CrossRef] [PubMed]

47. Lu, N.; Gao, A.R.; Dai, P.F.; Mao, H.J.; Zuo, X.L.; Fan, C.H.; Wang, Y.L.; Li, T. Ultrasensitive detection of dual cancer biomarkers with integrated cmos-compatible nanowire arrays. *Anal. Chem.* **2015**, *87*, 11203–11208. [CrossRef] [PubMed]

48. Huang, Y.W.; Wu, C.S.; Chuang, C.K.; Pang, S.T.; Pan, T.M.; Yang, Y.S.; Ko, F.H. Real-time and label-free detection of the prostate-specific antigen in human serum by a polycrystalline silicon nanowire field-effect transistor biosensor. *Anal. Chem.* **2013**, *85*, 7912–7918. [CrossRef] [PubMed]

49. Kim, D.J.; Sohn, I.Y.; Jung, J.H.; Yoon, O.J.; Lee, N.E.; Park, J.S. Reduced graphene oxide field-effect transistor for label-free femtomolar protein detection. *Biosens. Bioelectron.* **2013**, *41*, 621–626. [CrossRef] [PubMed]

50. Lee, J.; Dak, P.; Lee, Y.; Park, H.; Choi, W.; Alam, M.A.; Kim, S. Two-dimensional layered mos2 biosensors enable highly sensitive detection of biomolecules. *Sci. Rep.* **2014**, *4*, 7352. [CrossRef] [PubMed]

51. Wang, L.; Wang, Y.; Wong, J.I.; Palacios, T.; Kong, J.; Yang, H.Y. Functionalized mos2 nanosheet-based field-effect biosensor for label-free sensitive detection of cancer marker proteins in solution. *Small* **2014**, *10*, 1101–1105. [CrossRef] [PubMed]

52. Myung, S.; Solanki, A.; Kim, C.; Park, J.; Kim, K.S.; Lee, K.B. Graphene-encapsulated nanoparticle-based biosensor for the selective detection of cancer biomarkers. *Adv. Mater.* **2011**, *23*, 2221–2225. [CrossRef] [PubMed]

53. Reyes, P.I.; Ku, C.J.; Duan, Z.Q.; Lu, Y.C.; Solanki, A.; Lee, K.B. Zno thin film transistor immunosensor with high sensitivity and selectivity. *Appl. Phys. Lett.* **2011**, *98*, 173702. [CrossRef]

54. Zhu, K.Y.; Zhang, Y.; Li, Z.Y.; Zhou, F.; Feng, K.; Dou, H.Q.; Wang, T. Simultaneous detection of alpha-fetoprotein and carcinoembryonic antigen based on si nanowire field-effect transistors. *Sensors* **2015**, *15*, 19225–19236. [CrossRef] [PubMed]

55. Zhang, Y.L.; Chen, R.M.; Xu, L.; Ning, Y.; Xie, S.G.; Zhang, G.J. Silicon nanowire biosensor for highly sensitive and multiplexed detection of oral squamous cell carcinoma biomarkers in saliva. *Anal. Sci.* **2015**, *31*, 73–78. [CrossRef] [PubMed]

56. Jung, J.H.; Sohn, I.Y.; Kim, D.J.; Kim, B.Y.; Fang, M.; Lee, N.E. Enhancement of protein detection performance in field-effect transistors with polymer residue-free graphene channel. *Carbon* **2013**, *62*, 312–321. [CrossRef]

57. Chen, H.C.; Chen, Y.T.; Tsai, R.Y.; Chen, M.C.; Chen, S.L.; Xiao, M.C.; Chen, C.L.; Hua, M.Y. A sensitive and selective magnetic graphene composite-modified polycrystalline-silicon nanowire field-effect transistor for bladder cancer diagnosis. *Biosens. Bioelectron.* **2015**, *66*, 198–207. [CrossRef] [PubMed]

58. Justino, C.I.L.; Freitas, A.C.; Amaral, J.P.; Rocha-Santos, T.A.P.; CardosoC, S.; Duarte, A.C. Disposable immunosensors for c-reactive protein based on carbon nanotubes field effect transistors. *Talanta* **2013**, *108*, 165–170. [CrossRef] [PubMed]

59. Kim, K.; Park, C.; Kwon, D.; Kim, D.; Meyyappan, M.; Jeon, S.; Lee, J.S. Silicon nanowire biosensors for detection of cardiac troponin i (ctni) with high sensitivity. *Biosens. Bioelectron.* **2016**, *77*, 695–701. [CrossRef] [PubMed]

60. Kong, T.; Su, R.G.; Zhang, B.B.; Zhang, Q.; Cheng, G.S. Cmos-compatible, label-free silicon-nanowire biosensors to detect cardiac troponin i for acute myocardial infarction diagnosis. *Biosens. Bioelectron.* **2012**, *34*, 267–272. [CrossRef] [PubMed]

61. Lu, N.; Dai, P.F.; Gao, A.R.; Valiaho, J.; Kallio, P.; Wang, Y.L.; Li, T. Label-free and rapid electrical detection of htsh with cmos-compatible silicon nanowire transistor arrays. *ACS Appl. Mater. Interfaces* **2014**, *6*, 20378–20384. [CrossRef] [PubMed]

62. Zhang, C.; Xu, J.-Q.; Li, T.; Huang, L.; Pang, D.-W.; Ning, Y.; Huang, W.-H.; Zhang, Z.; Zhang, G.-J. Photocatalysis-induced renewable field-effect transistor for protein detection. *Anal. Chem.* **2016**, *88*, 4048–4054. [CrossRef] [PubMed]

63. Oh, J.; Yoo, G.; Chang, Y.W.; Kim, H.J.; Jose, J.; Kim, E.; Pyun, J.C.; Yoo, K.H. A carbon nanotube metal semiconductor field effect transistor-based biosensor for detection of amyloid-beta in human serum. *Biosens. Bioelectron.* **2013**, *50*, 345–350. [CrossRef] [PubMed]

64. Regonda, S.; Tian, R.H.; Gao, J.M.; Greene, S.; Ding, J.H.; Hu, W. Silicon multi-nanochannel fets to improve device uniformity/stability and femtomolar detection of insulin in serum. *Biosens. Bioelectron.* **2013**, *45*, 245–251. [CrossRef] [PubMed]

65. Xue, Q.N.; Bian, C.; Tong, J.H.; Sun, J.Z.; Zhang, H.; Xia, S.H. Fet immunosensor for hemoglobin a1c using a gold nanofilm grown by a seed-mediated technique and covered with mixed self-assembled monolayers. *Microchim. Acta* **2012**, *176*, 65–72. [CrossRef]

66. Lerner, M.B.; Goldsmith, B.R.; McMillon, R.; Dailey, J.; Pillai, S.; Singh, S.R.; Johnson, A.T.C. A carbon nanotube immunosensor for salmonella. *Aip Adv.* **2011**, *1*. [CrossRef]

67. Starodub, N.F.; Ogorodnijchuk, J.O. Immune biosensor based on the isfets for express determination of *Salmonella typhimurium*. *Electroanal* **2012**, *24*, 600–606. [CrossRef]

68. Chang, J.B.; Mao, S.; Zhang, Y.; Cui, S.M.; Zhou, G.H.; Wu, X.G.; Yang, C.H.; Chen, J.H. Ultrasonic-assisted self-assembly of monolayer graphene oxide for rapid detection of *Escherichia coli* bacteria. *Nanoscale* **2013**, *5*, 3620–3626. [CrossRef] [PubMed]

69. Garcia-Aljaro, C.; Cella, L.N.; Shirale, D.J.; Park, M.; Munoz, F.J.; Yates, M.V.; Mulchandani, A. Carbon nanotubes-based chemiresistive biosensors for detection of microorganisms. *Biosens. Bioelectron.* **2010**, *26*, 1437–1441. [CrossRef] [PubMed]

70. Huang, Y.X.; Dong, X.C.; Liu, Y.X.; Li, L.J.; Chen, P. Graphene-based biosensors for detection of bacteria and their metabolic activities. *J. Mater. Chem.* **2011**, *21*, 12358–12362. [CrossRef]

71. Yamada, K.; Kim, C.T.; Kim, J.H.; Chung, J.H.; Lee, H.G.; Jun, S. Single walled carbon nanotube-based junction biosensor for detection of *Escherichia coli*. *PLoS ONE* **2014**, *9*. [CrossRef] [PubMed]

72. Liu, F.; Kim, Y.H.; Cheon, D.S.; Seo, T.S. Micropatterned reduced graphene oxide based field-effect transistor for real-time virus detection. *Sens. Actuat. B Chem.* **2013**, *186*, 252–257. [CrossRef]

73. Wong, J.I.; Wang, L.; Shi, Y.M.; Palacios, T.; Kong, J.; Dong, X.C.; Yang, H.Y. Real-time, sensitive electrical detection of cryptosporidium parvum oocysts based on chemical vapor deposition-grown graphene. *Appl. Phys. Lett.* **2014**, *104*, 063705. [CrossRef]

74. Belkhamssa, N.; Justino, C.I.L.; Santos, P.S.M.; Cardoso, S.; Lopes, I.; Duarte, A.C.; Rocha-Santos, T.; Ksibi, M. Label-free disposable immunosensor for detection of atrazine. *Talanta* **2016**, *146*, 430–434. [CrossRef] [PubMed]

75. Wijaya, I.P.M.; Nie, T.J.; Gandhi, S.; Boro, R.; Palaniappan, A.; Hau, G.W.; Rodriguez, I.; Suri, C.R.; Mhaisalkar, S.G. Femtomolar detection of 2,4-dichlorophenoxyacetic acid herbicides via competitive immunoassays using microfluidic based carbon nanotube liquid gated transistor. *Lab Chip* **2010**, *10*, 634–638. [CrossRef] [PubMed]

76. Park, M.; Cella, L.N.; Chen, W.F.; Myung, N.V.; Mulchandani, A. Carbon nanotubes-based chemiresistive immunosensor for small molecules: Detection of nitroaromatic explosives. *Biosens. Bioelectron.* **2010**, *26*, 1297–1301. [CrossRef] [PubMed]

77. Moreau, A.L.D.; Janissen, R.; Santos, C.A.; Peroni, L.A.; Stach-Machado, D.R.; de Souza, A.A.; de Souza, A.P.; Cotta, M.A. Highly-sensitive and label-free indium phosphide biosensor for early phytopathogen diagnosis. *Biosens. Bioelectron.* **2012**, *36*, 62–68. [CrossRef] [PubMed]

78. Chiang, P.L.; Chou, T.C.; Wu, T.H.; Li, C.C.; Liao, C.D.; Lin, J.Y.; Tsai, M.H.; Tsai, C.C.; Sun, C.J.; Wang, C.H.; et al. Nanowire transistor-based ultrasensitive virus detection with reversible surface functionalization. *Chem. Asian J.* **2012**, *7*, 2073–2079. [CrossRef] [PubMed]

79. Ivanov, Y.D.; Pleshakova, T.O.; Kozlov, A.F.; Malsagova, K.A.; Krohin, N.V.; Shumyantseva, V.V.; Shumov, I.D.; Popov, V.P.; Naumova, O.V.; Fomin, B.I.; et al. Soi nanowire for the high-sensitive detection of hbsag and alpha-fetoprotein. *Lab Chip* **2012**, *12*, 5104–5111. [CrossRef] [PubMed]

80. Sharma, A.; Hong, S.; Singh, R.; Jang, J. Single-walled carbon nanotube based transparent immunosensor for detection of a prostate cancer biomarker osteopontin. *Anal. Chim. Acta* **2015**, *869*, 68–73. [CrossRef] [PubMed]

81. Chen, H.; Choo, T.K.; Huang, J.F.; Wang, Y.; Liu, Y.J.; Platt, M.; Palaniappan, A.; Liedberg, B.; Tok, A.I.Y. Label-free electronic detection of interleukin-6 using horizontally aligned carbon nanotubes. *Mater. Des.* **2016**, *90*, 852–857. [CrossRef]

82. Tlili, C.; Myung, N.V.; Shetty, V.; Mulchandani, A. Label-free, chemiresistor immunosensor for stress biomarker cortisol in saliva. *Biosens. Bioelectron.* **2011**, *26*, 4382–4386. [CrossRef] [PubMed]

83. Lerner, M.B.; Dailey, J.; Goldsmith, B.R.; Brisson, D.; Johnson, A.T.C. Detecting lyme disease using antibody-functionalized single-walled carbon nanotube transistors. *Biosens. Bioelectron.* **2013**, *45*, 163–167. [CrossRef] [PubMed]

84. Lerner, M.B.; D'Souza, J.; Pazina, T.; Dailey, J.; Goldsmith, B.R.; Robinson, M.K.; Johnson, A.T.C. Hybrids of a genetically engineered antibody and a carbon nanotube transistor for detection of prostate cancer biomarkers. *ACS Nano* **2012**, *6*, 5143–5149. [CrossRef] [PubMed]

85. Yeh, C.H.; Kumar, V.; Moyano, D.R.; Wen, S.H.; Parashar, V.; Hsiao, S.H.; Srivastava, A.; Saxena, P.S.; Huang, K.P.; Chang, C.C.; et al. High-performance and high-sensitivity applications of graphene transistors with self-assembled monolayers. *Biosens. Bioelectron.* **2016**, *77*, 1008–1015. [CrossRef] [PubMed]

86. Kwon, O.S.; Lee, S.H.; Park, S.J.; An, J.H.; Song, H.S.; Kim, T.; Oh, J.H.; Bae, J.; Yoon, H.; Park, T.H.; et al. Large-scale graphene micropattern nano-biohybrids: High-performance transducers for fet-type flexible fluidic HIV immunoassays. *Adv. Mater.* **2013**, *25*, 4177–4185. [CrossRef] [PubMed]

87. Kim, B.Y.; Sohn, I.Y.; Lee, D.; Han, G.S.; Lee, W.I.; Jung, H.S.; Lee, N.E. Ultrarapid and ultrasensitive electrical detection of proteins in a three-dimensional biosensor with high capture efficiency. *Nanoscale* **2015**, *7*, 9844–9851. [CrossRef] [PubMed]

88. Wu, C.C.; Pan, T.M.; Wu, C.S.; Yen, L.C.; Chuang, C.K.; Pang, S.T.; Yang, Y.S.; Ko, F.H. Label-free detection of prostate specific antigen using a silicon nanobelt field-effect transistor. *Int. J. Electrochem. Sci.* **2012**, *7*, 4432–4442.

89. Magliulo, M.; De Tullio, D.; Vikholm-Lundin, I.; Albers, W.; Munter, T.; Manoli, K.; Palazzo, G.; Torsi, L. Label-free c-reactive protein electronic detection with an electrolyte-gated organic field-effect transistor-based immunosensor. *Anal. Bioanal. Chem.* **2016**, *408*, 3943–3952. [CrossRef] [PubMed]

90. Basu, J.; Datta, S.; RoyChaudhuri, C. A graphene field effect capacitive immunosensor for sub-femtomolar food toxin detection. *Biosens. Bioelectron.* **2015**, *68*, 544–549. [CrossRef] [PubMed]

91. Tan, F.; Saucedo, N.M.; Ramnani, P.; Mulchandani, A. Label-free electrical immunosensor for highly sensitive and specific detection of microcystin-lr in water samples. *Environ. Sci. Technol.* **2015**, *49*, 9256–9263. [CrossRef] [PubMed]

92. Hu, J.T.; Odom, T.W.; Lieber, C.M. Chemistry and physics in one dimension: Synthesis and properties of nanowires and nanotubes. *Acc. Chem. Res.* **1999**, *32*, 435–445. [CrossRef]

93. Cui, Y.; Duan, X.F.; Hu, J.T.; Lieber, C.M. Doping and electrical transport in silicon nanowires. *J. Phys. Chem. B* **2000**, *104*, 5213–5216. [CrossRef]

94. Hasan, M.; Huq, M.F.; Mahmood, Z.H. A review on electronic and optical properties of silicon nanowire and its different growth techniques. *Springerplus* **2013**, *2*, 151. [CrossRef] [PubMed]

95. Chen, K.I.; Li, B.R.; Chen, Y.T. Silicon nanowire field-effect transistor-based biosensors for biomedical diagnosis and cellular recording investigation. *Nano Today* **2011**, *6*, 131–154. [CrossRef]

96. Noor, M.O.; Krull, U.J. Silicon nanowires as field-effect transducers for biosensor development: A review. *Anal. Chim. Acta* **2014**, *825*, 1–25. [CrossRef] [PubMed]

97. Cui, Y.; Wei, Q.Q.; Park, H.K.; Lieber, C.M. Nanowire nanosensors for highly sensitive and selective detection of biological and chemical species. *Science* **2001**, *293*, 1289–1292. [CrossRef] [PubMed]

98. Puppo, F.; Doucey, M.-A.; Moh, T.S.Y.; Pandraud, G.; Sarro, P.M.; De Micheli, G.; Carrara, S. Femto-molar sensitive field effect transistor biosensors based on silicon nanowires and antibodies. In Proceedings of the 2013 IEEE Sensors Proceedings, Baltimore, MD, USA, 4–6 November 2013.

99. Popov, V.N. Carbon nanotubes: Properties and application. *Mater. Sci. Eng. R* **2004**, *43*, 61–102. [CrossRef]

100. Iijima, S. Helical microtubules of graphitic carbon. *Nature* **1991**, *354*, 56–58. [CrossRef]

101. Iijima, S.; Ichihashi, T. Single-shell carbon nanotubes of 1-nm diameter. *Nature* **1993**, *363*, 603–605. [CrossRef]

102. Bethune, D.S.; Kiang, C.H.; Devries, M.S.; Gorman, G.; Savoy, R.; Vazquez, J.; Beyers, R. Cobalt-catalyzed growth of carbon nanotubes with single-atomic-layerwalls. *Nature* **1993**, *363*, 605–607. [CrossRef]

103. Sarkar, T.; Gao, Y.N.; Mulchandani, A. Carbon nanotubes-based label-free affinity sensors for environmental monitoring. *Appl. Biochem. Biotechnol.* **2013**, *170*, 1011–1025. [CrossRef] [PubMed]

104. Geim, A.K.; Novoselov, K.S. The rise of graphene. *Nat. Mater.* **2007**, *6*, 183–191. [CrossRef] [PubMed]

105. Novoselov, K.S.; Geim, A.K.; Morozov, S.V.; Jiang, D.; Zhang, Y.; Dubonos, S.V.; Grigorieva, I.V.; Firsov, A.A. Electric field effect in atomically thin carbon films. *Science* **2004**, *306*, 666–669. [CrossRef] [PubMed]

106. Park, S.; Ruoff, R.S. Chemical methods for the production of graphenes. *Nat. Nanotechnol.* **2009**, *4*, 217–224. [CrossRef] [PubMed]

107. Guo, S.J.; Dong, S.J. Graphene nanosheet: Synthesis, molecular engineering, thin film, hybrids, and energy and analytical applications. *Chem. Soc. Rev.* **2011**, *40*, 2644–2672. [CrossRef] [PubMed]

108. Di Bartolomeo, A. Graphene schottky diodes: An experimental review of the rectifying graphene/semiconductor heterojunction. *Phys. Rep.* **2016**, *606*, 1–58. [CrossRef]

109. Mas-Balleste, R.; Gomez-Navarro, C.; Gomez-Herrero, J.; Zamora, F. 2D materials: To graphene and beyond. *Nanoscale* **2011**, *3*, 20–30. [CrossRef] [PubMed]

110. Partoens, B.; Peeters, F.M. From graphene to graphite: Electronic structure around the k point. *Phys. Rev. B* **2006**, *74*, 075404. [CrossRef]

111. Schwierz, F. Graphene transistors. *Nat. Nanotechnol.* **2010**, *5*, 487–496. [CrossRef] [PubMed]

112. Di Bartolomeo, A.; Giubileo, F.; Santandrea, S.; Romeo, F.; Citro, R.; Schroeder, T.; Lupina, G. Charge transfer and partial pinning at the contacts as the origin of a double dip in the transfer characteristics of graphene-based field-effect transistors. *Nanotechnology* **2011**, *22*, 275702. [CrossRef] [PubMed]

113. Liu, Y.X.; Dong, X.C.; Chen, P. Biological and chemical sensors based on graphene materials. *Chem. Soc. Rev.* **2012**, *41*, 2283–2307. [CrossRef] [PubMed]

114. Okamoto, S.; Ohno, Y.; Maehashi, K.; Inoue, K.; Matsumoto, K. Immunosensors based on graphene field-effect transistors fabricated using antigen-binding fragment. *Jpn. J. Appl. Phys.* **2012**, *51*, 06FD08. [CrossRef]

115. Tong, X.; Ashalley, E.; Lin, F.; Li, H.D.; Wang, Z.M.M. Advances in mos2-based field effect transistors (fets). *Nano-Micro Lett.* **2015**, *7*, 203–218. [CrossRef]

116. Sarkar, D.; Liu, W.; Xie, X.J.; Anselmo, A.C.; Mitragotri, S.; Banerjee, K. Mos2 field-effect transistor for next-generation label-free biosensors. *ACS Nano* **2014**, *8*, 3992–4003. [CrossRef] [PubMed]

117. Sarkar, D.; Xie, X.J.; Kang, J.H.; Zhang, H.J.; Liu, W.; Navarrete, J.; Moskovits, M.; Banerjee, K. Functionalization of transition metal dichalcogenides with metallic nanoparticles: Implications for doping and gas-sensing. *Nano Lett.* **2015**, *15*, 2852–2862. [CrossRef] [PubMed]

118. Bai, J.; Zhou, B.X. Titanium dioxide nanomaterials for sensor applications. *Chem. Rev.* **2014**, *114*, 10131–10176. [CrossRef] [PubMed]

119. Wang, R.H.; Ruan, C.M.; Kanayeva, D.; Lassiter, K.; Li, Y.B. Tio2 nanowire bundle microelectrode based impedance immunosensor for rapid and sensitive detection of listeria monocytogenes. *Nano Lett.* **2008**, *8*, 2625–2631. [CrossRef] [PubMed]

120. Chu, Y.M.; Lin, C.C.; Chang, H.C.; Li, C.M.; Guo, C.X. Tio(2) nanowire fet device: Encapsulation of biomolecules by electro polymerized pyrrole propylic acid. *Biosens. Bioelectron.* **2011**, *26*, 2334–2340. [CrossRef] [PubMed]

121. Arya, S.K.; Saha, S.; Ramirez-Vick, J.E.; Gupta, V.; Bhansali, S.; Singh, S.P. Recent advances in zno nanostructures and thin films for biosensor applications: Review. *Anal. Chim. Acta* **2012**, *737*, 1–21. [CrossRef] [PubMed]

122. Yano, M.; Koike, K.; Mukai, K.; Onaka, T.; Hirofuji, Y.; Ogata, K.; Omatu, S.; Maemoto, T.; Sasa, S. Zinc oxide ion-sensitive field-effect transistors and biosensors. *Phys. Status Solidi A* **2014**, *211*, 2098–2104. [CrossRef]

123. Zhao, Z.W.; Lei, W.; Zhang, X.B.; Wang, B.P.; Jiang, H.L. Zno-based amperometric enzyme biosensors. *Sensors* **2010**, *10*, 1216–1231. [CrossRef] [PubMed]

124. Lee, K.; Weis, M.; Wei, O.Y.; Taguchi, D.; Manaka, T.; Iwamoto, M. Effects of gold nanoparticles on pentacene organic field-effect transistors. *Jpn. J. Appl. Phys.* **2011**, *50*, 041601.

125. Mao, S.; Lu, G.H.; Yu, K.H.; Chen, J.H. Specific biosensing using carbon nanotubes functionalized with gold nanoparticle-antibody conjugates. *Carbon* **2010**, *48*, 479–486. [CrossRef]

126. Mao, S.; Lu, G.H.; Yu, K.H.; Bo, Z.; Chen, J.H. Specific protein detection using thermally reduced graphene oxide sheet decorated with gold nanoparticle-antibody conjugates. *Adv. Mater.* **2010**, *22*, 3521–3526. [CrossRef] [PubMed]

127. Mao, S.; Yu, K.H.; Chang, J.B.; Steeber, D.A.; Ocola, L.E.; Chen, J.H. Direct growth of vertically-oriented graphene for field-effect transistor biosensor. *Sci. Rep.* **2013**, *3*, 1696. [CrossRef] [PubMed]

128. Cheng, S.S.; Hideshima, S.; Kuroiwa, S.; Nakanishi, T.; Osaka, T. Label-free detection of tumor markers using field effect transistor (fet)-based biosensors for lung cancer diagnosis. *Sens. Actuat. B Chem.* **2015**, *212*, 329–334. [CrossRef]

129. Jayasena, S.D. Aptamers: An emerging class of molecules that rival antibodies in diagnostics. *Clin. Chem.* **1999**, *45*, 1628–1650. [PubMed]

130. Tuerk, C.; Gold, L. Systematic evolution of ligands by exponential enrichment—RNA ligands to bacteriophage-t4 DNA-polymerase. *Science* **1990**, *249*, 505–510. [CrossRef] [PubMed]

131. Ellington, A.D.; Szostak, J.W. Invitro selection of rna molecules that bind specific ligands. *Nature* **1990**, *346*, 818–822. [CrossRef] [PubMed]

132. Kim, Y.S.; Raston, N.H.A.; Gu, M.B. Aptamer-based nanobiosensors. *Biosens. Bioelectron.* **2016**, *76*, 2–19. [PubMed]

133. Chiu, T.C.; Huang, C.C. Aptamer-functionalized nano-biosensors. *Sensors* **2009**, *9*, 10356–10388. [CrossRef] [PubMed]

134. Lee, J.O.; So, H.M.; Jeon, E.K.; Chang, H.; Won, K.; Kim, Y.H. Aptamers as molecular recognition elements for electrical nanobiosensors. *Anal. Bioanal. Chem.* **2008**, *390*, 1023–1032. [CrossRef] [PubMed]

135. Farid, S.; Meshik, X.; Choi, M.; Mukherjee, S.; Lan, Y.; Parikh, D.; Poduri, S.; Baterdene, U.; Huang, C.E.; Wang, Y.Y.; et al. Detection of interferon gamma using graphene and aptamer based fet-like electrochemical biosensor. *Biosens. Bioelectron.* **2015**, *71*, 294–299. [CrossRef] [PubMed]

136. Lee, H.S.; Kim, K.S.; Kim, C.J.; Hahn, S.K.; Jo, M.H. Electrical detection of vegfs for cancer diagnoses using anti-vascular endotherial growth factor aptamer-modified si nanowire fets. *Biosens. Bioelectron.* **2009**, *24*, 1801–1805. [CrossRef] [PubMed]

137. Maehashi, K.; Matsumoto, K.; Takamura, Y.; Tamiya, E. Aptamer-based label-free immunosensors using carbon nanotube field-effect transistors. *Electroanal* **2009**, *21*, 1285–1290. [CrossRef]

138. Ohno, Y.; Maehashi, K.; Matsumoto, K. Label-free biosensors based on aptamer-modified graphene field-effect transistors. *J. Am. Chem. Soc.* **2010**, *132*, 18012–18013. [CrossRef] [PubMed]

139. Park, J.W.; Na, W.; Jang, J. One-pot synthesis of multidimensional conducting polymer nanotubes for superior performance field-effect transistor-type carcinoembryonic antigen biosensors. *RSC Adv.* **2016**, *6*, 14335–14343. [CrossRef]

140. So, H.M.; Won, K.; Kim, Y.H.; Kim, B.K.; Ryu, B.H.; Na, P.S.; Kim, H.; Lee, J.O. Single-walled carbon nanotube biosensors using aptamers as molecular recognition elements. *J. Am. Chem. Soc.* **2005**, *127*, 11906–11907. [CrossRef] [PubMed]

141. Wang, C.; Kim, J.; Zhu, Y.; Yang, J.; Lee, G.-H.; Lee, S.; Yu, J.; Pei, R.; Liu, G.; Nuckolls, C.; et al. An aptameric graphene nanosensor for label-free detection of small-molecule biomarkers. *Biosens. Bioelectron.* **2015**, *71*, 222–229. [CrossRef] [PubMed]

142. Pacios, M.; Martin-Fernandez, I.; Borrise, X.; del Valle, M.; Bartroli, J.; Lora-Tamayo, E.; Godignon, P.; Perez-Murano, F.; Esplandiu, M.J. Real time protein recognition in a liquid-gated carbon nanotube field-effect transistor modified with aptamers. *Nanoscale* **2012**, *4*, 5917–5923. [CrossRef] [PubMed]

143. Grant, S.; Peter, W.; Tal, S.; Matthew, R.L.; Jenna, L.W.; Christopher, A.H.; Adeniyi, A.A.; Vincent, T.R.; Ethan, D.M. Scalable graphene field-effect sensors for specific protein detection. *Nanotechnology* **2013**, *24*, 355502.

chemosensors

MDPI

Review

Recent Advances in Electrochemical-Based Sensing Platforms for Aflatoxins Detection

Atul Sharma [1,2], Kotagiri Yugender Goud [2,3], Akhtar Hayat [2,4], Sunil Bhand [1] and Jean Louis Marty [2,*]

1   Biosensor Lab, Department of Chemistry, BITS, Pilani K. K. Birla Goa Campus, Zuarinagar, 403726 Goa, India; p2012407@goa.bits-pilani.ac.in (A.S.); sgbhand@gmail.com (S.B.)
2   BAE Laboratoire, Université de Perpignan Via Domitia, 52 Avenue Paul Alduy, 66860 Perpignan, France; yugenderkotagiri@gmail.com (K.Y.G.); akhtarloona@gmail.com (A.H.)
3   Department of Chemistry, National Institute of Technology, Warangal, 506004 Telangana, India
4   Interdisciplinary Research Centre in Biomedical Materials (IRCBM), COMSATS Institute of Information Technology (CIIT), Lahore 54000, Pakistan
*   Correspondence: jlmarty@univ-perp.fr; Tel.: +33-04-6866-2254; Fax: +33-04-6866-2223

Academic Editors: Paolo Ugo and Ligia Moretto
Received: 25 August 2016; Accepted: 20 December 2016; Published: 26 December 2016

Abstract: Mycotoxin are small (MW ~700 Da), toxic secondary metabolites produced by fungal species that readily colonize crops and contaminate them at both pre- and post-harvesting. Among all, aflatoxins (AFs) are mycotoxins of major significance due to their presence in common food commodities and the potential threat to human health worldwide. Based on the severity of illness and increased incidences of AFs poisoning, a broad range of conventional and analytical detection techniques that could be useful and practical have already been reported. However, due to the variety of structural analogous of these toxins, it is impossible to use one common technique for their analysis. Numerous recent research efforts have been directed to explore alternative detection technologies. Recently, immunosensors and aptasensors have gained promising potential in the area of sample preparation and detection systems. These sensors offer the advantages of disposability, portability, miniaturization, and on-site analysis. In a typical design of an aptasensor, an aptamer (ssDNA or RNA) is used as a bio-recognition element either integrated within or in intimate association with the transducer surface. This review paper is focused on the recent advances in electrochemical immuno- and aptasensing platforms for detection of AFs in real samples.

Keywords: aflatoxins; electrochemical techniques; aptasensor; biosensor; food

1. Introduction

With the increasing incidence and stubbornly high mycotoxin mortality around the world, the earlier diagnosis of mycotoxin contamination has drawn significant attention. The presence of mycotoxin in food and feed due to their associated toxic effects on human health and the environment has now became a primary concern [1]. Mycotoxins are the toxic fungal metabolites produced by fungi (micromycetes and macromycetes) under specific conditions of temperature and moisture [2]. The optimal condition of temperature for mycotoxin—producing molds ranging between 24 °C and 35 °C and a relative humidity of ≥70%. Toxicity of these metabolites in human and warm-blooded animals is commonly known as mycotoxicosis. More than 300 mycotoxins (aflatoxins, ochratoxins, trichothecane) commonly exist, but only some of them are practically important. Among all, the most commonly studied groups of mycotoxins are aflatoxins (AFs). Initially, AFs were isolated and identified after the death of young turkeys on poultry farms in England, which were found to be related due to the consumption of a Brazilian peanut meal. AFs are the difuranocoumarin derivatives mainly produced by *Aspergillus parasiticus*, *Aspergillus flavus*, and rarely by *Aspergillus nomius* [3]. AFs are often present in corn, peanuts, nuts, almonds, milk, cheese, and wide varieties of agriculture foodstuffs and beverages [4,5]. They have been classified based on their fluorescent properties under ultraviolet light (365 nm) and chromatographic mobility into different structural analogs, such as B-group (cyclopentane ring, blue fluorescence), G-group (lactone ring, yellow-green fluorescence), and a metabolite of B-group known as AFM1 and AFM2 (Figure 1). Among AFs, AFB1 is the most common and highly toxic contaminant responsible for more than 75% of all AF contamination in food and animal feed [6]. Subsequent exposure of AFB-contaminated feed to lactating animals leads to secretion of AFM1 and M2 in milk through the hydroxylation reaction mechanism. AFM1 and AFM2 are quite stable during milk pasteurization, as well as dairy product processing, which may persist to the final stage during human consumption [7–9].

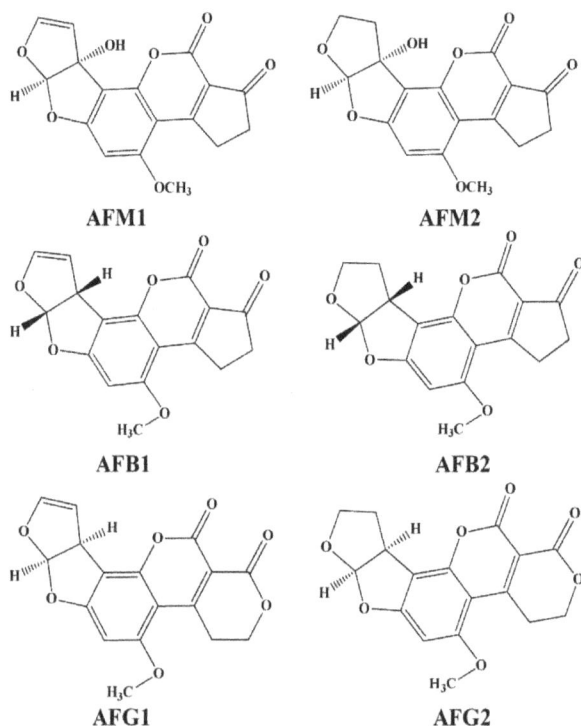

Figure 1. Structure of commonly found aflatoxins (AFs).

1.1. Toxicity of AFs

The biotransformation of AFB1 comprises the derivatives of AFM1, aflatoxin-exo-8,9-epoxide, AFQ1. Among them, the AFM1 and AFQ1 are less reactive and easily eliminated from the body through urination [10]. However, aflatoxin B1-8,9-exo-epoxide is a known mutagen, which is electrophilic in nature and capable of forming covalent bonds with nucleophilic sites of macromolecules, such as nucleic acids and proteins [11]. These covalent bond formations determines the formation of aflatoxin B1-DNA adducts, which results in mutagenic and carcinogenic effects of AFB1, such as G→T transversion mutation and attacks on the guanine base of DNA. This introduces the mutation in the normal cells and formation of various types of carcinomas in humans, especially in liver [12]. Typically, the AFB1 mutation can cause hepatocellular carcinoma, point mutation, inversion of base sequences (DNA and RNA), and destruction of protein structures [13]. The epoxide attacks at the position of seventh (7th) guanine nitrogen (both DNA and RNA), altering the hybridization of nucleic acids and transcription process. AFB1 has a negative impact on carbohydrate metabolism, which results in the reduction in hepatic glycogen and increased blood glucose levels. Although the toxic effects of AFM1 are less than that of AFB1, nevertheless, it causes the oxidative damage due to intracellular radical generation, DNA intercalation, base impairment, teratogenicity, birth defects, and genetic mutation [14].

According to the report of the Food and Agricultural Organization (FAO) on mycotoxin published in 2004, globally 99 countries had fixed the maximum stringent limits for mycotoxins in food and food products. In 2012, the Rapid Alert System for Food and Feed (RASFF) declared the AFs as one of the principle hazards in European Union [15–17]. To minimize the production losses and ensure the safety of human health, the European Commission (EC) has established the maximum stringent limits for most of the mycotoxins in food and food products as mentioned in the Commission Regulation (EC number 1881/2006), as well as through methods of sampling and analysis for their control (EC number 401/2006) [18,19]. Table 1 summarizes the permissible limits of aflatoxins in food by different agencies such as; European Union (EU), US Food and Drugs Administration (USFDA), the Codex Alimentous Commission (CAC), and the Food Safety and Standards Authority of India (FSSAI) [18,20].

Table 1. The regulatory standard for aflatoxins (AFs).

Matrix	Maximum Permissible Level			
	EU	USFDA	CAC	FSSAI
Milk and Milk based products	25 pg/mL (AFM1) 50 pg/mL (AFM1)	500 pg/mL (AFM1)	500 pg/mL (AFM1)	500 pg/mL (AFM1)
Nuts and dried food	5 µg/kg (AFB1) 10 µg/kg-Total	20 µg/kg-Total	Not specified	30 µg/kg
Groundnuts & dried fruits and their processed products	2 µg/kg (AFB1) 4 µg/kg-Total	20 µg/kg-Total	15 µg/kg-Total	30 µg/kg
Cereals	2 µg/kg (AFB1) 4 µg/kg-Total	20 µg/kg-Total	Not specified	30 µg/kg

1.2. Monitoring of Aflatoxins (AFs)

AF contamination seriously influences the quality of agricultural production, animal feeds, food quality, and other dietary products with potential threats to the human health and the environment, due to economic losses. Considering the above facts, the rapid, sensitive, and accurate detection of AF contamination in food and feed products, agriculture, and exposure levels in the human body require regular screening and risk monitoring. The reported classical methods for AFs detection are based on the chromatographic techniques, such as thin-layer chromatography [21–23], high-performance liquid chromatography [24,25], liquid chromatography coupled with mass spectroscopy [26–28], high-performance liquid chromatography coupled with fluorescence detection [29–31], and liquid chromatography/atmospheric pressure chemical ionization mass spectrometry (LC/APCI-MS) [32]. However, the inherent properties involved in the chromatographic techniques, such as long and complicated sample pre-treatment procedures, expensive instruments, and the requirement of trained technicians, limits their wider utility in high-throughput and on-site analysis of samples. The traditional immunoassays, mainly enzyme-linked immunosorbent assay (ELISA) are commonly used to detect AFs. However, the disadvantages involved in the immunoassay, such as long reaction time, difficulty in the automation of the process, in vivo production of antibodies and low sensitivity in different assays, decrease their involvement in real samples analysis. Some innovation and enhancement in the development of immunoassay such integration of nanomaterial, miniaturization have been reported. Meanwhile, the emergence of biosensing techniques has been witnessed as an alternative to the above problems. In the present review, we will discuss novel electrochemical biosensors and assay platforms for the detection of aflatoxins in different food matrices.

2. Biosensors (as an Alternative Tool)

A biosensor is a compact analytical device used for the detection of a target analyte based on optical, thermal, piezoelectric, and electrochemical signal generation, which are generated by the interaction between the recognition element and analyte of interest [33]. The molecular recognition elements are, consequently, the key for biosensors since their binding affinity and specificity greatly influences the sensor performance. Initially, the recognition elements (antibodies, enzymes, isolated receptors, etc.) were isolated from living organism i.e., goats, mice, horses. The antibodies were generated by animal immunization when it responds to the different antigens such as toxins, drugs, chemicals, virus particles, spores, and other foreign substrates [34]. Currently, synthetic or bio-engineered recognition elements are available in the laboratory, including antibodies, enzymes, molecularly-imprinted polymers, and lectins with the improved features of selectivity and specificity in biosensing. The specific and selective interaction between a particular antibody and an antigen is the basic principle involved in immunoassays. The results obtained from these immunoassays must be reproducible and repeatable in order to enable proper detection of analytes. Depending upon the technique used, immunosensors can be optical [35], mass-sensitive [36], and electrochemical [37]. Among these, electrochemical immunosensors are widely used, since they involve comparatively inexpensive, simple, and easy to use instruments. Immunosensors based on screen-printed electrodes (SPEs) are very convenient to use as they are easy to fabricate, portable, suitable for mass production, and provide inexpensive kits for the rapid and accurate detection and quantification of antigens and antibodies in a sample matrix. However, this is possible only when such a system is thoroughly characterized, well optimized, and immobilized on the surface of the electrodes.

Unfortunately, these recognition elements exhibit certain limitations. For instance, the antibodies and enzymes are sensitive to working pH and temperature, which is reflected in the short shelf life and irreversible denaturation [38]. The need of animal immunization for antibody production, which often involves the animal suffering, batch to batch variation, and difficulty in labeling of the specific recognition site, decreases the wide utility of antibodies. Finally, due to the requirement of immobilization and extensive washing in antibody-based affinity assays, it is difficult to carry out the homogeneous assays [39]. Therefore, it is highly desirable to seek the alternative ligands or recognition elements as a new platform for biosensing and analytical applications.

Aptamer

In the last decades, aptamers have attracted tremendous interest in therapeutic and bioanalytical applications, either used as an active separation material in chromatography or as recognition material in biosensing applications to replace commonly-used bio-receptors [40–45]. Aptamers are synthetic oligonucleotides sequences (30–100 nucleotides) with high affinity and specificity to recognize their cognate target molecules, ranging from small ions to large peptides. Upon target recognition, the aptamer folds into a specific 3D structure known as the antiparallel G-quadruplex aptamer complex form. Most of the aptamers are obtained through a combinatorial process called a systemic evolution of ligands by exponential enrichment (SELEX) from vast populations of random sequences. In SELEX, a random oligonucleotides library is exposed to the specific analyte of interest under a set of pH, ionic strength, and temperature conditions. However, it is difficult to optimize the SELEX parameters and select the aptamer sequence with high binding efficiency but, once optimized, it will reflect the sensing environment for the detection of target molecules [46]. The effect of structural analogs or interference against target molecules might hinder the aptamer synthesis. The selection of complex real matrices such as extracts, food, or bacterial samples for testing, ensure that the synthesized aptamer has the potential to work in real samples and detect mycotoxins. Considering these factors, there remain a number of mycotoxins for which aptamers could be selected. Another potential advantage of aptamer technology over the antibody is that the selection of oligonucleotides sequences can be rationally determined and altered to optimize molecular recognition performance. The high binding sequences can be partitioned from the sequences lacking affinity against the target. For small molecules, such as mycotoxins, the SELEX is often achieved by the immobilizing of target molecules to a solid phase or beads, allowing the easy removal of unwanted sequences through multiple washing steps. As recognition elements, aptamers offer many advantages over conventional antibodies. Due to their small size, high affinity, specificity, ease of denaturation, high stability (especially DNA aptamers), ease of modification, and labeling, aptamers have gained significant potential for developing practical, inexpensive and robust biosensing platforms [47,48]. Even though the promising potential of aptamer-based sensing strategies exist in the food industry, therapeutics, and clinical diagnostics, only a few aptamer-based products are commercializedThis review surveys the recent literature dealing with immuno- and aptasensors for AF detection. These studies can open the way to novel analytical devices of commercial interest with several advantages, such as miniaturization, portability, disposability, low sample requirement, and suitability for practical and on-site applications.

3. Electrochemical Immunosensors for AFs Detections

In the existing literature, several electrochemical immunosensing platforms have been reported for AFB1 and AFM1 detections. Firstly, an indirect competitive ELISA was performed on SPE electrodes using DPV analysis for AFB1 detection. The presented method was successfully applied for detection of AFB1 in barley samples with high sensitivity and good recoveries. The SPE-based ELISA showed better analytical performance than spectrophotometric ELISA with a LOD of 30 pg/mL [49]. After one year, a direct HRP-linked chronoamperometric immunosensor was developed for detection of AFM1 in milk samples [50]. Obtained results showed that, using SPEs, AFM1 can be measured up to 25 pg/mL with a dynamic working range between 30 and 160 pg/mL. Meanwhile, in a study by Parker et al., it was concluded that the presence of divalent ion (calcium) is highly recommended to stabilize the milk samples on metal electrodes [51]. For multi-analyte determinations, a competitive ELISA combined with 96-well screen-printed microplate-based multichannel electrochemical detection was developed for AFB1 detection in corn samples using intermittent pulse amperometry (IPA) [52]. The author reported a LOD of 30 pg/mL with the high throughput ELISA procedure. In the last decades, nanomaterial-based signal amplification strategies for conventional ELISA and electrochemical detection have been applied for AFM1 detection [53,54]. For AFB1 detection, an impedimetric immunosensor based on colloidal gold and silver electrodeposition for AFM1 detection was developed by Vig et al. [55]. The signal amplification was carried out by silver electrodeposition using chronoamperometry. The results of calculating charge transfer resistance (EIS signal) correspond to the amount of AFM1 present. Obtained results were further compared with linear sweep voltammetry (LSV) measurements. In the same context, an indirect enzymatic immunosensor for AFB1 detection was fabricated on gold electrodes using signal amplification strategies based on silver electrodeposition. LSV measurements were carried out to quantify the metal silver, which corresponds to the amount of AFB1 in rice samples [37]. Bacher et al., reported a label-free impedimetric immunosensor based on antibody-coupled silver wire for detection of AFM1 in milk samples [56]. The anti-AFM1 mAb and AFM1 interaction were quantified on the basis of impedance change at 10 mV potential. The developed sensor has the highest sensitivity with a LOD of 1 pg/mL with an analysis time of 20 min. In order to miniaturize and improve the sensitivity and selectivity of a conventional electrode system, a gold microelectrode array was used as a platform for AF analysis. Parker et al., reported the development of direct competitive ELISA based on the gold microelectrode array for direct analysis of AFM1 in milk samples [57].

Integration of nanotechnology or nanostructures in biosensing applications improves the analytical performance of the electrochemical biosensing methods owing to their high surface area impact ratio, excellent surface catalytic activity, ease of preparation, and bioconjugation [58].

Glassy carbon electrodes (GCEs) or modified GCE surfaces have been widely used for preparing electrochemical immunosensors for AF detection. After capturing the analyte on the sensor surface, the electrochemical signal change is measured by DPV, EIS, and amperometry measurements [59–62]. Similarly, single-walled carbon nanotubes (SWNTs) have also been widely used in the development of electrochemical immunosensing platforms. For AFB1 detection in corn powder, an AFB1-BSA immobilized conjugate on SWNTs/chitosan-modified GCE surfaces was used for development of an electrochemical-based indirect competitive immunoassay [63]. Graphene oxide (GO), has been used for fabricating electrochemical immunosensors for mycotoxin detection. Recently, GO based electrochemiluminescent and EIS immunosensors have been developed for AFM1 and AFB1 detection [64,65]. Immunosensing platforms reported for AF detection based on electrochemical signal generation have been summarized in Table 2.

Table 2. Summary of literature reports describing electrochemical immunosensors for aflatoxins detection.

Analyte	Method	LOD	Matrix	Reference
AFB1	ELISA with DPV	30 pg/mL	Barley	[49]
AFM1	Amperomertic	25 pg/mL	Milk	[50]
AFB1	intermittent pulse amperometry (IPA)	30 pg/mL	Corn	[52]
AFM1	EIS	15 ng/L	Milk	[55]
AFB1	LSV	0.06 ng/mL	Rice	[37]
AFM1	EIS	1 pg/mL	Milk	[56]
AFM1	EIS	8 ng/L	Milk	[57]
AFB1	DPV	0.07 ng/mL	-	[58]
AFM1	DPV	0.2 ng/mL	PBS	[59]
		0.7 ng/mL	-	[60]
AFB1	EIS	0.01 ng/mL	Bee pollen	[61]
AFB1	Amperometric	0.05 ng/mL	Human serum and grape samples	[62]
AFB1	DPV	3.5 pg/mL	Corn powder	[63]
AFM1	Electrochemiluminescent (ECL)	0.3 pg/mL	Milk	[64]
AFB1	EIS	0.5 pg/mL	Corn samples	[65]

4. Aptamer-Based Sensing Strategies for Aflatoxins (AFs) Detection

Recently, aptamer sequences possessing a high affinity for AFs (different structural analogs) have been reported in either publications or patents (Table 3) [66–70]. In the present scenario, the problems associated with the analysis of complex samples, such as blood, serum, and cellular extracts, have been solved using electrochemical biosensors [71,72].

Table 3. Aptamer sequences for aflatoxins (AFs).

Target	Aptamer Length	Base Pair Sequences (No.)	Sequences	Kd (nM)	Ref.
AFB1	50	16	GTTGGGCACGTGTTGTCTCTCTGTGTCTC GTGCCCTTCGCTAGGCCCACA	N.R.	[66]
	80	26	AGCAGCACAGAGGTCAGATGGTGCTAT CATGCGCTCAATGGGAGACTTTAGCTG CCCCCACCTATGCGTGCTACCGTGAA	11.29 ± 1.27	[67]
AFB2	80	26	AGCAGCACAGAGGTCAGATGCTGACA CCCTGGACCTTGGGATTCCGGAAGTT TTCCGGTACCTATGCGTGCTACCGTGAA	9.83 ± 0.99	[68]
AFM1	21	7	ACTGCTAGAGATTTTCCACAT	N.R.	[69]
	72	24	ATCCGTCACACCTGCTCTGACGCTG GGGTCGACCCGGAGAAATGCATTCC CCTGTGGTGTTGGCTCCCGTAT	35.6 ± 2.6	[70]

In the last decades, the development of electrochemical aptasensing platforms has gained considerable attention in the analysis of target analytes. Immobilization of aptamers on the electrode surface is highly important. Several immobilization techniques, such as thiolation, diazonium coupling, and click chemistry, are reported [73–75]. Among all, diazotization coupling provides better immobilization impact due to the lack of leakage of bio-recognition elements on storage. The generation of the electrochemical signal corresponds to the amount of analyte present. The electrochemical aptasensors can be easily modified and offer the advantages of high sensitivity, selectivity, stability, compatibility with microfabrication, disposability, portability, high detection speed, and the requirement of low sample volume. Based on these advantages, electrochemical sensors appear to be well suited for practical applications. In the last decade, a large number of papers and reviews have been published in this field.

Electrochemical Aptasensors for AFs Detection

In recent years, the development of aptasensors for detection of toxins and environmental pollutants has gained significant attention. The merging of aptamer capabilities and the versatility of nanomaterials has opened new strategies for the amplified detection of mycotoxins. The various developed electrochemical aptasensors for AFs detection has been summarized in Table 4. Nguyen et al. have reported the label-free electrochemical aptasensor for detection of AFM1 [69]. For the construction of electrochemical aptasensor, a Fe_3O_4 polyaniline (Fe_3O_4/PANi) film was polymerized on the interdigitated electrode (IDE) for AFM1 detection. Immobilized aptamers as affinity capture reagents, and magnetic nanoparticles for signal amplification were employed in construction of sensing platform. For AFM1 quantification, label-free and direct measurements of the AFM1 aptamer on the Fe_3O_4/PANi interface were carried out using cyclic (CV) and square wave voltammetry (SWV). The developed aptasensor showed a LOD of 1.98 ng/L with a good sensitivity in the range 6–60 ng/L. Later, the aptasensor performance was successfully demonstrated in milk samples. A DNA biosensor based on the interaction of AFM1 and a self-assembled metal supported lipid bilayer membrane (s-BLMs) and its effect on DNA hybridization was reported [76]. The interactions of AFM1 with s-BLMs was composed of egg phosphatidylcholine, responsible for an increase in ion current, which corresponds to the concentration of toxin. The presence of ssDNA causes an increase in ion current across s-BLMs, whereas the current decrease is due to the formation of double-stranded DNA (dsDNA). The captured signal decreases in the presence of the toxin and increases the time to reach equilibrium. This aptasensor provided the rapid (<1 min) detection and the low detection limit (0.5 nM) of AFM1 based on the measurements of the initial rate of hybridization. Dinckaya et al. reported the development of an impedimetric DNA biosensor for detection of AFM1 in milk and dairy products [77]. The DNA biosensor was constructed by covalent immobilization of the thio-modified single-stranded DNA (ss-HSDNA) on the gold surface using self-assembled monolayer. Using impedance spectroscopy, a detection limit of 1–14 ng/mL was obtained.

Very recently, an impedimetric aptasensor for detection of AFM1 was reported by Istamboulie et al. [78]. In this work, the hexaethyleneglycol-modified oligonucleotides (seven base pair sequences) of anti-AFM1 aptamer were immobilized on the diazotized screen-printed carbon electrode (SPCEs) via a carbodiimide coupling reaction. The fabricated aptasensor was characterized at each step using CV and EIS using ferri/ferrocyanide as a redox probe. A dynamic range of 2–150 ng/L AFM1 was obtained with a LOD of 1.15 ng/L. For real sample analysis, a simple filtration through a 0.2 mm polytetrafluoroethylene (PTFE) membrane was carried out to allow the determination of AFM1 in milk samples.

Table 4. The reported literature based on electrochemical aptasensors for detection of Aflatoxins.

Target	Method Used	Limit of Detection (LOD)	Matrix	Ref.
AFM1	Cyclic (CV) and square wave voltammetry (SWV)	1.98 ng/L	Milk	[69]
AFM1	Amperomertic	0.5 nM	Milk	[76]
AFM1	Electrochemical impedance spectroscopy (EIS)	N.R.	Milk	[77]
AFM1	CV and SWV	1.98 ng/L	Milk	[78]
AFB1	CV and EIS	0.40 ± 0.03 nM	peanuts-corn snacks	[79]
AFB1	CV	0.10 nM	peanuts, cashew nuts, white wine and soy sauce	[80]
	EIS	0.05 nM		
AFB1	EIS	0.12 ng/mL (seqA)	Alcoholic beverages	[81]
		0.25 ng/mL (seqB)		
AFB1	SWV	0.6×10^{-4} ng/L	Corn	[82]

Castillo et al. reported the development of an electrochemical aptasensor using polyamidoamine PAMAM dendrimers for AFB1 detection [79]. For sensor fabrication, a single-stranded (ss) amino-modified DNA aptamer highly specific to AFB1 was immobilized on the assembly of a multilayer framework of immobilized PAMAM dendrimers on the gold electrode. The CV and EIS measurements were performed to capture the signal response by means of redox indicators: $K[Fe(CN)_6]^{3-}/^{4-}$. The aptasensor allowed AFB1 determination in the range of 0.1–10 nM AFB1. The sensor possesses the LOD of 0.40 ± 0.03 nM, with a stability of 60 h at 4 °C. In previous years, the use of mediators, such as methylene blue, ferrocene, ferri-ferrocyanide, and methylene green in the electrochemical sensor has been successfully reported for a decrease in potential and amplification of signals [83–86]. Previously, an electrochemical aptasensor based on the electropolymerization of neutral red on the electrode surface for detection of AFB1 has been reported by Evtugyn et al. [80]. The aptasensor was prepared using covalent immobilization of anti-AFB1 DNA aptamer to the polycarboxylated macrocyclic ligand immobilized (Thiacalix arene A) on an electropolymerised layer of neutral red, which acts as a redox probe (Figure 2). For quantitative measurements of AFB1, CV and EIS measurements were carried out. The developed aptasensor showed a LOD of 0.05 nM with EIS in 4-(2-hydroxyethyl)-1-piperazineethanesulfonic acid (HEPES) binding buffer. It was reported that the developed protocol provides the enhancement in stability of the surface layer and improved reproducibility of the voltammetric signal in multiple food matrices, such as peanuts, cashew nuts, white wine, and soy sauce.

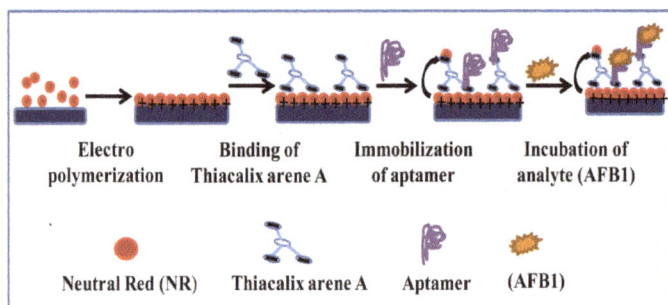

Figure 2. Schematic representation of an electrochemical aptasensor used for determination of AFB1 using electropolymerized modified electrodes (scheme illustration from [80]).

Very recently, a label-free EIS aptasensor for the detection of AFB1 in alcoholic beverages [81]. An EIS aptasensor was fabricated over SPCEs via immobilization of anti-AFB1 aptamer using a diazonium coupling mechanism (Figure 3). In this work, the two different sequences of an anti-AFB1 aptamer were used and compared for their analytical performance. On incubation of AFB1, a dynamic detection range from 0.125 to 16 ng/mL was obtained with a LOD of 0.12 and 0.25 ng/mL for seqA and seqB. The performance of the EIS aptasensor was successfully demonstrated in alcoholic beverages (beer and wine samples) with recoveries between 92% and 102%. The developed aptasensor offers the advantages of disposability and portability for on-site analysis. Among the reported electrochemical techniques, one important strategy is the designing of a switchable on-off electrochemical aptasensor, which results in a signal upon target recognition depending upon conformational changes. However, for the "signal-off" electrochemical sensor, it is generally recognized that the suppression of the signal alters the sensitivity and specificity of the developed platforms [87]. Presently, to effectively avoid the inherent drawbacks of signal-off biosensors, the integration of different amplification approaches became attractive, such as in vitro DNA amplification. Based on the above facts, an enzyme-based signal amplification electrochemical aptasensing platform for ultrasensitive detection of AFB1 has been reported by Zheng et al. [82]. In this work, a heteroenzyme-based two-round signal amplification electrochemical aptasensor approach was designed for AFB1 detection. In the first round, the telomerase-based amplification led to the generation of a high current signal, which increases the detection range (Figure 4), whereas in the second round the EXO III-based amplification led to the generation of an observable signal response corresponding to the trace concentrations of AFB1. Based on the advantage of the two-round signal amplification strategy, the sensitivity and detection range of proposed electrochemical aptasensors were greatly improved.

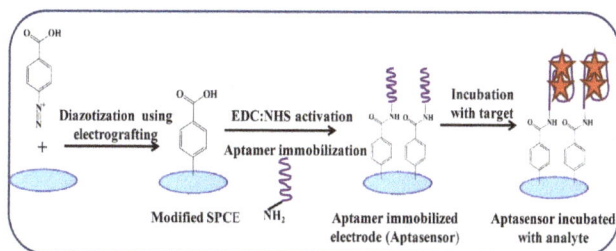

Figure 3. Schematic representation of an electrochemical aptasensor used for determination of AFB1 using diazotized SPCEs (scheme illustration from [81]).

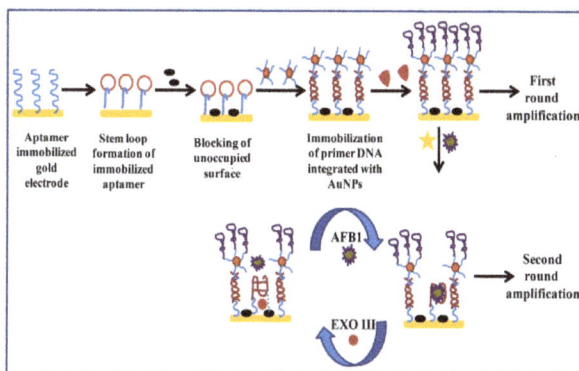

Figure 4. Schematic representation of two signals amplified signal on an electrochemical aptasensor for AFB1 detection (scheme illustration from [82]).

5. Safety Notes

AFs are highly carcinogenic and should be handled with extreme care. After use, the AF-contaminated labwares must be decontaminated with an aqueous solution of sodium hypochlorite (5%). The AFs are subject to light degradation; therefore, the samples must be protected from daylight and standards must be stored in amber-colored vials. For aqueous solutions of AFs, the use of non-acid-washed glassware may result in the loss of AF, thus, special attention and precautions should be paid in cleaning new glassware, which should be soaked in dilute acid (10% sulfuric acid) for several hours and then thoroughly rinsed with distilled water to remove all of the traces of acid [88,89].

6. Conclusions and Future Perspectives

Various analytical methods employed in the analysis of aflatoxins in agricultural, food, crops, and feeds have been reported. Over the well-established antigen-antibody-based (immunosensor) detection systems, the aptamer-based (aptasensors) strategies have been explored due to their inherent practical benefits over the antibodies as recognition elements. Preferably, a detection method should be able to detect the target analyte at very low levels with high specificity. In this context, immunosensors with a very high level of analytical performance; lower LODs, high stability with high precision and accuracy, has been reported. Despite of their numerous advantages, the immunosensors still require some improvements for better analysis of food and environmental samples, whereas the in vitro design and selection of the aptamer sequences allow the unparalleled control over binding conditions and possible cross-reactivity. The SELEX experiments can be carefully designed, including the counter selections against toxins or other possible interferences. Additionally, the selection of aptamer directly in complex matrices, such as extracts from the crops or food, could help to ensure their reliable performances in real-world samples. Considering these factors, there is scope to explore the SELEX process for selection of aptamer against a series of mycotoxins for which aptamers are not known.

It is worth noting that although many sensitive methods have been described for the analysis of AFs, based on electrochemical signal generation. The EIS aptasensors offers the advantages of disposability, portability, miniaturization, and on-site analysis. Therefore, the development of simple, label-free, rapid, and sensitive tools that are based on electrochemical responses can provide versatile, portable, sensitive, and accurate devises for AFs on-site detection. The discussed signal amplification strategies possess the significant potential to overcome bottleneck in the traditional signal-off biosensor. One of the major breakthrough studies could be the integration of signal amplification strategies with sensing platforms based on screen printed electrodes.

Acknowledgments: Atul Sharma (A.S.) and Kotagiri Yugender Goud (K.Y.G.) would like to thanks EUPHRATES (ERASMUS Mundus) Doctoral Fellowship program. A.S. would also acknowledge NFBSFARA (Project No. PHT/4007/2013-14), ICAR, New Delhi, India for Senior Research Fellowship. Authors would like to thank the reviewers and academic editors for their valuable inputs to improve the quality of manuscript.

Conflicts of Interest: The authors declare no conflict of interest.

References

1. Guo, X.; Wen, F.; Zheng, N.; Luo, Q.; Wang, H.; Wang, H.; Li, S.; Wang, J. Development of an ultrasensitive aptasensor for the detection of aflatoxin B1. *Biosens. Bioelectron.* **2014**, *56*, 340–344. [CrossRef] [PubMed]
2. Boonen, J.; Malysheva, S.V.; Taevernier, L.; Diana Di Mavungu, J.; De Saeger, S.; De Spiegeleer, B. Human skin penetration of selected model mycotoxins. *Toxicology* **2012**, *301*, 21–32. [CrossRef] [PubMed]
3. Yao, H.; Hruska, Z.; Mavungu, J.D.D. Developments in detection and determination of aflatoxins. *World Mycotoxin J.* **2015**, *8*, 181–191. [CrossRef]
4. Bakirci, I. A study on the occurrence of aflatoxin M1 in milk and milk products produced in van province of turkey. *Food Control* **2001**, *12*, 47–51. [CrossRef]
5. Sharma, A.; Catanante, G.; Hayat, A.; Istamboulie, G.; Ben Rejeb, I.; Bhand, S.; Marty, J.L. Development of structure switching aptamer assay for detection of aflatoxin M1 in milk sample. *Talanta* **2016**, *158*, 35–41. [CrossRef] [PubMed]

6.	Ayub, M.; Sachan, D. Dietary factors affecting aflatoxin B1 carcinogenicity. *Malays. J. Nutr.* **1997**, *3*, 161–197.
7.	Codex Committee on Food Additives and Contaminants. *Comments Submitted on the Draft Maximum Level for Aflatoxin M1 in Milk, CL CX/FAC 01/20*; Codex Alimentarious Commission Netherlands: Hague, The Netherlands, 2001; pp. 1–9.
8.	Stroka, J.; Anklam, E. New strategies for the screening and determination of aflatoxins and the detection of aflatoxin-producing moulds in food and feed. *TrAC Trends Anal. Chem.* **2002**, *21*, 90–95. [CrossRef]
9.	Badea, M.; Micheli, L.; Messia, M.C.; Candigliota, T.; Marconi, E.; Mottram, T.; Velasco-Garcia, M.; Moscone, D.; Palleschi, G. Aflatoxin M1 determination in raw milk using a flow-injection immunoassay system. *Anal. Chim. Acta* **2004**, *520*, 141–148. [CrossRef]
10.	Wild, C.P.; Turner, P.C. The toxicology of aflatoxins as a basis for public health decisions. *Mutagenesis* **2002**, *17*, 471–481. [CrossRef] [PubMed]
11.	Wacoo, A.P.; Wendiro, D.; Vuzi, P.C.; Hawumba, J.F. Methods for detection of aflatoxins in agricultural food crops. *J. Appl. Chem.* **2014**, *2014*, 706291. [CrossRef]
12.	De Oliveira, C.A.; Germano, P.M. Aflatoxins: Current concepts on mechanisms of toxicity and their involvement in the etiology of hepatocellular carcinoma. *Rev. Saude Publ.* **1997**, *31*, 417–424.
13.	Levy, D.D.; Groopman, J.D.; Lim, S.E.; Seidman, M.M.; Kraemer, K.H. Sequence specificity of aflatoxin B1-induced mutations in a plasmid replicated in xeroderma pigmentosum and DNA repair proficient human cells. *Cancer Res.* **1992**, *52*, 5668–5673. [PubMed]
14.	Hamid, A.S.; Tesfamariam, I.G.; Zhang, Y.; Zhang, Z.G. Aflatoxin B1-induced hepatocellular carcinoma in developing countries: Geographical distribution, mechanism of action and prevention. *Oncol. Lett.* **2013**, *5*, 1087–1092. [PubMed]
15.	Marin, S.; Ramos, A.J.; Cano-Sancho, G.; Sanchis, V. Mycotoxins: Occurrence, toxicology, and exposure assessment. *Food Chem. Toxicol.* **2013**, *60*, 218–237. [CrossRef] [PubMed]
16.	Official Journal of the European Union. *Commission Regulation (EU) No 165/2010 of 26 February 2010 Amending Regulation (EC) No 1881/2006 Setting Maximum Levels for Certain Contaminants in Foodstuffs as Regards Aflatoxins*; OJ L 50; Official Journal of the European Union: Brusseles, Belgium, 2010; pp. 8–126.
17.	Official Journal of the European Union. *Commission Regulation (EU) No 1058/2012 of 12 November 2012 Amending Regulation (EC) No 1881/2006 as Regards Maximum Levels for Aflatoxins in Dried Figs Text with EEA Relevance*; OJ L 313; Official Journal of the European Union: Brusseles, Belgium, 2012; pp. 14–15.
18.	Official Journal of the European Union. *Commission Regulation (EC) No. 1881/2006 of 19 December 2006 Setting Maximum Levels for Certain Contaminants in Foodstuffs*; Official Journal of the European Union: Brusseles, Belgium, 2006; pp. L364/5–L364/24.
19.	Official Journal of the European Union. *Commission Regulation (EC) No. 401/2006 Laying down the Methods of Sampling and Analysis for the Official Control of the Levels of Mycotoxins in Foodstuffs*; Official Journal of the European Union: Brusseles, Belgium, 2006; Volume L70, p. 1234.
20.	Dorsm, G.C.; Caldas, S.C.; Feddern, V.; Bemvenuti, R.; Hackbart, H.C.D.S.; De Souza, M.M.; Oliveira, M.D.S.; Buffon, J.G.; Primel, E.G.; Furlong, E.B. *Aflatoxins: Contamination, Analysis and Control, Aflatoxins-Biochemistry and Molecular Biology*; Guevara-Gonzalez, R.G., Ed.; InTech: Rijeka, Croatia, 2011.
21.	Soares, L.M.; Rodriguez-Amaya, D.B. Survey of aflatoxins, ochratoxin a, zearalenone, and sterigmatocystin in some brazilian foods by using multi-toxin thin-layer chromatographic method. *J Assoc. Off. Anal. Chem.* **1989**, *72*, 22–26. [PubMed]
22.	Stroka, J.; Otterdijk, R.V.; Anklam, E. Immunoaffinity column clean-up prior to thin-layer chromatography for the determination of aflatoxins in various food matrices. *J. Chromatogr. A* **2000**, *904*, 251–256. [CrossRef]
23.	Stroka, J.; Anklam, E. Development of a simplified densitometer for the determination of aflatoxins by thin-layer chromatography. *J. Chromatogr. A* **2000**, *904*, 263–268. [CrossRef]
24.	Rodríguez Velasco, M.L.; Calonge Delso, M.M.; Ordóñez Escudero, D. ELISA and HPLC determination of the occurrence of aflatoxin M1 in raw cow's milk. *Food Addit. Contam.* **2003**, *20*, 276–280. [CrossRef] [PubMed]
25.	Turner, N.W.; Subrahmanyam, S.; Piletsky, S.A. Analytical methods for determination of mycotoxins: A review. *Anal. Chim. Acta* **2009**, *632*, 168–180. [CrossRef] [PubMed]
26.	Spanjer, M.C.; Rensen, P.M.; Scholten, J.M. LC–MS/MS multi-method for mycotoxins after single extraction, with validation data for peanut, pistachio, wheat, maize, cornflakes, raisins and figs. *Food Addit. Contam. Part A* **2008**, *25*, 472–489. [CrossRef] [PubMed]

27. Vahl, M.; Jørgensen, K. Determination of aflatoxins in food using LC/MS/MS. *Zeitschrift für Lebensmitteluntersuchung und Forschung A* **1998**, *206*, 243–245. (In German) [CrossRef]

28. Monbaliu, S.; Van Poucke, C.; Detavernier, C.L.; Dumoulin, F.; Van De Velde, M.; Schoeters, E.; Van Dyck, S.; Averkieva, O.; Van Peteghem, C.; De Saeger, S. Occurrence of mycotoxins in feed as analyzed by a multi-mycotoxin lc-ms/ms method. *J. Agric. Food Chem.* **2010**, *58*, 66–71. [CrossRef] [PubMed]

29. Braga, S.M.; de Medeiros, F.D.; de Oliveira, E.J.; Macedo, R.O. Development and validation of a method for the quantitative determination of aflatoxin contaminants in maytenus ilicifolia by HPLC with fluorescence detection. *Phytochem. Anal.* **2005**, *16*, 267–271. [CrossRef] [PubMed]

30. Navas, S.A.; Sabino, M.; Rodriguez-Amaya, D.B. Aflatoxin M1 and ochratoxin a in a human milk bank in the city of São Paulo, Brazil. *Food Addit. Contam.* **2005**, *22*, 457–462. [CrossRef] [PubMed]

31. Iha, M.H.; Barbosa, C.B.; Favaro, R.M.; Trucksess, M.W. Chromatographic method for the determination of aflatoxin M1 in cheese, yogurt, and dairy beverages. *J. AOAC Int.* **2011**, *94*, 1513–1518. [CrossRef] [PubMed]

32. Tanaka, H.; Takino, M.; Sugita-Konishi, Y.; Tanaka, T. Development of a liquid chromatography/time-of-flight mass spectrometric method for the simultaneous determination of trichothecenes, zearalenone and aflatoxins in foodstuffs. *Rapid Commun. Mass Spectrom.* **2006**, *20*, 1422–1428. [CrossRef] [PubMed]

33. Thevenot, D.R.; Tóth, K.; Durst, R.A.; Wilson, G.S. Electrochemical biosensors: Recommended definitions and classification. *Pure Appl. Chem.* **1999**, *71*, 2333–2338. [CrossRef]

34. Radi, A.-E. Electrochemical aptamer-based biosensors: Recent advances and perspectives. *Int. J. Electrochem.* **2011**, *2011*, 863196. [CrossRef]

35. Sassolas, A.; Catanante, G.; Hayat, A.; Marty, J.-L. Development of an efficient protein phosphatase-based colorimetric test for okadaic acid detection. *Anal. Chim. Acta* **2011**, *702*, 262–268. [CrossRef] [PubMed]

36. Mishra, G.K.; Sharma, A.; Bhand, S. Ultrasensitive detection of streptomycin using flow injection analysis-electrochemical quartz crystal nanobalance (FIA-EQCN) biosensor. *Biosens. Bioelectron.* **2015**, *67*, 532–539. [CrossRef] [PubMed]

37. Tan, Y.; Chu, X.; Shen, G.-L.; Yu, R.-Q. A signal-amplified electrochemical immunosensor for aflatoxin B1 determination in rice. *Anal. Biochem.* **2009**, *387*, 82–86. [CrossRef] [PubMed]

38. Jayasena, S.D. Aptamers: An emerging class of molecules that rival antibodies in diagnostics. *Clin. Chem.* **1999**, *45*, 1628–1650. [PubMed]

39. Zhou, W.; Huang, P.-J.J.; Ding, J.; Liu, J. Aptamer-based biosensors for biomedical diagnostics. *Analyst* **2014**, *139*, 2627–2640. [CrossRef] [PubMed]

40. Ellington, A.D.; Szostak, J.W. In vitro selection of RNA molecules that bind specific ligands. *Nature* **1990**, *346*, 818–822. [CrossRef] [PubMed]

41. Stojanovic, M.N.; de Prada, P.; Landry, D.W. Fluorescent sensors based on aptamer self-assembly. *J. Am. Chem. Soc.* **2000**, *122*, 11547–11548. [CrossRef]

42. Bunka, D.H.J.; Stockley, P.G. Aptamers come of age—At last. *Nat. Rev. Microbiol.* **2006**, *4*, 588–596. [CrossRef] [PubMed]

43. Keefe, A.D.; Pai, S.; Ellington, A. Aptamers as therapeutics. *Nat. Rev. Drug Discov.* **2010**, *9*, 537–550. [CrossRef] [PubMed]

44. Hayat, A.; Marty, J.L. Aptamer based electrochemical sensors for emerging environmental pollutants. *Front. Chem.* **2014**, *2*, 41. [CrossRef] [PubMed]

45. Sharma, A.; Hayat, A.; Mishra, R.K.; Catanante, G.; Shahid, S.A.; Bhand, S.; Marty, J.L. Design of a fluorescence aptaswitch based on the aptamer modulated nano-surface impact on the fluorescence particles. *RSC Adv.* **2016**, *6*, 65579–65587. [CrossRef]

46. McKeague, M.; McConnell, E.M.; Cruz-Toledo, J.; Bernard, E.D.; Pach, A.; Mastronardi, E.; Zhang, X.; Beking, M.; Francis, T.; Giamberardino, A.; et al. Analysis of in vitro aptamer selection parameters. *J. Mol. Evol.* **2015**, *81*, 150–161. [CrossRef] [PubMed]

47. Mairal, T.; Ozalp, V.C.; Lozano Sanchez, P.; Mir, M.; Katakis, I.; O'Sullivan, C.K. Aptamers: Molecular tools for analytical applications. *Anal. Bioanal. Chem.* **2008**, *390*, 989–1007. [CrossRef] [PubMed]

48. Wang, J.; Lv, R.; Xu, J.; Xu, D.; Chen, H. Characterizing the interaction between aptamers and human IgE by use of surface plasmon resonance. *Anal. Bioanal. Chem.* **2008**, *390*, 1059–1065. [CrossRef] [PubMed]

49. Ammida, N.H.S.; Micheli, L.; Palleschi, G. Electrochemical immunosensor for determination of aflatoxin B1 in barley. *Anal. Chim. Acta* **2004**, *520*, 159–164. [CrossRef]

50. Micheli, L.; Grecco, R.; Badea, M.; Moscone, D.; Palleschi, G. An electrochemical immunosensor for aflatoxin m1 determination in milk using screen-printed electrodes. *Biosens. Bioelectron.* **2005**, *21*, 588–596. [CrossRef] [PubMed]

51. Parker, C.O.; Tothill, I.E. Development of an electrochemical immunosensor for aflatoxin M1 in milk with focus on matrix interference. *Biosens. Bioelectron.* **2009**, *24*, 2452–2457. [CrossRef] [PubMed]

52. Piermarini, S.; Micheli, L.; Ammida, N.H.S.; Palleschi, G.; Moscone, D. Electrochemical immunosensor array using a 96-well screen-printed microplate for aflatoxin B1 detection. *Biosens. Bioelectron.* **2007**, *22*, 1434–1440. [CrossRef] [PubMed]

53. Kanungo, L.; Pal, S.; Bhand, S. Miniaturised hybrid immunoassay for high sensitivity analysis of aflatoxin M1 in milk. *Biosens. Bioelectron.* **2011**, *26*, 2601–2606. [CrossRef] [PubMed]

54. Paniel, N.; Radoi, A.; Marty, J.-L. Development of an electrochemical biosensor for the detection of aflatoxin M1 in milk. *Sensors* **2010**, *10*, 9439–9448. [CrossRef] [PubMed]

55. Vig, A.; Radoi, A.; Muñoz-Berbel, X.; Gyemant, G.; Marty, J.-L. Impedimetric aflatoxin M1 immunosensor based on colloidal gold and silver electrodeposition. *Sens. Actuators B Chem.* **2009**, *138*, 214–220. [CrossRef]

56. Bacher, G.; Pal, S.; Kanungo, L.; Bhand, S. A label-free silver wire based impedimetric immunosensor for detection of aflatoxin M1 in milk. *Sens. Actuators B Chem.* **2012**, *168*, 223–230. [CrossRef]

57. Parker, C.O.; Lanyon, Y.H.; Manning, M.; Arrigan, D.W.; Tothill, I.E. Electrochemical immunochip sensor for aflatoxin M1 detection. *Anal. Chem.* **2009**, *81*, 5291–5298. [CrossRef] [PubMed]

58. Wang, X.; Niessner, R.; Tang, D.; Knopp, D. Nanoparticle-based immunosensors and immunoassays for aflatoxins. *Anal. Chim. Acta* **2016**, *912*, 10–23. [CrossRef] [PubMed]

59. Masoomi, L.; Sadeghi, O.; Banitaba, M.H.; Shahrjerdi, A.; Davarani, S.S.H. A non-enzymatic nanomagnetic electro-immunosensor for determination of aflatoxin b1 as a model antigen. *Sens. Actuators B Chem.* **2013**, *177*, 1122–1127. [CrossRef]

60. Owino, J.; Arotiba, O.; Hendricks, N.; Songa, E.; Jahed, N.; Waryo, T.; Ngece, R.; Baker, P.; Iwuoha, E. Electrochemical immunosensor based on polythionine/gold nanoparticles for the determination of aflatoxin B1. *Sensors* **2008**, *8*, 8262–8274. [CrossRef] [PubMed]

61. Zaijun, L.; Zhongyun, W.; Xiulan, S.; Yinjun, F.; Peipei, C. A sensitive and highly stable electrochemical impedance immunosensor based on the formation of silica gel-ionic liquid biocompatible film on the glassy carbon electrode for the determination of aflatoxin B1 in bee pollen. *Talanta* **2010**, *80*, 1632–1637. [CrossRef] [PubMed]

62. Sun, A.-L.; Qi, Q.-A.; Dong, Z.-L.; Liang, K.Z. An electrochemical enzyme immunoassay for aflatoxin B1 based on bio-electrocatalytic reaction with room-temperature ionic liquid and nanoparticle-modified electrodes. *Sens. Instrum. Food Qual. Saf.* **2008**, *2*, 43–50. [CrossRef]

63. Zhang, X.; Li, C.-R.; Wang, W.-C.; Xue, J.; Huang, Y.-L.; Yang, X.-X.; Tan, B.; Zhou, X.-P.; Shao, C.; Ding, S.-J.; et al. A novel electrochemical immunosensor for highly sensitive detection of aflatoxin B1 in corn using single-walled carbon nanotubes/chitosan. *Food Chem.* **2016**, *192*, 197–202. [CrossRef] [PubMed]

64. Gan, N.; Zhou, J.; Xiong, P.; Hu, F.; Cao, Y.; Li, T.; Jiang, Q. An ultrasensitive electrochemiluminescent immunoassay for aflatoxin M1 in milk, based on extraction by magnetic graphene and detection by antibody-labeled cdte quantumn dots-carbon nanotubes nanocomposite. *Toxins* **2013**, *5*, 865–883. [CrossRef] [PubMed]

65. Wang, D.; Hu, W.; Xiong, Y.; Xu, Y.; Li, M.C. Multifunctionalized reduced graphene oxide-doped polypyrrole/pyrrolepropylic acid nanocomposite impedimetric immunosensor to ultra-sensitively detect small molecular aflatoxin B1. *Biosens. Bioelectron.* **2015**, *63*, 185–189. [CrossRef] [PubMed]

66. Le, C.L.; Cruz-Aguado, J.; Penner, G.A. DNA Ligands for Aflatoxin and Zearalenone. US Patent No. US20120225494 A1, 6 September 2012.

67. Ma, X.; Wang, W.; Chen, X.; Xia, Y.; Wu, S.; Duan, N.; Wang, Z. Selection, identification, and application of aflatoxin B1 aptamer. *Eur. Food Res. Technol.* **2014**, *238*, 919–925. [CrossRef]

68. Ma, X.; Wang, W.; Chen, X.; Xia, Y.; Duan, N.; Wu, S.; Wang, Z. Selection, characterization and application of aptamers targeted to aflatoxin B2. *Food Control* **2015**, *47*, 545–551. [CrossRef]

69. Nguyen, B.H.; Tran, L.D.; Do, Q.P.; Nguyen, H.L.; Tran, N.H.; Nguyen, P.X. Label-free detection of aflatoxin M1 with electrochemical Fe_3O_4/polyaniline-based aptasensor. *Mater. Sci. Eng. C* **2013**, *33*, 2229–2234. [CrossRef] [PubMed]

70. Malhotra, S.; Pandey, A.K.; Rajput, Y.S.; Sharma, R. Selection of aptamers for aflatoxin M1 and their characterization. *J. Mol. Recognit.* **2014**, *27*, 493–500. [CrossRef] [PubMed]

71. Zuo, X.; Song, S.; Zhang, J.; Pan, D.; Wang, L.; Fan, C. A target-responsive electrochemical aptamer switch (TREAS) for reagentless detection of nanomolar ATP. *J. Am. Chem. Soc.* **2007**, *129*, 1042–1043. [CrossRef] [PubMed]

72. Zuo, X.; Xiao, Y.; Plaxco, K.W. High specificity, electrochemical sandwich assays based on single aptamer sequences and suitable for the direct detection of small-molecule targets in blood and other complex matrices. *J. Am. Chem. Soc.* **2009**, *131*, 6944–6945. [CrossRef] [PubMed]

73. Adams, N.M.; Jackson, S.R.; Haselton, F.R.; Wright, D.W. Design, synthesis, and characterization of nucleic-acid-functionalized gold surfaces for biomarker detection. *Langmuir* **2012**, *28*, 1068–1082. [CrossRef] [PubMed]

74. Goud, K.Y.; Sharma, A.; Hayat, A.; Catanante, G.; Gobi, K.V.; Gurban, A.M.; Marty, J.L. Tetramethyl-6-carboxyrhodamine quenching-based aptasensing platform for aflatoxin B1: Analytical performance comparison of two aptamers. *Anal. Biochem.* **2016**, *508*, 19–24. [CrossRef] [PubMed]

75. Hayat, A.; Sassolas, A.; Marty, J.-L.; Radi, A.-E. Highly sensitive ochratoxin A impedimetric aptasensor based on the immobilization of azido-aptamer onto electrografted binary film via click chemistry. *Talanta* **2013**, *103*, 14–19. [CrossRef] [PubMed]

76. Siontorou, C.G.; Nikolelis, D.P.; Miernik, A.; Krull, U.J. Rapid methods for detection of aflatoxin M1 based on electrochemical transduction by self-assembled metal-supported bilayer lipid membranes (s-BLMs) and on interferences with transduction of DNA hybridization. *Electrochim. Acta* **1998**, *43*, 3611–3617. [CrossRef]

77. Dinçkaya, E.; Kınık, Ö.; Sezgintürk, M.K.; Altuğ, Ç.; Akkoca, A. Development of an impedimetric aflatoxin M1 biosensor based on a DNA probe and gold nanoparticles. *Biosens. Bioelectron.* **2011**, *26*, 3806–3811. [CrossRef] [PubMed]

78. Istamboulié, G.; Paniel, N.; Zara, L.; Granados, L.R.; Barthelmebs, L.; Noguer, T. Development of an impedimetric aptasensor for the determination of aflatoxin M1 in milk. *Talanta* **2016**, *146*, 464–469. [CrossRef] [PubMed]

79. Castillo, G.; Spinella, K.; Poturnayová, A.; Šnejdárková, M.; Mosiello, L.; Hianik, T. Detection of aflatoxin B1 by aptamer-based biosensor using pamam dendrimers as immobilization platform. *Food Control* **2015**, *52*, 9–18. [CrossRef]

80. Evtugyn, G.; Porfireva, A.; Stepanova, V.; Sitdikov, R.; Stoikov, I.; Nikolelis, D.; Hianik, T. Electrochemical aptasensor based on polycarboxylic macrocycle modified with neutral red for aflatoxin B1 detection. *Electroanalysis* **2014**, *26*, 2100–2109. [CrossRef]

81. Yugender Goud, K.; Catanante, G.; Hayat, A.; Satyanarayana, M.; Vengatajalabathy Gobi, K.; Marty, J.L. Disposable and portable electrochemical aptasensor for label free detection of aflatoxin B1 in alcoholic beverages. *Sens. Actuators B Chem.* **2016**, *235*, 466–473. [CrossRef]

82. Zheng, W.; Teng, J.; Cheng, L.; Ye, Y.; Pan, D.; Wu, J.; Xue, F.; Liu, G.; Chen, W. Hetero-enzyme-based two-round signal amplification strategy for trace detection of aflatoxin B1 using an electrochemical aptasensor. *Biosens. Bioelectron.* **2016**, *80*, 574–581. [CrossRef] [PubMed]

83. Wu, J.; Chu, H.; Mei, Z.; Deng, Y.; Xue, F.; Zheng, L.; Chen, W. Ultrasensitive one-step rapid detection of ochratoxin a by the folding-based electrochemical aptasensor. *Anal. Chim. Acta* **2012**, *753*, 27–31. [CrossRef] [PubMed]

84. Evtugyn, G.; Porfireva, A.; Ivanov, A.; Konovalova, O.; Hianik, T. Molecularly imprinted polymerized methylene green as a platform for electrochemical sensing of aptamer–thrombin interactions. *Electroanalysis* **2009**, *21*, 1272–1277. [CrossRef]

85. Wang, Y.; He, X.; Wang, K.; Ni, X.; Su, J.; Chen, Z. Electrochemical detection of thrombin based on aptamer and ferrocenylhexanethiol loaded silica nanocapsules. *Biosens. Bioelectron.* **2011**, *26*, 3536–3541. [CrossRef] [PubMed]

86. Kwon, D.; Jeong, H.; Chung, B.H. Label-free electrochemical detection of human α-thrombin in blood serum using ferrocene-coated gold nanoparticles. *Biosens. Bioelectron.* **2011**, *28*, 454–458. [CrossRef] [PubMed]

87. Yin, H.; Zhou, Y.; Chen, C.; Zhu, L.; Ai, S. An electrochemical signal 'off-on' sensing platform for microrna detection. *Analyst* **2012**, *137*, 1389–1395. [CrossRef] [PubMed]
88. Dragacci, S.; Grosso, F. Immunoaffinity column cleanup with liquid chromatography for determination of aflatoxinm1 in liquid milk: Collaborative study. *J. AOAC Int.* **2001**, *84*, 437–443. [PubMed]
89. Kanungo, L.; Bhand, S. Fluorimetric immunoassay for multianalysis of aflatoxins. *J. Anal. Methods Chem.* **2013**, *2013*, 584964. [CrossRef] [PubMed]

Chapter 3:
Technical Note

chemosensors

MDPI

Technical Note

A Low-Cost Label-Free AFB$_1$ Impedimetric Immunosensor Based on Functionalized CD-Trodes

Marcos Vinicius Foguel [1,*], Gabriela Furlan Giordano [1], Célia Maria de Sylos [2], Iracilda Zeppone Carlos [3], Antonio Aparecido Pupim Ferreira [1], Assis Vicente Benedetti [1] and Hideko Yamanaka [1,*]

1   Department of Analytical Chemistry, Institute of Chemistry, UNESP-Univ Estadual Paulista, Rua Francisco Degni, 55, Quitandinha, 14800-060 Araraquara, SP, Brazil; gabigiordano@uol.com.br (G.F.G.); antoferr@iq.unesp.br (A.A.P.F.); benedeti@iq.unesp.br (A.V.B.)
2   Department of Food and Nutrition, School of Pharmaceutical Sciences, UNESP-Univ Estadual Paulista, Rodovia Araraquara Jaú, Km 01, s/n, Campos Ville, 14800-903 Araraquara, SP, Brazil; syloscm@fcfar.unesp.br
3   Department of Clinical Analysis, School of Pharmaceutical Sciences, UNESP-Univ Estadual Paulista, Rodovia Araraquara Jaú, Km 01, s/n, Campos Ville, 14800-903 Araraquara, SP, Brazil; carlosiz@fcfar.unesp.br
*   Correspondence: mvfoguel@gmail.com (M.V.F.); hidekoy@iq.unesp.br (H.Y.); Tel.: +55-16-3301-9622

Academic Editors: Paolo Ugo and Ligia Moretto
Received: 29 April 2016; Accepted: 19 August 2016; Published: 1 September 2016

Abstract: This work describes the investigation of a label-free immunosensor for the detection of aflatoxin B$_1$ (AFB$_1$). CD-trodes (electrodes obtained from recordable compact disks) were used as low-cost and disposable transducers after modification with a self-assembled monolayer (SAM) of lipoic acid. The anti-aflatoxin B$_1$ antibody was immobilized via EDC/NHS activation, followed by blocking with bovine serum albumin and immunoassays with AFB$_1$. The optimization of analytical parameters and the detection were carried out using electrochemical impedance measurements. Using chemometric tools, the best conditions for the immunosensor development were defined as: anti-AFB$_1$ antibody at 1:2000 dilution and surface blocking with 0.5% bovine serum albumin, both incubated for 1 h, and antibody–antigen immunoreaction for 30 min. The impedimetric immunosensor showed a linear range from 5×10^{-9} to 1×10^{-7} mol·L^{-1} (1.56–31.2 ng·mL^{-1}), limit of detection and limit of quantification, respectively, 3.6×10^{-10} and 1.1×10^{-9} mol·L^{-1} (0.11 and 0.34 ng·mL^{-1}). The proposed immunosensor was applied to analyze peanut samples.

Keywords: aflatoxin B$_1$; immunosensor; electrochemical impedance spectroscopy

1. Introduction

Aflatoxin B$_1$ (AFB$_1$) is a toxic metabolite produced by fungi *Aspergillus flavus* and *A. parasiticus* that is characterized by acute toxicity, teratogenicity, mutagenicity, and carcinogenicity [1]. The acute effects are observed mainly in the liver; they can result in necrosis, hemorrhage, injury, fibrosis, cirrhosis, and cancer. This toxin infects a wide range of agricultural products [2], especially peanuts, corn, wheat, rice, nuts, dried fruits, among others.

The detection of aflatoxin in food and feed is usually performed by instrumental methodologies based on synchronous fluorescence spectrometry [3], high-performance liquid chromatography with amperometric detection [4], fluorescence detection [5], thin layer chromatography [6], or immunochromatographic assay [7]. An alternative is offered by the use of immunosensors, due to their sensibility, stability, and ease of handling. In this case, the antibodies are immobilized on an electrode and must maintain their biological activity on the transducer [8]. One of the procedures is based on the formation of a self-assembled monolayer (SAM) on the electrode surface [9], which consists of the interaction of a highly-organized organic molecule layer on the surface and one of the functional groups of the selected organic molecule having the function of providing the biological material's immobilization (e.g., enzymes, proteins, nucleic acid, antibody, etc.) via the free functional group of SAM [10]. Organic monolayers with a sulfur group on the electrode surface are of great interest, because the sulfur binds strongly to the gold surface and the reactive functional group is maintained free for the immobilization of biological molecules [11]. Surface Plasmon resonance devices [12], conductometric [13], fluorometric [14], and amperometric [15,16] biosensors indicated good performance. Related to the impedimetric immunosensor for AFB$_1$, the literature registered the immobilization of antibody on Pt, glassy carbon, and gold electrode, as displayed in Table 1.

Table 1. Comparison of the performances of different impedimetric immunosensors for the determination of aflatoxin B$_1$ (AFB$_1$).

Matrix	Dynamic Range	LOD [1]	Ref.
Pt electrodes modified with polyaniline and polystyrene sulphonic acid	0–6 mg·L^{-1}	0.1 mg·L^{-1}	[17]
Silica gel–ionic liquid biocompatible film on the glassy carbon electrode	0.1–10 ng·mL^{-1}	0.01 ng·mL^{-1}	[18]
1,6-hexanedithiol, colloidal Au, and aflatoxin B$_1$—bovine serum albumin conjugate on a gold electrode	0.08–100 ng·mL^{-1}	0.05 ng·mL^{-1}	[19]
Graphene oxide on Au electrode	0.5–5 ng·mL^{-1}	0.23 ng·mL^{-1}	[20]
Graphene/polypyrrole/pyrrolepropylic acid composite film on glassy carbon electrode	10 fg·mL^{-1}–10 pg·mL^{-1}	10 fg·mL^{-1}	[21]
Poly(amidoamine) dendrimers of fourth generation immobilized on gold electrode covered by cystamine	0.03–3.1 ng·mL^{-1}	0.12 ng·mL^{-1}	[22] *
MWCNT [2]/ionic liquid composite films on glassy carbon electrode	0.1–10 ng·mL^{-1}	0.03 ng·mL^{-1}	[23]
Poly(ophenylenediamine) electropolymerized film modified gold three-dimensional nanoelectrode ensembles	0.04–8.0 ng·mL^{-1}	0.019 ng·mL^{-1}	[24]
Screen-printed interdigitated microelectrodes modified with 3-Dithiobis-(sulfosuccinimidyl-propionate) and Protein G	5–20 ng·mL^{-1}	5 ng·mL^{-1}	[25]

[1] LOD = Limit of detection, [2] MWCNT: multi-walled carbon nanotubes. * Original value of dynamic range: 0.1–10 nmol·L^{-1} and LOD: 0.4 nmol·L^{-1}.

This work reports the development of an impedimetric immunosensor for the determination of aflatoxin B$_1$, through gold CD-trode (electrode obtained from recordable compact disks) surface modification with a self-assembled monolayer (SAM) of lipoic acid activated via EDC (1-ethyl-3-(3-(dimethylamino)-propyl)carbodiimide)/NHS (N-hydroxy succinimide) for the immobilization of anti-aflatoxin B$_1$ antibody. CD-trodes can be easily obtained by simple treatment of wasted gold CDs to obtain cheap but efficient electrochemical sensors [26,27]. The optimization of the experimental parameters involved in the development of the CD-trode sensor was performed chemometrically, by means of full factorial design. The proposed biosensor was applied to determine the antigen in peanut samples.

2. Materials and Methods

2.1. Reagents

69%–70% HNO$_3$, 95%–98% H$_2$SO$_4$, methanol, and chloroform were purchased from J. T. Baker (Phillipsburg, NJ, USA). 1-ethyl-3-(3-(dimethylamino)-propyl)carbodiimide (EDC) was obtained from Fluka (Buchs, Switzerland). Lipoic acid (C$_8$H$_{14}$O$_2$S$_2$), N-hydroxy succinimide (NHS), aflatoxin B$_1$ (AFB$_1$) from *Aspergillus flavus*, anti-AFB$_1$ antibody from rabbit, bovine serum albumin (BSA),

NaH$_2$PO$_4$·H$_2$O, Na$_2$HPO$_4$, K$_3$Fe(CN)$_6$, K$_4$Fe(CN)$_6$, CuSO$_4$·5H$_2$O, and KCl were purchased from Sigma (St. Louis, MO, USA). Ultra-pure water (resistivity 18.2 MΩ·cm) was used to prepare the solutions.

2.2. Apparatus and Electrochemical Cell

The electrochemical measurements were carried out using a potentiostat/galvanostat AUTOLAB PGSTAT 302 with impedance module FRA 2 (frequency response analyzer) and software version 4.9.005. The experiments were carried out using a one-compartment electrochemical cell with a volume of 5 mL, and a three-electrode system: gold CD-trode (A$_{geom}$ = 0.071 cm^2 and active area of 0.091 cm^2, estimated from by Randles–Sevik equation [28]), Ag | AgCl | KCl$_{(sat)}$ and platinum wire (A$_{geom}$ = 4 cm^2) as work, reference, and auxiliary electrodes, respectively.

2.3. Construction of Gold Electrode (AuCD-Trode)

The working electrode was constructed from a recordable compact disc (Mitsui Archive Gold CD-R 100) containing a gold film with thickness between 50 and 100 nm, using a previously reported procedure [26]. Briefly, to access the metal layer of the CD-R, concentrated HNO$_3$ was added to the surface; after 5–10 min, the protective layers were totally removed and the gold surface was washed thoroughly with distilled water [26]. The CD-R was cut and the working area of the electrode (E$_w$) was pre-set with perforated galvanoplastic tape. The electrical contact of the CD-trode was a laminated copper wire fixed and insulated with polytetrafluoroethylene (PTFE) tape. The characterization of the AuCD-trode as an electrochemical transducer was done previously [27]. CD-trodes are disposable, so they were used only once. A scheme of CD-trode is shown in Figure 1.

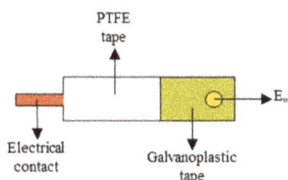

Figure 1. Scheme of modification of the gold CD-trode surface. PTFE: polytetrafluoroethylene.

2.4. Immunosensor Development

Scheme 1 shows the schemes of the different steps of modification of CD-trode to obtain the impedimetric immunosensor [29,30].

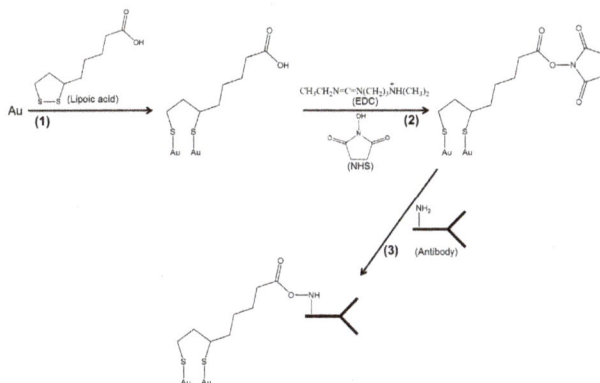

Scheme 1. Scheme of the gold CD-trode surface modification.

The first step in the development of an impedimetric immunosensor (Scheme 1-1) is the addition of 10 μL of 1×10^{-3} mol·L^{-1} lipoic acid prepared in ethanol/water solution (1:10) to the gold CD-trode surface and incubation for 2 h to form a SAM [31]. According to literature, both thiol and disulfide groups interact with gold [11]. For the immobilization of biological material on the SAM, it was necessary to activate the carboxyl group of the lipoic acid with 10 μL of 0.4/0.1 mmol·L^{-1} EDC/NHS prepared in deionized water that was added on the surface of the modified electrode, and the incubation time was 60 min (Scheme 1-2) [32]. The immobilization of the anti-AFB$_1$ antibody (Scheme 1-3) was performed by adding 10 μL of antibody solution prepared in 0.01 mol·L^{-1} phosphate buffer solution (pH 7.4) with different concentrations and incubation times on the modified electrode. In order to avoid unspecific interaction, the electrode surface was blocked with 10 μL of 0.5% bovine serum albumin (BSA) prepared in 0.01 mol·L^{-1} phosphate buffer solution (pH 7.4) containing tween 20 (PBS-T) for 60 min. Finally, 10 μL of aflatoxin B$_1$, prepared in 0.5% BSA solution, was added on the immunosensor surface and incubated for 30 min. The incubation steps were performed at 25 °C. After each incubation time at different steps of modification, the electrode was washed by immersion in ultra-pure water three times for 10 s under stirring.

The analysis of factorial designs results was conducted with statistical and graphical analysis software MINITAB® Release 15, developed by Minitab Inc., State College, PA, USA. All experiments were carried out in triplicate and the reproducibility was evaluated statistically by the MINITAB® software.

After the construction of the immunosensor, a standard solution of AFB$_1$—prepared in 0.01 mol·L^{-1} phosphate buffer solution at pH 7.4 containing 0.5% BSA—was added to the electrode, and the affinity reaction was monitored.

2.5. Electrochemical Measurements

In order to clean and homogenize the surface, the CD-trodes were submitted to pretreatments with 10 voltammetric cycles in 0.5 mol·L^{-1} sulfuric acid in the potential range between +0.2 and +1.5 V at 100 mV·s^{-1}.

The impedance spectra were obtained by applying a sine wave of 10 mV (rms) on E$_{ocp}$ from 100 kHz to 100 mHz and recording 10 points/frequency decade. Measurements were performed in 0.1 mol·L^{-1} phosphate buffer solution pH 7.0 containing 1.0×10^{-3} mol·L^{-1} Fe(CN)$_6^{3-/4-}$ redox pair. All electrochemical measurements were done in triplicate at 25 ± 2 °C in a Faraday cage. The value of charge transfer resistance (R$_{ct}$), calculated from the Nyquist plot, was used as parameter related to the response of the immunosensor. These values were obtained for each experiment. The real impedance (Z$_{re}$) of the frequency in the maximum of the semicircle was taken; it is the solution resistance (R$_{\Omega}$) plus half of the charge transfer resistance. Therefore, the R$_{ct}$ can be defined by Equation (1) [33].

$$R_{ct} = 2\,Z_{re} - 2\,R_{\Omega} \tag{1}$$

2.6. Extraction of AFB$_1$ from Peanut Samples

The immunosensor was applied to AFB$_1$ analysis in peanut samples. The toxin extraction from the sample was carried out as follows: 50 g of ground raw peanuts were mixed with 270 mL of methanol and 30 mL of 4% KCl in a blender for 5 min, followed by filtration through a qualitative filter paper. Then, the filtrate was mixed with 150 mL of 10% CuSO$_4$·5H$_2$O and celite for 5 min, and then filtered again on filter paper. The filtrate was mixed with water (1:1), and the aflatoxin extracted with 10 mL of chloroform. Another aliquot of chloroform was added to the aqueous solution. The organic phases were mixed and dried in a water bath at 40 °C [34]. The dried aliquots were re-suspended with 1.0 mL methanol and diluted in 0.01 mol·L^{-1} phosphate buffer solution (pH 7.4, containing 0.5% BSA) and analyzed.

3. Results and Discussion

3.1. Chemometric Study of Antibody, BSA, and AFB₁ Concentrations and Incubation Time

The first step in the development of the immunosensor concerned the optimization of the concentration of anti-AFB$_1$ (1:2000 and 1:500, or 0.0017 and 0.0067 $\mu g \cdot \mu L^{-1}$) and its incubation time (1 and 12 h). When the antibodies were incubated for 1 or 12 h, the results indicated almost the same value of R$_{ct}$, and the capacitive circle presented no inductive loop [35]. This means that one hour of incubation is enough.

The concentration of the anti-AFB$_1$ antibody and incubation time optimization were performed by full factorial design of type 2^2. The Pareto plot in Figure 2a shows that the most important parameter is the antibody dilution (D$_{Ab}$), while the incubation time of the solution (t$_{inc}$) and interactions between factors (D$_{Ab}$ and t$_{inc}$) on the modified CD-trode do not have great influence on the impedance measurements. The influences of the high and low levels of each variable were also determined. The lines shown in Figure 2b express the trend of the R$_{ct}$ values when changing the parameter from low to high level. The line with higher slope indicates the most influential parameter of the system; thus, the dilution presents the greater influence to the system. The 1:2000 dilution of the antibody with 1 h of incubation tended to result in higher impedance value.

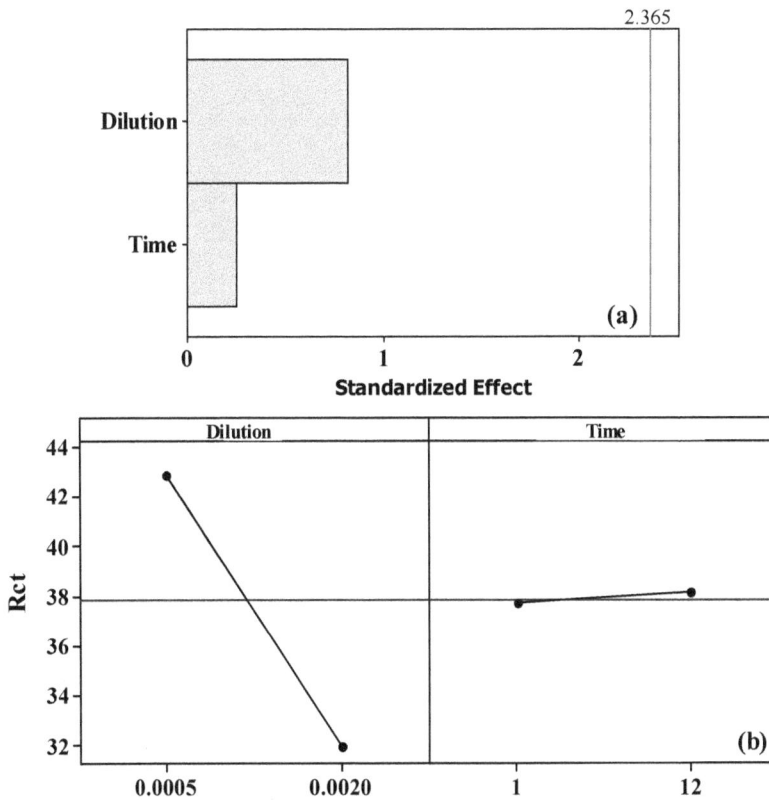

Figure 2. (a) Pareto plot: the influence of parameters on immobilization of Ab-AFB$_1$ on modified CD-trode with a self-assembled monolayer (SAM). D$_{Ab}$ = dilution of anti-AFB$_1$ antibody and t$_{incub.}$ = incubation time; (b) Trend of R$_{ct}$ to high and low levels of each variable.

The optimization of bovine serum albumin and aflatoxin B_1 concentration and incubation time of antigen were investigated by means of full factorial design of type 2^3. In Figure 3a, the Pareto plot shows that the most important parameter is the AFB_1 concentration, followed by incubation time of the antigen. The BSA concentration and interactions between the factors have little influence on the response of the system. The plot in Figure 3b indicates that at high AFB_1 concentration, R_{ct} is higher when the incubation time is 30 min, because these were the parameters that tended to higher values. As the variation of the BSA concentration does not present influence, the 0.5% concentration was adopted. To evaluate the incubation time of BSA, different times were studied: 30, 40, and 60 min (data not shown). The R_{ct} values for 30 and 40 min of incubation were quite similar to the R_{ct} of the previous modification step (Au-SAM-Ab). However, with 60 min of incubation, the R_{ct} value was higher than the previous step, and the reproducibility of the measurements was better. Thereby, the incubation time of 60 min was used for the blocking step.

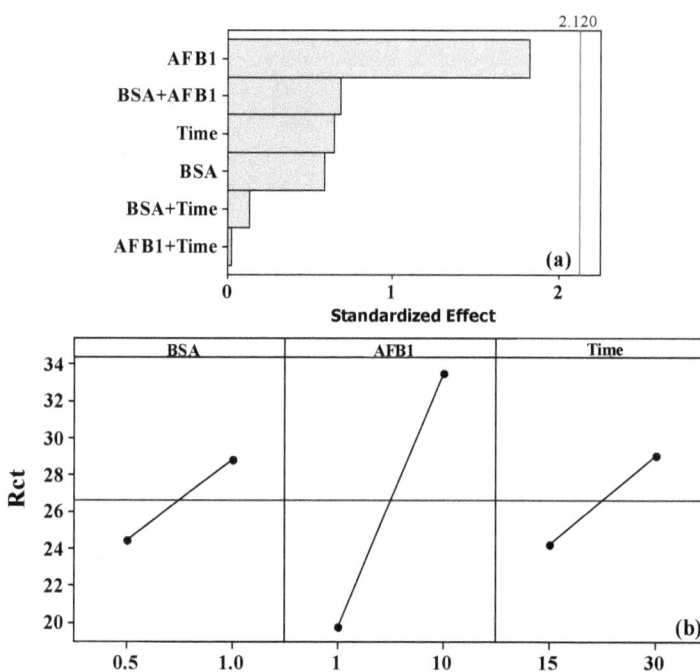

Figure 3. (a) Pareto plot: the influence of parameters on immobilization of AFB_1 antigen on the immobilized antibody on the SAM; (b) Trend of R_{ct} to high and low levels of each variable. C_{BSA} = bovine serum albumin concentration, C_{AFB1} = aflatoxin B_1 concentration, and $t_{incub.}$ = incubation time of aflatoxin B_1.

Figure 4 presents the Nyquist diagram for the various steps of the construction of the optimized immunosensor: SAM layer, active SAM layer, antibody layer, blocking with BSA, and interaction of antigen with the immunosensor. The figures indicate that each layer deposition on the immunosensor surface increases the impedance of the system. This increase is due to the changes of the electrical characteristics of the gold/electrolyte interface.

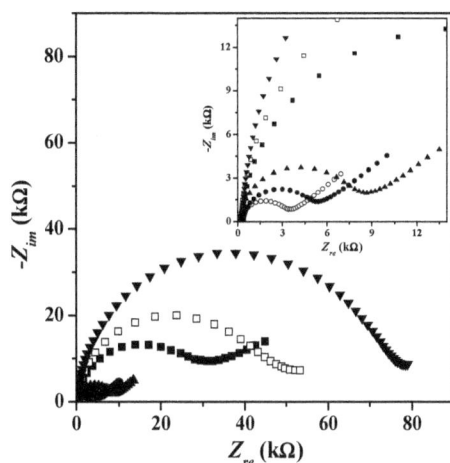

Figure 4. Nyquist plot in 0.1 mol·L^{-1} phosphate buffer solution pH 7.0 containing 1.0×10^{-3} mol·L^{-1} Fe(CN)$_6^{3/-4-}$ for unmodified CD-trode ($\circ$), modified CD-trode with 1.0×10^{-3} mol·L^{-1} lipoic acid ($\bullet$), CD-trode with carboxyl group of SAM activated with EDC/NHS ($\blacktriangle$), CD-trode with anti-AFB$_1$ antibody immobilized on active SAM ($\blacksquare$), Ab immobilized and blocked with 0.5% BSA and 60 min of incubation ($\square$), and CD-trode with 1.0×10^{-7} mol·L^{-1} AFB$_1$ on the Ab with 30 min of incubation ($\blacktriangledown$). E$_{ocp}$ vs. Ag | AgCl | KCl$_{(sat)}$. Inset: Zoom of high frequency data.

3.2. Analytical Curve

After optimization of all steps for the development of the impedimetric immunosensor, an analytical curve was constructed from 5.0×10^{-9} to 1.0×10^{-7} mol·L^{-1} (1.56–31.2 ng·mL^{-1}) as shown in Figure 5. The curve presented a linear range with a correlation coefficient of 0.99858, and the limits of detection and quantification of 3.6×10^{-10} and 1.1×10^{-9} mol·L^{-1} (0.11 and 0.34 ng·mL^{-1}), respectively.

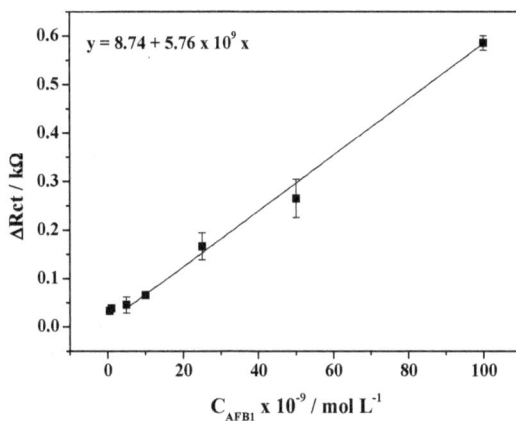

Figure 5. Analytical curve of AFB$_1$ using impedimetric immunosensor ($n = 3$).

3.3. Application in Peanut Samples

Impedimetric immunosensors were applied to three different samples of raw peanuts, provided by a food industry that performs quality analysis to verify the eventual contamination of the products with aflatoxin. According to the supplier industry, two of these samples were contaminated with AFB_1, and in the third one the concentration level was unknown, but below the limit established by the Brazilian National Agency of Sanitary Surveillance (20 µg of total aflatoxin in 1 kg of peanut).

Table 2 shows the AFB_1 concentration for each sample, analyzed by interpolating in the analytical curve. Samples A and B presented levels around three times above the limit set by Brazilian legislation. In sample C, a concentration of 25 µg·kg^{-1} was detected by the sensor, which is slightly above the legal limit. This contrasts the datum determined previously by the producer, using classical analytical procedures. This could indicate a better sensitivity of the sensor with respect to standard methods, however we cannot exclude a possible growth of *Aspergillus* sp. and subsequent production of aflatoxin during storage, because of the long time interval elapsed between the two analyses (around 7 months).

Table 2. AFB_1 concentration values found in peanuts samples ($n = 3$).

Sample	Concentration $\times$ 10^{-9}/mol·L^{-1}	Concentration/µg·kg^{-1}
A	13 ± 1	65 ± 6
B	13.3 ± 0.6	67 ± 3
C	5.0 ± 0.3	25 ± 1

4. Conclusions

On the proposed methodology, the dynamic range is 0.16 to 3 ng·mL^{-1} and LOD 0.35 ng·mL^{-1}. In comparison with data reported in previous literature, displayed in Table 1, the performances of the CD-trode sensor are comparable with those obtained with some of the immunosensors previously described in the literature [20,22,25], but with the advantage of being based on the use of disposable and low-cost CD-trodes. These preliminary results are encouraging in order to progress with an in-depth validation of the sensor, in particular concerning matrix effects and recovery tests.

Acknowledgments: The authors thank FAPESP (Proc. 2008/07729-8), FP7–PEOPLE–IRSES (Proc. 2008-230849) and CNPq (Proc. 309275/2012-1) for financial support.

Conflicts of Interest: The authors declare no conflict of interest.

Abbreviations

The following abbreviations were used in this manuscript:

AFB_1	Aflatoxin B_1
CD-trode	Electrode obtained from recordable compact disk
EDC	1-Ethyl-3-(3-(dimethylamino)-propyl)carbodiimide
NHS	N-hydroxy succinimide
SAM	Self-assembled monolayer

References

1. Nayak, S.; Sashidhar, R.B. Metabolic intervention of aflatoxin B_1 toxicity by curcumin. *J. Ethnopharmacol.* **2010**, *127*, 641–644. [CrossRef] [PubMed]
2. Ye, Y.; Zhou, Y.; Mo, Z.; Cheng, W.; Yang, S.; Wang, X.; Chen, F. Rapid detection of aflatoxin B_1 on membrane by dot-immunogold filtration assay. *Talanta* **2010**, *81*, 792–798. [CrossRef] [PubMed]
3. Aghamohammadi, M.; Hashemi, J.; Kram, G.A.; Alizadeh, N. Enhanced synchronous spectrofluorimetric determination of aflatoxin B_1 in pistachio samples using multivariate analysis. *Anal. Chim. Acta* **2007**, *582*, 288–294. [CrossRef] [PubMed]
4. Elizalde-González, M.P.; Mattusch, J.; Wennrich, R. Stability and determination of aflatoxins by high-performance liquid chromatography with amperometric detection. *J. Chromatogr. A* **1998**, *828*, 439–444. [CrossRef]

5. Khayoon, W.S.; Saad, B.; Yan, C.B.; Hashim, N.H.; Ali, A.S.M.; Salleh, M.I.; Salleh, B. Determination of aflatoxins in animal feeds by HPLC with multifunctional column clean-up. *Food Chem.* **2010**, *118*, 882–886. [CrossRef]

6. Gimeno, A.; Martins, M.L. Rapid thin layer chromatographic determination of patulin, citrinin, and aflatoxin in apples and pears, and their juices and jams. *J. Assoc. Off. Anal. Chem.* **1983**, *66*, 85–91. [PubMed]

7. Xiulan, S.; Xiaolian, Z.; Jian, T.; Xiaohong, G.; Jun, Z.; Chu, F.S. Development of an immunochromatographic assay for detection of aflatoxin B_1 in foods. *Food Control* **2006**, *17*, 256–262. [CrossRef]

8. Ansari, A.A.; Kaushik, A.; Solanki, P.R.; Malhotra, B.D. Nanostructured zinc oxide platform for mycotoxin detection. *Bioelectrochemistry* **2010**, *77*, 75–81. [CrossRef] [PubMed]

9. Moccelini, S.K.; Fernandes, S.C.; Vieira, I.C. Bean sprout peroxidase biosensor based on l-cysteine self-assembled monolayer for the determination of dopamine. *Sens. Actuators B* **2008**, *133*, 364–369. [CrossRef]

10. Asav, E.; Akyilmaz, E. Preparation and optimization of a bienzymic biosensor based on self-assembled monolayer modified gold electrode for alcohol and glucose detection. *Biosens. Bioelectron.* **2010**, *25*, 1014–1018. [CrossRef] [PubMed]

11. Biebuyck, H.A.; Bain, C.D.; Whitesides, G.M. Comparison of organic monolayers on polycrystalline gold spontaneously assembled from solutions containing dialkyl disulfides or alkanethiols. *Langmuir* **1994**, *10*, 1825–1831. [CrossRef]

12. Van der Gaag, B.; Spath, S.; Dietrich, H.; Stigter, E.; Boonzaaijer, G.; van Osenbruggen, T.; Koopal, K. Biosensors and multiple mycotoxin analysis. *Food Control* **2003**, *14*, 251–254. [CrossRef]

13. Liu, Y.; Qin, Z.; Wu, X.; Jiang, H. Immune-biosensor for aflatoxin B_1 based bio-electrocatalytic reaction on micro-comb electrode. *Biochem. Eng. J.* **2006**, *32*, 211–217. [CrossRef]

14. Carlson, M.A.; Bargeron, C.B.; Benson, R.C.; Fraser, A.B.; Phillips, T.E.; Velky, J.T.; Groopman, J.D.; Strickland, P.T.; Ko, H.W. An automated, handheld biosensor for aflatoxin. *Biosens. Bioelectron.* **2000**, *14*, 841–848. [CrossRef]

15. Piermarini, S.; Micheli, L.; Ammida, N.H.S.; Palleschi, G.; Moscone, D. Electrochemical immunosensor array using a 96-well screen-printed microplate for aflatoxin B_1 detection. *Biosens. Bioelectron.* **2007**, *22*, 1434–1440. [CrossRef] [PubMed]

16. Sun, A.L.; Qi, Q.A.; Dong, Z.L.; Liang, K.Z. An electrochemical enzyme immunoassay for aflatoxin B1 based on bio-electrocatalytic reaction with room-temperature ionic liquid and nanoparticle-modified electrodes. *Sens. Instrum. Food Qual. Saf.* **2008**, *2*, 43–50. [CrossRef]

17. Owino, J.H.O.; Ignaszak, A.; Al-Ahmed, A.; Baker, P.G.L.; Alemu, H.; Ngila, J.C.; Iwuoha, E.I. Modelling of the impedimetric responses of an aflatoxin B_1 immunosensor prepared on an electrosynthetic polyaniline platform. *Anal. Bioanal. Chem.* **2007**, *388*, 1069–1074. [CrossRef] [PubMed]

18. Zaijun, L.; Zhongyun, W.; Xiulan, S.; Yinjun, F.; Peipei, C. A sensitive and highly stable electrochemical impedance immunosensor based on the formation of silica gel-ionic liquid biocompatible film on the glassy carbon electrode for the determination of aflatoxin B_1 in bee pollen. *Talanta* **2010**, *80*, 1632–1637. [CrossRef] [PubMed]

19. Chen, L.; Jiang, J.; Shen, G.; Yu, R. A label-free electrochemical impedance immunosensor for the sensitive detection of aflatoxin B_1. *Anal. Methods* **2015**, *7*, 2354–2359. [CrossRef]

20. Srivastava, S.; Ali, M.A.; Umrao, S.; Parashar, U.K.; Srivastava, A.; Sumana, G.; Malhotra, B.D.; Pandey, S.S.; Hayase, S. Graphene oxide-based biosensor for food toxin detection. *Appl. Biochem. Biotechnol.* **2014**, *174*, 960–970. [CrossRef] [PubMed]

21. Wang, D.; Hu, W.; Xiong, Y.; Xu, Y.; Li, C.M. Multifunctionalized reduced graphene oxide-doped polypyrrole/pyrrolepropylic acid nanocomposite impedimetric immunosensor to ultra-sensitively detect small molecular aflatoxin B_1. *Biosens. Bioelectron.* **2015**, *63*, 185–189. [CrossRef] [PubMed]

22. Castillo, G.; Spinella, K.; Poturnayová, A.; Šnejdárková, M.; Mosiello, L.; Hianik, T. Detection of aflatoxin B_1 by aptamer-based biosensor using PAMAM dendrimers as immobilization platform. *Food Control* **2015**, *52*, 9–18. [CrossRef]

23. Yu, L.; Zhang, Y.; Hu, C.; Wu, H.; Yang, Y.; Huang, C.; Jia, N. Highly sensitive electrochemical impedance spectroscopy immunosensor for the detection of AFB_1 in olive oil. *Food Chem.* **2015**, *176*, 22–26. [CrossRef] [PubMed]

24. Hu, H.; Cao, L.; Li, Q.; Ma, K.; Yan, P.; Kirk, D.W. Fabrication and modeling of an ultrasensitive label free impedimetric immunosensor for Aflatoxin B_1 based on poly(o-phenylenediamine) modified gold 3D nano electrode ensembles. *RSC Adv.* **2015**, *5*, 55209–55217. [CrossRef]

25. Li, Z.; Ye, Z.; Fu, Y.; Xiong, Y.; Li, Y. A portable electrochemical immunosensor for rapid detection of trace aflatoxin B 1 in rice. *Anal. Methods* **2016**, *8*, 548–553. [CrossRef]

26. Angnes, L.; Richter, E.M.; Augelli, M.A.; Kume, G.H. Gold electrodes from recordable CDs. *Anal. Chem.* **2000**, *72*, 5503–5506. [CrossRef] [PubMed]

27. Foguel, M.V.; Santos, G.P.; Ferreira, A.A.P.; Magnani, M.; Mascini, M.; Skladal, P.; Benedetti, A.V.; Yamanaka, H. Comparison of gold CD-R types as electrochemical device and as platform for biosensors. *J. Braz. Chem. Soc.* **2016**, *27*, 650–662. [CrossRef]

28. Zoski, C.G. Preface. In *Handbook of Electrochemistry*; Elsevier: Oxford, UK, 2007.

29. Su, X.L.; Li, Y. A self-assembled monolayer-based piezoelectric immunosensor for rapid detection of Escherichia coli O157:H7. *Biosens. Bioelectron.* **2004**, *19*, 563–574. [CrossRef]

30. Hermanson, G.T. *Bioconjugate Techniques*; Academic Press: San Diego, CA, USA, 2013.

31. Uliana, C.V.; Tognolli, J.O.; Yamanaka, H. Application of factorial design experiments to the development of a disposable amperometric DNA biosensor. *Electroanalysis* **2011**, *23*, 2607–2615. [CrossRef]

32. Hnaien, M.; Diouani, M.F.; Helali, S.; Hafaid, I.; Hassen, W.M.; Renault, N.J.; Ghram, A.; Abdelghani, A. Immobilization of specific antibody on SAM functionalized gold electrode for rabies virus detection by electrochemical impedance spectroscopy. *Biochem. Eng. J.* **2008**, *39*, 443–449. [CrossRef]

33. Brett, C.; Brett, A. *Electrochemistry: Principles, Methods, and Applications*; Oxford University Press: Oxford, UK, 1993.

34. Ensminger, L.G. The association of official analytical chemists. *Clin. Toxicol.* **1976**, *9*, 471. [CrossRef]

35. Pupim Ferreira, A.A.; Alves, M.J.M.; Barrozo, S.; Yamanaka, H.; Benedetti, A.V. Optimization of incubation time of protein Tc85 in the construction of biosensor: Is the EIS a good tool? *J. Electroanal. Chem.* **2010**, *643*, 1–8. [CrossRef]

chemosensors

MDPI

Editorial

Electrochemical Immunosensors and Aptasensors

Paolo Ugo * and Ligia M. Moretto *

Department of Molecular Sciences and Nanosystems, University Ca' Foscari of Venice, via Torino 155, 30172 Mestre-Venice, Italy
* Correspondence: ugo@unive.it (P.U.); moretto@unive.it (L.M.M.);
    Tel.: +39-041-234-8503 (P.U.); +39-041-234-8585 (L.M.M.)

Academic Editor: Igor Medintz
Received: 30 March 2017; Accepted: 30 March 2017; Published: 2 April 2017

1. Introduction

Since the first electrochemical biosensor for glucose detection, pioneered in 1962 by Clark and Lyons [1], research and application in the field has grown at an impressive rate and we are still witnessing a continuing evolution of research on this topic [2].

According to the definition recommended by IUPAC some years later [3], a biosensor is a self-contained device that is capable of providing specific quantitative or semi-quantitative analytical information using a biological recognition element (biochemical receptor) that is in direct spatial contact with the transduction element.

The easiest and perhaps most efficient way to favor the exchange of chemical information between a biomolecular recognition layer and a signal transducer is to immobilize the former directly on the surface of an electrochemical transducer, namely an electrode.

Initially, major research and application efforts were devoted to developing biocatalytic electrochemical sensors [4,5], aimed at exploiting the specificity of the reaction between an enzyme and its substrate to typically detect (in the simplest case) the latter as the analyte [6]. In the 1980s–1990s, the first examples of immunochemical reactions being used for electrochemical sensing were proposed by Heineman and Halsall [7], to be followed by the first proposal of electrochemical immunosensors where antibodies were immobilized on the electrode surface [8].

These biosensors exploit the specificity of the antigen (Ag)–antibody (Ab) interaction in order to detect one of the two partners as the analyte. For instance, the Ag can be immobilized on the electrode to capture the Ab or vice-versa. In order to achieve electrochemical detection by this approach, it is necessary to use a label which should be electroactive itself or able to generate or consume an electroactive molecule. By exploiting the long-lasting know-how developed for ELISA assays, the most commonly used labels in electrochemical immunosensors are enzyme labels. Following this approach, many sensors have been developed to detect a number of disease markers [9].

Recent research trends in the field of affinity biosensors are indeed moving beyond the above described ELISA-based approach, and new directions are being explored.

On one side, many studies are aimed at developing and applying novel capture agents, not necessarily belonging to the antibody category. Aptamers represent a successful example of efficient capture molecules as an alternative to antibodies. Aptamers are single-stranded oligonucleotides which act as biorecognition elements for proteins and other molecules. An aptamer for a specific target is selected from a large pool of random RNA or DNA sequences by the so-called SELEX (systematic evolution of ligands by exponential enrichment) methodology and is amplified by the polymerase chain reaction [10,11].

On the other hand, the possibility to avoid the use of enzyme labels appears very attractive in order to simplify detection schemes, avoiding complex functionalization procedures.

All this recent progress in the field of affinity biosensors has also led to the applications of affinity biosensors being widened to areas not necessarily of biomedical interest, extending to environmental [12] and food control [13].

2. The Special Issue

This Special Issue compiles eleven contributions and offers a platform on which to present recent developments in the field of electrochemical immunosensors and aptasensors, outlining future prospects and research trends.

The use of advanced capturing agents as an alternative to "classical" antibodies constitutes the focus of three original research articles. Marrazza and coworkers [14] propose the use of affibodies immobilized on magnetic particles to detect, with screen printed electrodes, the human epidermal growth factor receptor 2 (HER2) which is overexpressed by breast cancer cells. De Wael et al. [15] study the immobilization of aptamers on core-shell nanoparticles for improving the electrochemical detection of bacteria. Aptamers are also studied by Casalis and coworkers [16] who exploit an approach based on atomic force microscopy to control the bioaffinity properties of the sensor at the nanoscale. A miniaturized electrochemical cell for electrochemical impedance spectroscopy is developed and tested for thrombin detection.

Three review papers set the state-of-the-art on the use of aptamers for different biosensing purposes. The review by Oliveira-Brett and Chiorcea-Paquim [17] deals with the electrochemical characterization of four-stranded guanine structures (G-quadruplex), discussing the development of G-quadruplex aptasensors and hemin/G-quadruplex biosensors with peroxidase activity. Szunerits, Vasilescu et al. [18] review recent research on the development of aptamers-based electrochemical sensors suitable to detect lysozyme, an ubiquitous protein with antibacterial activity; this protein is also a potential biomarker for several diseases and can act as an allergen in foods. The review by Qi, Xu and coworkers [19] presents and discusses recent research aimed at combining aptasensing with stripping voltammetry. This combination offers interesting analytical prospects for the sensitive detection of a variety of analytes, from small biomolecules to proteins and cancer cells.

Electrochemical detection methods, as alternatives to voltammetry, amperometry or even electrochemical impedance spectroscopy, are discussed by de Moraes and Kubota [20], who reviewed the recent literature on immunosensors based on field effect transistors (FETs). The electrical signal of FET immunosensors is generated as a result of the antibody–antigen conjugation. FET biosensors allow real-time and rapid response, exhibiting high sensitivity and selectivity for analysis in small sample volume.

The applications of affinity sensors to food analyses and control are also a hot topic and, in this Special Issue, four papers deal with the exploitation of the analytical potentialities of immunosensors for this purpose. The review by Marty and coworkers [21] presents the state-of-the-art in the field of aflatoxin (AF) detection with affinity biosensors. Mycotoxins are toxic secondary metabolites produced by fungal species that can colonize and contaminate crops. Among them, aflatoxins (AFs) are of major significance because of their possible presence in many food commodities, thus constituting a potential threat to human health. The paper discusses and compares the characteristics and analytical performances of sensors for AF detection based on antibodies or aptamers as capture reagents.

Three research papers deal with the development of immunosensors for food control, trying to push the frontiers by using non-conventional electrode systems.

Pingarron, Campuzano and coworkers [22] present a novel magnetic beads-based immunosensing approach for the rapid and simultaneous determination of the main peanut allergenic proteins, namely, Ara h 1 and Ara h 2. The use of dual screen-printed carbon electrodes and the H2O2/hydroquinone system allowed interestingly low detection limits to be reached, suitable for detecting the two allergens in food extracts and wheat flour.

Yamanaka, Foguel et al. [23] propose the use of the so-called CD-trodes as transducers for the label-free immunosensing of aflatoxin B1 (AFB1). The CD-trodes are obtained by recycling

Au-compact discs (CDs), by the suitable removal of the polycarbonate outer layer. The electrode is then functionalized with anti-AFB1 using a self assemble monolayer of lipoic acid as an anchoring agent.

Moretto, Ugo et al. [24] apply ensembles of nanoelectrodes (NEEs), prepared by template Au deposition in nanoporous membranes, to detect, in food integrators, immunoglobulin IgY from hen-eggs yolk. Anti-IgY is immobilized on the polycarbonate templating membrane which surrounds the Au nanoelectrodes, achieving high antibody loadings while keeping unaltered the extremely low detection limits typical of NEEs.

Acknowledgments: We would like to thank all authors who contributed with their excellent papers to this Special Issue (SI). We thank the anonymous reviewers for their fundamental help in assuring the high quality of the SI. We are grateful to the *Chemosensors* Editorial Office for giving us this opportunity and, in particular, to Linda Wang for her continuous support in managing and organizing this SI.

References

1. Clark, L.C., Jr.; Lyons, C. Electrode systems for continuous monitoring in cardiovascular surgery. *Ann. NY Acad. Sci.* **1962**, *102*, 29–45. [CrossRef] [PubMed]
2. Ronkainen, N.J.; Halsall, H.B.; Heineman, W.R. Electrochemical biosensors. *Chem. Soc. Rev.* **2010**, *39*, 1747–1763. [CrossRef] [PubMed]
3. Thevenot, D.R.; Toth, K.; Durst, R.A.; Wilson, G.S. Electrochemical biosensors: Recommended definitions and classifications. *Pure Appl. Chem.* **1999**, *71*, 2333–2348. [CrossRef]
4. Turner, A.; Karube, I.; Wilson, G.S. *Biosensors: Fundamentals and Applications*; Oxford University Press: Oxford, UK, 1987.
5. Cunningham, A.J. *Introduction to Bioanalytical Sensors*; John Wiley & Sons: West Sussex, UK, 1998.
6. Bartlett, P.N. *Bioelectrochemistry, Fundamentals, Experimental Techniques and Applications*; John Wiley & Sons: West Sussex, UK, 2008.
7. Heineman, W.R.; Halsall, H.B. Antibodies-production, functions and applications in biosensors strategies for electrochemical immunoassay. *Anal. Chem.* **1985**, *57*, 1321A–1331A. [CrossRef] [PubMed]
8. Killard, A.J.; Deasy, B.; O'Kennedy, R.; Smyth, M.R. Antibodies: Production, functions and applications in biosensors. *TrAC Trends Anal. Chem.* **1995**, *14*, 257–266. [CrossRef]
9. Rusling, J.F. Nanomaterials-based electrochemical immunosensors for proteins. *Chem Rec.* **2012**, *12*, 164–176. [CrossRef] [PubMed]
10. Ellington, A.D.; Szostak, J.W. In vitro selection of RNA molecules that bind specific ligands. *Nature* **1990**, *346*, 818–822. [CrossRef] [PubMed]
11. Tierk, C.; Gold, L. Systematic evolution of ligands by exponential enrichment: RNA ligands to bacteriophage T4 DNA polymerase. *Science* **1990**, *249*, 505–510. [CrossRef]
12. Mascini, M. Affinity electrochemical biosensors for pollution control. *Pure Appl. Chem.* **2001**, *73*, 23–30. [CrossRef]
13. Scognamiglio, V.; Arduini, F.G.; Palleschi, G.; Rea, G. Biosensing technology for sustainable food safety. *TrAC Trends Anal. Chem.* **2014**, *62*, 1–10. [CrossRef]
14. Ilkhani, H.; Ravalli, A.; Marrazza, G. Design of an affibody-based recognition strategy for human epidermal growth factor receptor 2 (HER2) detection by electrochemical biosensors. *Chemosensors* **2016**, *4*, 23. [CrossRef]
15. Hamidi-Asl, E.; Dardenne, V.; Pilehvar, S.; Blust, R.; De Wael, V. Unique properties of core shell Ag@Au nanoparticles for the aptasensing of bacterial cells. *Chemosensors* **2016**, *4*, 16. [CrossRef]
16. Bosco, A.; Ambrosetti, E.; Mavri, J.; Capaldo, P.; Casalis, L. Miniaturized aptamer-based assays for protein detection. *Chemosensors* **2016**, *4*, 18. [CrossRef]
17. Chiorcea-Paquim, A.M.; Oliveira-Brett, A.M. Guanine quadruplex electrochemical aptasensors. *Chemosensors* **2016**, *4*, 13. [CrossRef]
18. Vasilescu, A.; Wang, Q.; Li, M.; Boukherroub, R.; Szunerits, S. Aptamer-based electrochemical sensing of lysozyme. *Chemosensors* **2016**, *4*, 10. [CrossRef]
19. Qi, W.; Wu, D.; Xu, G.; Nsabimana, J.; Nsabimana, A. Aptasensors based on stripping voltammetry. *Chemosensors* **2016**, *4*, 12. [CrossRef]
20. De Moraes, A.C.M.; Kubota, L.T. Recent trends in field-effect transistors-based immunosensors. *Chemosensors* **2016**, *4*, 20. [CrossRef]

21. Sharma, A.; Goud, K.Y.; Hayat, A.; Bhand, S.; Marty, J.L. Recent Advances in electrochemical-based sensing platforms for aflatoxins detection. *Chemosensors* **2017**, *5*, 1. [CrossRef]

22. Ruiz-Valdepenas Montiel, V.; Torrente-Rodríguez, R.M.; Campuzano, S.; Pellicanò, A.; Reviejo, Á.J.; Cosio, M.S.; Pingarrón, J.M. Simultaneous determination of the main peanut allergens in foods using disposable amperometric magnetic beads-based immunosensing platforms. *Chemosensors* **2016**, *4*, 11. [CrossRef]

23. Foguel, M.V.; Giordano, G.F.; de Sylos, C.M.; Carlos, I.Z.; Ferreira, A.A.P.; Benedetti, A.V.; Yamanaka, H. A low-cost label-free AFB$_1$ impedimetric immunosensor based on functionalized CD-trodes. *Chemosensors* **2016**, *4*, 17. [CrossRef]

24. Gaetani, C.; Ambrosi, E.; Ugo, P.; Moretto, L.M. Electrochemical immunosensor for detection of IgY in food and food supplements. *Chemosensors* **2017**, *5*, 10. [CrossRef]

MDPI AG

St. Alban-Anlage 66

4052 Basel, Switzerland

Tel. +41 61 683 77 34

Fax +41 61 302 89 18

http://www.mdpi.com

Chemosensors Editorial Office

E-mail: chemosensors@mdpi.com

http://www.mdpi.com/journal/chemosensors